Integrated Improvement of Food Legumes

Aditya Pratap • Chandra Mohan Singh
Deepak Kumar Verma • Awdhesh Kumar Mishra
Editors

Integrated Improvement of Food Legumes

Springer

Editors
Aditya Pratap
Crop Improvement Division
ICAR-Indian Institute of Pulses Research
Kanpur, India

Chandra Mohan Singh
Department Genetics and Plant Breeding
Banda University of Agriculture
and Technology
Banda, Uttar Pradesh, India

Deepak Kumar Verma
Agricultural and Food Engineering
Department
Indian Institute of Technology
Kharagpur, West Bengal, India

Awdhesh Kumar Mishra
Department of Biotechnology
Yeungnam University
Yeungnam, Korea (Republic of)

ISBN 978-3-031-82628-3 ISBN 978-3-031-82629-0 (eBook)
https://doi.org/10.1007/978-3-031-82629-0

This Springer imprint is published by the registered company Springer Nature Switzerland AG
The registered company address is: Gewerbestrasse 11, 6330 Cham, Switzerland

Preface

Grain legumes provide an excellent source of dietary protein, carbohydrates, iron, and zinc. Besides, they also contribute appreciably to enhancing soil nutrition status through N fixation and offer proven benefits to agriculture through reduced water consumption, less fertilizer and pesticide use, lowered greenhouse gas emissions, and improving physical properties. More than two dozen grain legumes are grown globally which ensure sustainability of the predominantly cereal-based cropping systems. Besides, these also ensure unparalleled food and nutritional security in the primarily vegetarian diets. The role of pulses in promoting human health and nutrition and management of noncommunicable diseases such as obesity, diabetes, and cardiovascular problems is amply appreciated by the global scientific communities and nutritionists. Subsequently, the last two decades have especially seen a focused interest of the global communities toward increasing the role of grain legumes in human nutrition. Of late, ushered by the technology revolution, country-specific policy initiatives and farmer-centric approach, an unprecedented awareness has arisen among the pulse producers, consumers, and researchers and a revolution has arisen in their production and productivity. This is contrary to their situation until the end of the twentieth century when pulses were mostly considered as orphan crops. Gradually, a shift in the research focus increased consciousness toward personal health by the enlightened masses, and these became the mainstream crops.

Tremendous progress has been made on research front including development of improved varieties, matching production and protection technologies, and development and deployment of omics-based resources toward genetic amelioration of grain legumes. A whole lot of pulses have been sequenced in the last one decade and ample scientific literature has been generated, which would be helpful toward accelerating the genetic improvement of grain legumes through translational research. In these efforts voluminous literature and knowledge resources have been generated. Through this book, the editors aimed to bring all relevant information about grain legumes on a single platform which is easily accessible to the readers. The book is proposed with 13 chapters on different grain legume covering an array of research aspects. Each chapter has been contributed by acknowledged pulses researchers. All of them have put their best efforts and deserve our sincere acknowledgments for their diligence and hard work.

We are highly thankful to Dr. Himanshu Pathak and Dr. M.L. Jat, the former and the present Director Generals of ICAR and Secretary, DARE, Ministry of Agriculture and Farmers' Welfare, Government of India; Dr. D.K. Yadav and Dr. T. R. Sharma, the present and the former Deputy Director Generals (Crop Science), ICAR; Dr. Sanjeev Gupta, Additional Director General (O&P), ICAR; Dr. N. P. Singh, former Vice Chancellor, BUAT, Banda; and Prof. Mukul Kumar, Dean (PGS), BUAT, Banda, for their consistent support and encouragement. We also wish to place on records the unconditional support of our families who despite their own professional and family commitments helped us to complete this endeavor. The research scholars and students working with us, namely, Chirag, Ruchi, and Priyanka Singh, also deserve appreciations for their kind help in cross checking the references. The team at Springer Nature, especially the project managers dealing with this volume, are also acknowledged for their involvement and support to make this publication a reality. We hope that this book will give an interesting reading to its readers.

Kanpur, India
Banda, Uttar Pradesh, India
Kharagpur, West Bengal, India
Yeungnam, Korea (Republic of)
30 April 2025

Aditya Pratap
Chandra Mohan Singh
Deepak Kumar Verma
Awdhesh Kumar Mishra

Contents

Editors and Contributors

About the Editors

Aditya Pratap has made contributions of eminence toward genetic improvement of crop plants, especially food legumes, deploying translational genetics. Amalgamating conventional and molecular tools he developed 16 varieties, most of them as the team lead. He has to his credit the first pan India extra early mungbean variety, Virat, which revolutionized summer mungbean cultivation in India besides one of the first marker assisted back crossing (MABC)-derived Fusarium wilt resistant chickpea varieties, Samriddhi. He established useful marker-trait associations in mungbean for several traits including grain micronutrients and anti-nutritional characters besides pyramiding genes for Fusarium wilt resistance and accomplishing genetic mapping of QTLs for drought tolerance in chickpea. He elucidated the role and expression of *VrLEA* encoding genes for heat tolerance, *NBS-LRR*-encoding genes for MYMIV resistance, and auxin-encoding gene family for water-logging tolerance in different *Vigna* species. He also established a *Vigna* wide hybridization garden and paved a way toward the utilization of wild accessions in prebreeding and distant hybridization. Dr. Aditya also contributed in the development of country-wide map of YMD-causing viruses and their molecular detection. He has been a key partner in establishing "International Mungbean Improvement Network." He has >300 publications to his credit including >120 peer-reviewed research articles in high impact national and international journals, 8 books, and several recognitions and awards including the Fellow, National Academy of Agricultural Sciences; ICAR-Lal Bahadur Shastri Outstanding Young Scientist Award; Norman E. Borlaug International Agricultural Science and Technology Fellowship; Member, National Academy of Sciences (NASI); and Best Research Team Award of WorldVeg as well as AICRP, MULLaRP. He has also been the Organizing Secretary of International Conference on Pulses (ICPulses2023). Dr. Aditya is presently working as the Project Coordinator, AICRP on Kharif Pulses, and his current assignment includes overall coordination, monitoring, and appraisal of research and development of Kharif pulses in India. His focus is on developing commercially viable hybrid cultivars with >30% exploitable heterosis in pigeonpea, deployment of MABC approach for Fusarium wilt, SMD resistance in pigeonpea and YMD resistance in mungbean and urdbean, and development and release of biofortified and specialty varieties in kharif pulses.

Chandra Mohan Singh is an Assistant Professor at Department of Genetics and Plant Breeding, Banda University of Agriculture and Technology, Banda, Uttar Pradesh, India. Dr. Singh has expertise in Molecular Breeding and Biotechnology. He did Post Doctorate in Plant Biotechnology from Center for Plant Biotechnology and Genomics, Spain. He served as Scientist (SERB) at ICAR-IIPR, Kanpur. He has published about 50 research papers in various reputed national and international journals. His research interest is breeding of legumes and oilseeds, plant pathogen interaction, and functional genomics. He has developed one variety of sesame and three varieties of linseed, i.e., BUAT Til-1 (CVRC), BUAT Alsi-3 (SVRC), BUAT Alsi-4 (CVRC), and BUAT Alsi-5 (CVRC). In collaboration, he screened the whole linseed germplasm of National Gene Bank for budfly and submitted the 04 unique genetic stock for budfly resistance and registered as potential donors. He generated the genomics/transcriptomics resources and submitted the data in NCBI. He is the recipient of various awards like SERB Young Scientist Award by SERB-DST; Fellow of ISPRD, Kanpur; Young Scientist Award by Government of Uttar Pradesh, India.

Deepak Kumar Verma is a proficient specialist in Agri-food and BioScience domains. He currently serves as an Independent Researcher and a National Consultant for CERVA-IRRI-SARC (Center of Excellence in Rice Value Addition, International Rice Research Institute-South Asia Regional Centre, India). Additionally, he has previously worked as a Scientific Consultant for Director Professor (Dr.) Cristobal Noe Aguilar at the Office of Research and Postgraduate Programs, Universidad Autonoma de Coahuila, México. He received the "Young Scientist Award 2021" for his outstanding contribution and recognition in the field of Biochemistry by the jury committee at the 5th International Conference on Advances in Smart Agriculture and Biodiversity Conservation for Sustainable Development conducted by the Agricultural Technology Development Society (ATDS). In the year 2020, he successfully obtained a Doctor of Philosophy (PhD) degree in the field of food processing engineering from the esteemed Agricultural and Food Engineering Department at the Indian Institute of Technology Kharagpur, located in India. In the year 2012, he was awarded a DST-INSPIRE Fellowship for pursuing a doctoral degree by the Department of Science and Technology (DST), which operates under the Ministry of Science and Technology, under the governance of the Government of India. Verma was tasked with conducting a study titled "Extraction and Identification of Volatile Aroma Compounds from Eight Indian Rice Cultivars (*Oryza sativa* L.)." This research expands upon his previous work during his master's degree, which focused on investigating the "physico-chemical and cooking characteristics of Azad Basmati (CSAR 839-3), a recently developed variety of Basmati Rice (*Oryza sativa* L.)." Verma obtained his Bachelor of Science degree in agricultural science in 2009 from the Faculty of Agriculture at Gorakhpur University, Gorakhpur. He further pursued his Master of Science (Agriculture) in Agricultural Biochemistry in 2011, achieving the top rank in his program. He was also honored with the Department of Agricultural Biochemistry's top award at CSAUA&T in Kanpur, India. In addition to his specialized expertise in Plant Biochemistry, he has also acquired a comprehensive knowledge base in the fields of Plant Physiology, Plant Breeding, Plant Biotechnology, Molecular Biology and Genetic Engineering, Seed Science and Technology, Agricultural Microbiology, Food Science & Technology, etc. Furthermore, he holds membership in various professional bodies, and his endeavors encompass participation in conferences, seminars, workshops, and training sessions, as well as the dissemination of scientific research through the publication of articles, books, and book chapters.

Awdhesh Kumar Mishra completed his PhD in the field of Plant Biotechnology from the National Institute of Plant Genome Research (degree awarded by Jawaharlal Nehru University) in New Delhi, India, in 2014. His PhD research focused on the characterization of a different functional aspect of WD40 protein in foxtail millet (*Setaria italica*) and their evolutionary history. After this, he joined the Center for Plant Biotechnology and Genomics, UPM, Madrid, Spain, as an Erasmus Postdoctoral Fellow (2016). Then, moved to another Postdoctoral position, at the Department of Biotechnology, Yeungnam University, Republic of Korea (2017–2018). Afterward, He was promoted to his current position, as an International Research Professor. He has published more than 35 articles, having an h-index of 21.

Contributors

B. Daniel Adewale Department of Crop Science and Horticulture, Federal University Oye-Ekiti, Ikole-Ekiti Campus, Ikole-Ekiti, Nigeria

Cristobal Noe Aguilar Bioprocesses and Bioproducts Group, Food Research Department, School of Chemistry, Autonomous University of Coahuila, Coahuila, Mexico

C. O. Aremu Department of Crop and Soil Science, Landmark University, Omu-Aran, Nigeria

Mamoni Banerjee Bio-Research Laboratory, Rajendra Mishra School of Engineering Entrepreneurship, Indian Institute of Technology Kharagpur, Kharagpur, West Bengal, India

P. S. Basu ICAR-Indian Institute of Pulses Research, Kanpur, UP, India

Mónica L. Chávez González Bioprocesses and Bioproducts Group, Food Research Department, School of Chemistry, Autonomous University of Coahuila, Coahuila, Mexico

Chirag Banda University of Agriculture and Technology, Banda, Uttar Pradesh, India

Vishal Chugh Department of Basic and Social Sciences, College of Horticulture, Banda University of Agriculture and Technology, Banda, Uttar Pradesh, India

Man Mohan Deo Division of Crop Production, ICAR- Indian Institute of Pulses Research, Kanpur, Uttar Pradesh, India

F. T. Fayose Department of Agriculture and Bio-resources Engineering, Federal University Oye-Ekiti, Ikole-Ekiti Campus, Ikole-Ekiti, Nigeria

Prem Chand Gyani Division of Genetics, ICAR-Indian Agriculture Research Institute, New Delhi, India

B. Ikhajiagbe Department of Plant Biology and Biotechnology, University of Benin, Benin City, Nigeria

Sudhir Kumar Jha Division of Plant Biotechnology, ICAR-Indian Institute of Pulses Research, Kanpur, Uttar Pradesh, India

Vikash Kumar Jha Department of Molecular Biology and Genetic Engineering, Bihar Agricultural University, Sabour, Bhagalpur, Bihar, India

Shashi Kashyap Division of Molecular Biology and Biotechnology, CSIR-NBRI, Lucknow, Uttar Pradesh, India

Chandan Kishore Department of Plant Breeding and Genetics, Bihar Agricultural University, Sabour, Bhagalpur, Bihar, India

T. E. Konyeme Department of Biological Sciences, University of Delta, Agbor, Nigeria

Bal Krishna Bhola Paswan Shastri Agricultural College, Purnea, Bihar, India

Bihar Agricultural University, Bhagalpur, Bihar, India

Anand Kumar Department of Plant Breeding and Genetics, Bihar Agriculture University, Sabour, Bhagalpur, Bihar, India

Anil Kumar Department of Plant Breeding and Genetics, Bihar Agricultural University, Sabour, Bhagalpur, Bihar, India

Jitesh Kumar Department of Molecular Biology and Genetic Engineering, Bihar Agriculture University, Sabour, Bhagalpur, Bihar, India

Mukul Kumar Banda University of Agriculture and Technology, Banda, Uttar Pradesh, India

Ravi Ranjan Kumar Department of Molecular Biology and Genetic Engineering, Bihar Agricultural University, Bhagalpur, Bihar, India

Sanjay Kumar Department of Plant Breeding and Genetics, Bihar Agricultural University, Bhagalpur, Bihar, India

Vaibhav Kumar ICAR-Indian Institute of Pulses Research, Kanpur, UP, India

Vikas Kumar Department of Seed Science and Technology, Bihar Agricultural University, Bhagalpur, Bihar, India

Vinod Kumar Department of Molecular Biology and Genetic Engineering, Bihar Agricultural University, Bhagalpur, Bihar, India

Yogesh Kumar ICAR-Indian Institute of Pulses Research, Kanpur, UP, India

Juli Kumari Department of Molecular Biology and Genetic Engineering, Bihar Agricultural University, Bhagalpur, Bihar, India

Seema Kumari Department of Molecular Biology and Genetic Engineering, Bihar Agricultural University, Sabour, Bhagalpur, Bihar, India

Priyanka Kaundal Banda University of Agriculture and Technology, Banda, Uttar Pradesh, India

Archana Kushwaha College of Home Science, G.B.P.U.A. &T., Pant Nagar, Uttarakhand, India

Awdhesh Kumar Mishra Department of Biotechnology, Yeungnam University, Gyeongsan, South Korea

Vigya Mishra Department of Post Harvest Technology, College of Horticulture, Banda University of Agriculture and Technology, Banda, Uttar Pradesh, India

Mehul Misra ICAR-Indian Institute of Pulses Research, Kanpur, UP, India

Biswajit Mondal ICAR-Indian Institute of Pulses Research, Kanpur, UP, India

C. V. Nnamani Plant Systematics and Conservation Biology Research Unit, Department of Applied Biology, Faculty of Science, Ebonyi State University, Abakaliki, Nigeria

A. A. Odetoye Department of Crop Science and Horticulture, Federal University Oye-Ekiti, Ikole-Ekiti Campus, Ikole-Ekiti, Nigeria

B. O. Ojuederie Department of Biological Sciences, Kings University, Ode-Omu, Nigeria

Brijesh Pandey Mahatma Gandhi Central University, Marihari, Bihar, India

Sunita Pandey Department of Genetics and Plant Breeding, Agriculture University, Jodhpur, Rajasthan, India

J. O. Popoola Department of Biological Sciences, Bowen University, Iwo, Nigeria

Pawan Prabhakar Bio-Research Laboratory, Rajendra Mishra School of Engineering Entrepreneurship, Indian Institute of Technology Kharagpur, Kharagpur, West Bengal, India

Seema Prajapati Department of Agronomy, Bihar Agricultural University, Sabour, Bhagalpur, Bihar, India

Jai Prakash Prasad Bhola Paswan Shastri Agricultural College, Purnea, Bihar, India

Bihar Agricultural University, Bhagalpur, Bihar, India

Aditya Pratap ICAR-Indian Institute of Pulses Research, Kanpur, India

Shalini Purwar Department of Basic and Social Sciences, College of Forestry, Banda University of Agriculture and Technology, Banda, India

Kumari Rajani Bhola Paswan Shastri Agricultural College, Purnea, Bihar, India

Department of Seed Science and Technology, Bihar Agricultural University, Sabour, Bhagalpur, Bihar, India

Tushar Ranjan Department of Molecular Biology and Genetic Engineering, Bihar Agricultural University, Bhagalpur, Bihar, India

Meenal Rathore Division of Plant Biotechnology, ICAR-Indian Institute of Pulses Research, Kanpur, Uttar Pradesh, India

Ruchi Banda University of Agriculture and Technology, Banda, Uttar Pradesh, India

Ankita Sagar ICAR-Indian Institute of Pulses Research, Kanpur, UP, India

Akhilesh Kumar Singh Department of Entomology, College of Horticulture, Banda University of Agriculture and Technology, Banda, India

Chandra Mohan Singh Banda University of Agriculture and Technology, Banda, Uttar Pradesh, India

Farindra Singh ICAR-Indian Institute of Pulses Research, Kanpur, UP, India

N. P. Singh ICAR-Indian Institute of Pulses Research, Kanpur, UP, India

Nandan Singh ICAR-Indian Institute of Pulses Research, Kanpur, UP, India

Narendra Pratap Singh Banda University of Agriculture and Technology, Banda, Uttar Pradesh, India

Poornima Singh Mahatma Gandhi Central University, Marihari, Bihar, India

Ravi Shankar Singh Department of Molecular Biology and Genetic Engineering, Bihar Agricultural University, Sabour, Bhagalpur, Bihar, India

Smita Singh Department of Allied Health Sciences, Chitkara School of Health Sciences, Chitkara University, Rajpura, Punjab, India

Prem Prakash Srivastav Agricultural and Food Engineering Department, Indian Institute of Technology Kharagpur, Kharagpur, West Bengal, India

Dharamsheela Thakur Department of Molecular Biology and Genetic Engineering, Bihar Agricultural University, Sabour, Bhagalpur, Bihar, India

Ayushi Tripathi Amity University Noida, Noida, Uttar Pradesh, India

Soubhagya Tripathy Agricultural and Food Engineering Department, Indian Institute of Technology Kharagpur, Kharagpur, West Bengal, India

Gemilang Lara Utama Faculty of Agro-Industrial Technology, Universitas Padjadjaran, Sumedang, Indonesia

Center for Environment and Sustainability Science, Universitas Padjadjaran, Bandung, Indonesia

Deepak Kumar Verma Agricultural and Food Engineering Department, Indian Institute of Technology Kharagpur, Kharagpur, West Bengal, India

Prasoon Verma ICAR-Indian Institute of Pulses Research, Kanpur, UP, India

Division of Crop Production, ICAR- Indian Institute of Pulses Research, Kanpur, Uttar Pradesh, India

Aneeta Yadav Faculty of Agricultural Sciences and Allied Industries, RAMA University, Kanpur, UP, India

Pooja Yadav Department of Plant Breeding and Genetics, Bihar Agriculture University, Sabour, Bihar, India

Biofortification of Food Legumes as a Means of Combating Malnutrition

Deepak Kumar Verma, Soubhagya Tripathy, Smita Singh, Mónica L. Chávez González, Prem Prakash Srivastav, Gemilang Lara Utama, and Cristobal Noe Aguilar

Abstract Due to the high concentration of dietary proteins that may be found in legumes, they constitute an essential component of the world's food supply. In addition to having a large amount of plant protein, legumes also have a high concentration of a wide variety of other nutrients as well as helpful phytochemical compounds. In order to improve the nutritional value of legumes and so contribute to the fight against malnutrition on a global scale, biofortification has been shown to be the most effective method currently available. Researchers and scientists are putting forth their best effort to make use of a wide variety of biofortifying strategies in order to accomplish their mission and compose a compelling narrative. As a result, the topic of food legume biofortification and how it may function as a nutritional powerhouse to avoid malnutrition will be the primary focus of this chapter, which will also include a critical discussion on the topic. In addition, this chapter looked at a number of different ways to biofortification, some of which are more current while others are more conventional. In addition, we investigated our current situation and the work that we have completed, as well as the challenges that lie ahead and opportunities that lie ahead for the biofortification of food legumes. Our goal is to meet the demands of the rapidly expanding global population for food and animal feed in this way.

D. K. Verma (✉) · S. Tripathy · P. P. Srivastav
Agricultural and Food Engineering Department, Indian Institute of Technology Kharagpur, Kharagpur, West Bengal, India

S. Singh
Department of Allied Health Sciences, Chitkara School of Health Sciences, Chitkara University, Rajpura, Punjab, India

M. L. Chávez González · C. N. Aguilar
Bioprocesses and Bioproducts Group, Food Research Department, School of Chemistry, Autonomous University of Coahuila, Coahuila, Mexico

G. L. Utama
Faculty of Agro-Industrial Technology, Universitas Padjadjaran, Sumedang, Indonesia

Center for Environment and Sustainability Science, Universitas Padjadjaran, Bandung, Indonesia

A. Pratap et al. (eds.), *Integrated Improvement of Food Legumes*,
https://doi.org/10.1007/978-3-031-82629-0_1

Keywords Biofortification · Dietary proteins · Phytochemical compounds · Transgenic approaches · Mutagenesis · Molecular breeding · Quantitative trait loci (QTLs)

1 Introduction

Legumes are a crucial component of the global food supply because they have a high concentration of dietary proteins (Singh & Pratap, 2016; Jimenez-Lopez et al., 2020; Semba et al., 2021; Verma et al., 2021; Affrifah et al., 2023). In addition to being high in plant protein, legumes are also rich in other nutrients and beneficial substances called phytochemicals. Although legumes are a good source of protein and fiber and include 15 different vitamins and minerals, their bioavailability is limited because of antinutritional effects (Jimenez-Lopez et al., 2020; Semba et al., 2021; Verma et al., 2021; Affrifah et al., 2023). Since then, the rate at which the population of the globe is growing has been one that is quite concerning. Though the agricultural industry is currently dealing with the concurrent challenge of boosting productivity in order to provide a growing population with a greater quantity of micronutrients. This challenge can be overcome by increasing the resource use efficiency of agricultural operations. Micronutrient malnutrition, known as "hidden hunger," has emerged as one of the most critical issues confronting the global community, particularly among neglected and impoverished rural populations (Rehman et al., 2019; Jaiswal et al., 2022; Odesina et al., 2022; Dwivedi et al., 2023; Huertas et al., 2023). Food legumes might be an excellent target for boosting nutritional quality, particularly in countries that are still in the development stage (Roriz et al., 2020).

Legume foods offer a wide variety of energy, comparable to that of cereals; this energy comes mostly from carbohydrates, contributing to around 60% of total energy, and legumes are also richer in iron (Fe), zinc (Zn), magnesium (Mn), phosphorus (P), and calcium (Ca) than cereals, with a high potassium (K) concentration. Food legumes include large quantities of Mn and copper (Cu) as well (Venkidasamy et al., 2019; Jimenez-Lopez et al., 2020; Semba et al., 2021; Verma et al., 2021; Affrifah et al., 2023; Dwivedi et al., 2023; Huertas et al., 2023; Lagoriya et al., 2023; Tererai & Mcebisi, 2023). Beans and lentils, on the other hand, have the maximum Fe concentration, and lentils have the highest Se concentration (Singh & Pratap, 2016; Huertas et al., 2023). For there to be an increase in the consumption of food legumes, there must first be an increase in their availability, an improvement in their economic access, and a high degree of awareness regarding their nutritious worth (Jha & Warkentin, 2020). Malnutrition and other significant nutritional problems, which afflict almost half of the world's population due to inadequate recommended dietary allowance (RDA) consumption during meals, have grown in recent decades (Majeed et al., 2020; UNICEF/WHO/WB, 2021; Dwivedi et al., 2023). Table 1 elaborates on the positive health advantages of consuming legumes.

Table 1 Importance of consuming legumes for human health against a wide range of diseases and conditions

Various health concerns		Food legume crop/ product	Scientific name	Remarks	References
Cancer	Breast	Kidney bean	*Ph. vulgaris*	Consumption of kidney beans was shown to inhibit the multiplication of human breast cancer cells.	Chan et al. (2013)
	Colon	Bean extracts	–	Bean extract can prevent abnormal foci crypts from developing in rat colons.	Feregrino-Pérez et al. (2008)
Cardiovascular		Dry beans	–	Dry beans, peanuts, and peas were three legume foods that, when consumed regularly, have the potential to lower the risk of coronary heart disease and cardiovascular disease.	Bazzano et al. (2001)
		Peanuts	*Arachis hypogaea*		
		Peas	*P. sativum*		
Cardiovascular and diabetic		Beans	*Phaseolus* spp.	There is evidence that chickpeas, lentils, and beans, three different types of food legumes, have the ability to reduce levels of hemoglobin A1c (HbA1c), as well as the risk of developing coronary heart disease.	Winham and Hutchins (2007)
		Chickpeas	*C. arietinum*		
		Lentils	*L. culinaris* or *L. esculenta*		
Glycaemia and obesity		Beans with a low-GI diet	–	It has been claimed that consuming beans as part of a diet with a low glycemic index can enhance metabolic regulation in type 2 obese-diabetic individuals and lead to weight loss.	Jimenez-Cruz et al. (2003)
Hypercholesterolemia		Baked beans	–	The intake of baked beans has the potential to bring about a decrease in both total cholesterol and LDL cholesterol.	Winham and Hutchins (2007)
Hypertriglyceridemia		Adzuki beans	*V. angularis*	Triglyceride levels have been observed to decrease as a direct result of the capacity of adzuki beans to inhibit lipase inhibitory action.	Maruyama et al. (2008)
Metabolic syndrome		Yellow peas	–	Food legumes such as yellow peas, chickpeas, navy beans, and lentils have all been identified as having the potential to lower the risk factors associated with metabolic syndrome.	Mollard et al. (2012)
		Chickpeas	*C. arietinum*		
		Navy beans	*Ph. vulgaris*		
		Lentils	*L. culinaris* or *L. esculenta*		

In general, the issues that burden poor nations may be traced back to a shortage of fundamental infrastructure, such as adequate educational and medical facilities. Even though numerous initiatives have been started and are being maintained on a global scale to reduce hunger, poverty, and other forms of deprivation, including malnutrition, these problems persist (Sharma et al., 2016; UNICEF/WHO/WB, 2021; Kumar, 2022; Wang, 2022; Dwivedi et al., 2023; Lagoriya et al., 2023; Mandal et al., 2023; Tererai & Mcebisi, 2023). People, especially children, are more likely to be stunted or wasted as a direct outcome of geographical differences that exacerbated malnutrition, according to research published by UNICEF, WHO, and the World Bank in 2020 (UNICEF/WHO/WB, 2021; Dwivedi et al., 2023). Therefore, preventative approaches in the realm of nutrition have been the preferred route to take when attempting to reduce the prevalence of nutritional problems in humans, and for good reason (Pérez-Escamilla, 2017). The "Millennium Development Goals (MDGs)" in the area of malnutrition are international and national priorities that immediately need to be addressed as a public-private strategy in low-income nations (Athar et al., 2020). Nowadays, fortifying certain food crops is necessary, but this practice is too expensive for the economically disadvantaged and is typically only available in metropolitan areas with affluent populations (Glover & Poole, 2019; Athar et al., 2020; Van Der Straeten et al., 2020; Mishra et al., 2022; Dwivedi et al., 2023; Tererai & Mcebisi, 2023). Fortification and supplementation are doable and successful in metropolitan areas because of the availability of necessary technology and infrastructure; nevertheless, if supplementation is not done adequately in foods, overloading can result in many ailments rather than prevent them. Beans and other legumes can give an energy-nutrient-rich food that helps in the reduction and prevention of malnutrition. This is possible due to the high protein, fiber, and mineral content of beans and other legumes, as well as the low dietary fat content of these foods (Singh & Pratap, 2016; Ojiewo et al., 2019; Rehman et al., 2019; Semba et al., 2021; Verma et al., 2021; Odesina et al., 2022; Affrifah et al., 2023; Dwivedi et al., 2023).

The purpose of this chapter is to concentrate on and have a critical discussion on food legume biofortification and how it might work as a nutritional powerhouse to prevent malnutrition. In addition, this chapter discussed a variety of biofortification approaches, both traditional and contemporary. Furthermore, we have reviewed where we are today and what we have done, as well as future difficulties and prospects for the biofortification of food legumes in order to fulfill the food and feed demands of the fast-rising population throughout the world.

2 Food Legumes Biofortification: A Brief Overview

Consumption of food legumes may not be adequate on its own to solve concerns about nutritional security. Food legumes, on the other hand, provide communities that are vegetarian either by choice or because they cannot afford meat or other animal products with an inexpensive and easily accessible supply of proteins that are rich in quantity. In so many Asian and African nations, cereals are traditionally eaten alongside these foods as part of the traditional diet (Madodé et al., 2011; Verma et al., 2021). They are also a high source of dietary fiber as well as the important amino acid lysine, making them an excellent addition to diets based on cereals, which frequently have low lysine concentrations (Jimenez-Lopez et al., 2020; Semba et al., 2021; Odesina et al., 2022; Affrifah et al., 2023). Moreover, chickpeas, lentils, and other legumes are rich in β-carotene, lutein, and zeaxanthin, all of which are precursors to vitamin A (Bar-El Dadon et al., 2017; Margier et al., 2018; Venkidasamy et al., 2019; Verma et al., 2021). In humans, however, these chemicals are first to be extracted from the plant matrix, placed in micelles, ingested, and then separated to yield retinol (Reboul, 2019; Nkhata et al., 2020). Research on these phenomena in legumes used for human consumption is in its infancy. Some common food sources, including corn, cassava (*Morchella esculenta*), sweet potatoes, and others, have undergone biofortification to increase their β-carotene content. Yet, to the greatest of our understanding, biofortification projects have not yet begun to focus on increasing the carotenoid concentration of food legumes.

Some of the potential solutions to the problem of "hidden hunger" include diversifying diets, fortifying food with vitamins and other micronutrients, and supplementing the diet with vitamins and other micronutrients (Sharma et al., 2016; Gödecke et al., 2018; Rehman et al., 2019; Bamji et al., 2021; Huertas et al., 2023; Tererai & Mcebisi, 2023). Another method for reaching those who don't have convenient access to fortified food is called biofortification, which refers to the agronomic or genetic enhancement of food crops to increase their nutritional content (Venkidasamy et al., 2019; Joshi-Saha et al., 2022; Lagoriya et al., 2023; Tererai & Mcebisi, 2023). Genetic biofortification, which is achieved through traditional breeding, is the technique that has the widest acceptance and is the most cost-effective and sustainable way of creating nutritionally enriched crops with long-term viability (Saltzman et al., 2017; Ojiewo et al., 2019; Dhanasekar et al., 2021; Gupta et al., 2021; Joshi-Saha et al., 2022; Tererai & Mcebisi, 2023). Rice (*Oryza sativa*) (Ludwig & Slamet-Loedin, 2019; De Steur et al., 2022), wheat (*Triticum aestivum*) (Ludwig & Slamet-Loedin, 2019), maize (*Zea mays*) (Nkhata et al., 2020; Obaid et al., 2022), potato (*Solanum tuberosum*) (D'Amato et al., 2020), and cassava (*M. esculenta*) (Sayre, 2022) are considered to be the top five most important staple food crops in all of Africa, Asia, and South America; as a result, these five crops have been the primary focus of biofortification efforts (Madodé et al., 2011; Saltzman et al., 2017; Rehman et al., 2019; Gaikwad et al., 2020; Verma et al., 2021; Monika et al., 2022; Barman et al., 2023). In addition to the crops described previously, food legumes are a key component of the diet of many civilizations throughout Asia and Africa (Verma et al., 2021). Food legumes are cultivated all over the world, with Asia and Africa being the two continents that contribute the most in terms of both total area and production (Table 2 and Fig. 1). India is responsible for around 70% and 55% of the area and production under food legumes across Asian nations (FAOSTAT, 2019). According to the reports published by different groups of researchers and scientists, surprisingly, these are the continents that have the largest burden of chronic hunger as well as hidden hunger (Yadava et al., 2018; Gödecke et al., 2018; GNR, 2020; Mandal et al., 2023).

Table 2 Areas under cultivation and total production of legumes across the various continents

Continent	Cultivated area (%)	Production (%)
Asian	49.1	44.3
African	30.7	24.1
American	12.8	18.7
European	5.5	10.8
Oceania	1.9	2.2

Source: FAOSTAT (2019)

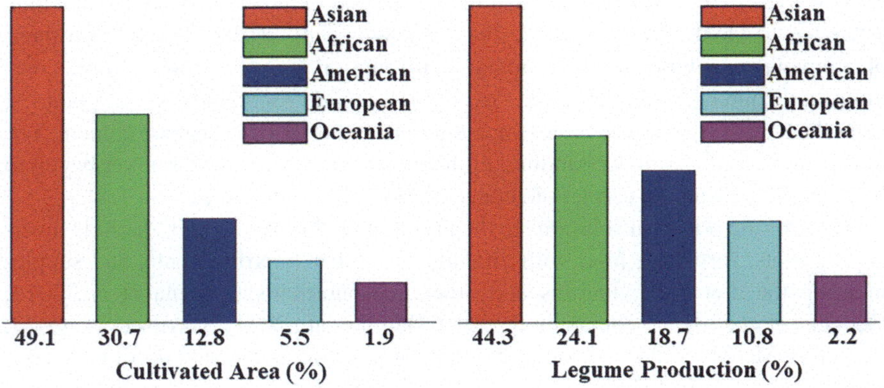

Fig. 1 Graphical demonstration of the area and production of various continents of the world that are involved in producing food legumes (FAOSTAT, 2019)

3 Food Legumes Biofortification as a Nutritional Powerhouse to Combat Malnutrition

Traditional plant breeding, agronomy, microbial-controlled activities, and transgenic approaches can all be used in biofortification to increase the nutrient density of a crop's edible parts (Saltzman et al., 2017; Garg et al., 2018; Tan et al., 2018; Ojiewo et al., 2019; Malik & Maqbool, 2020; Dhanasekar et al., 2021; Gupta et al., 2021; Tererai & Mcebisi, 2023). Consistent use of biofortified food products made from staple crops will lead to significant enhancements in human health and nutrition (Pérez-Escamilla, 2017; Semba et al., 2021; Jaiswal et al., 2022; Obaid et al., 2022; Dwivedi et al., 2023). As a result, biofortification of staple crops has emerged as a leading approach for combating the issue of micronutrient deficiencies and increasing the availability of vitamins and minerals in diets that are mostly composed of foods with lower nutrient densities (Bouis & Saltzman, 2017; Connorton & Balk, 2019; Athar et al., 2020; D'Amato et al., 2020; Dhanasekar et al., 2021; Mishra et al., 2022; Sayre, 2022; Barman et al., 2023; Dwivedi et al., 2023). Legumes consumed as food are directly responsible for the healthy development and growth of newborns, children, and even adults. Children who don't get enough of them can suffer from issues like growth retardation, obesity, brain damage, and protein-energy malnutrition (PEM). In order to maintain muscle mass, repair damaged tissues, and generate new cells, a healthy adult diet must include a sufficient amount of protein. Food legumes have a protein level of around two-to three-times higher than that of wheat (*T. aestivum*) and rice (*O. sativa*), respectively (Bar-El Dadon et al., 2017; Margier et al., 2018; Yadava et al., 2018; Venkidasamy et al., 2019; Semba et al., 2021; Affrifah et al., 2023; Dwivedi et al., 2023).

The uncontrollable rise in the world's population, which is expected to expand by 25% above its current level, is putting enormous strain on the world's food supply at the moment (UN, 2019). According to the most recent estimates, more than three billion individuals do not have the financial means to maintain a diet that is nutritionally sound and well-balanced, despite the fact that 57% of the world's population lives in sub-Saharan Africa and southern Asia, in addition to parts of North America and Europe (King et al., 2015; UNICEF/WHO/WB, 2021). As a result, putting a stop to the problem of malnutrition will not be simple (Dwivedi et al., 2023). The global output of food legumes was expected in an official report of the FAOSTAT to be 92.4 million MT in 2018, which represented a significant growth of almost 36 million MT, or 63%, between 1998 and 2018 (FAOSTAT, 2018). The output of some important food legumes has increased significantly in recent years, which include chickpeas (*Cicer arietinum*), common beans (*Phaseolus vulgaris*), cowpeas (*Vigna unguiculata*), lentils (*Lens culinaris* or *Lens esculenta*), pigeon peas (*Cajanus cajan*), etc. (Singh & Pratap, 2016; PULSEPOD, 2020; Affrifah et al., 2023). The total quantity of food legumes produced throughout the world was 92 million tonnes, of which 42 million tonnes have been provided in Asia alone (FAOSTAT, 2018). India is the country that produces and consumes the most food legumes, accounting for 25% of the world's total production, 27% of the world's total consumption, and 14% of the world's total imports, respectively (FAO, 2023).

4 Conventional and Modern Tools of Biofortification

4.1 Agronomic Biofortification in Legumes

In agronomic biofortification, the amount of micronutrients in the crop's edible section is increased by using inorganic fertilizers. This is an approach that is extensively used (Pal et al., 2019a, b, 2021; López-Morales et al., 2020; Silva et al., 2021; Lagoriya et al., 2023; Barman et al., 2023). It is necessary to apply these micronutrients agronomically during crop cultivation in order to nutritionally enhance food legumes with Fe and/or Zn. Yield loss may be avoided while increasing micronutrient concentration (Shivay et al., 2016; Pal et al., 2019a, b, 2021; López-Morales et al., 2020; Silva et al., 2021). There are a number of different agronomic biofortification strategies, such as seed priming and seed coating, as well as soil and/or foliar fertilization, which have the potential to raise the levels of micronutrients in grains. The viability of agronomic biofortification has been demonstrated by a number of experiments conducted on cereals and fewer studies conducted on food legumes, including chickpeas, peas, and common beans. These studies were recently evaluated by Garg et al. (2018). We have gone through some of the most significant

Table 3 A summary of the key findings from some of the most significant research on food legumes linked with agronomic biofortification

Food legumes				
Common name	Scientific name	Agronomic interventions for biofortification	Significant results and remarks	References
Chickpea	*C. arietinum*	Foliar applications of minerals, chelated, and nanoform fertilizers included 0.5% Zn and 0.5% iron; 0.3% Zn and 0.3% Fe; and 0.5% Zn and 0.5% Fe, respectively.	As compared to two separate foliar treatments of mineral and chelated forms, the results of a single foliar application of nano-fertilizers (0.5% ZnO NPs + 0.5% Fe_2O_3 NPs (ZFN)) during the pre-flowering stage showed the greatest ability to improve grain yield, Zn and Fe content, and their absorption.	Dhaliwal et al. (2021)
		In addition to the control treatments, which did not include the addition of any nutrients, individual and tank mix administrations of 0.5%, 0.5%, and 2% concentrations of $FeSO_4$, $ZnSO_4$, and urea were applied throughout the blooming and pod development phases, respectively.	When applied as a tank mix, the three components had the same beneficial effect as when applied separately, resulting in higher Zn content compared to the unsprayed control.	Pal et al. (2021)
		At the blooming and pod-forming stages, apply 25 kg per hectare of $ZnSO_4$ with a 0.5% foliar spray, with or without a 2% urea foliar spray.	The use of Zn and urea together resulted in higher levels of Zn and Fe in the chickpea compared to the use of Zn alone.	Pal et al. (2019a)
		Using 0.5% $FeSO_4$ as a foliar spray during the blooming and pod-forming phase, spraying solely during the flowering phase, or not applying any foliar fertilizer at all were compared with and without a 2% urea foliar spray during the flowering and pod-forming phase.	The use of Fe and urea together resulted in higher levels of Zn and Fe in the chickpea compared to the use of Fe alone.	Pal et al. (2019b)
		Seed priming (soaking in 0.05% ZnSHH) or soil application (25 kg/ha) of ZnSHH, or foliar application (0.5%) of a ZnSHH solution at the beginning of the vegetative stage, the beginning of flowering, and the beginning of pod filling.	Zn applied by foliar spray boosted the grain's Zn content by 21% and 22%, respectively, as compared to Zn applied via soil spray and seed priming. Additionally, the content varied depending on the genotype that was used.	Hidoto et al. (2017)

Cowpea	*V. unguiculata*	The soil was treated with $ZnSO_4$ at a rate of 25 kilograms per hectare.	There was no impact on yield, but there was a genotype-dependent increase in the Zn and protein content of the grain, as well as a reduction in the amount of phytic acid.	Silva et al. (2021)
		Application to the soil or as a foliar spray at different concentrations and stages of the plant's growth.	The amount of Zn in the grain and the size of the grain both increased across all doses.	López-Morales et al. (2020)
Lentil	*L. culinaris* or *L. esculenta*	The use of $ZnSO_4$ at a rate of 2 kg/ha, boric acid at a rate of 1.5 kg/ha, and sodium molybdate at a rate of 1 kg/ha as soil amendments, either individually or in combinations.	Both the micronutrients, such as Zn and boron, and the macronutrients, such as nitrogen and phosphorus, had their quantities raised in the seeds.	Islam et al. (2018)
Mungbean	*V. radiata*	$ZnSO_4$ was applied to the soil at a rate of 10 kg/ha and control (no application).	Zn fertilization enhanced all of the allometric and yield-related characteristics, as well as the Zn content of the grain, in a way that was dependent on the genotype.	Haider et al. (2021)
		Used a foliar spray of $ZnSO_4$ at concentrations of 0%, 0.5%, 1.0%, and 1.5%.	When compared to circumstances in which no Zn was applied, applications of Zn at concentrations of 0.5% and 1% resulted in increases of 86–38% and 78–156%, respectively, in grain yield and grain Zn content.	Haider et al. (2018)
Pea	*Pisum sativum*	Application of 0 mg, 4 mg, or 8 mg of ZnSHH per kilogram of soil, as well as foliar sprays with 0, 0.25%, or 0.5% of ZnSHH.7 H_2O, before flowering and throughout the early grain-filling period.	Grain Zn concentration was raised by both soil and foliar treatment. An important relationship between soil Zn and Zn in foliage was observed. Increases in total Zn content in raw grains following foliar Zn treatment were attenuated when soil Zn was present.	Poblaciones and Rengel (2016)

research linked with agronomic biofortification, and a summary of the most noteworthy findings from these studies can be seen in Table 3.

The success of agronomic biofortification can be attributed to a number of variables. These include the additional nutrient's chemical make-up, the administration technique (soil vs. foliar), the timing of the application, and the impact on the soil's existing mineral composition (Shivay et al., 2016; Pal et al., 2019a, 2021; Silva et al., 2021; Barman et al., 2023; Lagoriya et al., 2023). Foliar treatment of Zn is more effective than soil application of Zn that has been demonstrated that zinc ethylenediamine tetra acetic acid, also known as Zn-EDTA, is more effective than zinc sulfate heptahydrate, also known as ZnSHH, in the cultivation of chickpea (Shivay et al., 2016). Two sprays of Zn-EDTA were used in this research, one during the vegetative stage and another during the flowering stage, and both showed the highest agronomic effectiveness (grain yield increase in kilograms per gram of Zn administered). Crop production and micronutrient concentrations in mature grains may both benefit from the application of fertilizers, including zinc sulfate ($ZnSO_4$), zinc oxide (ZnO), iron sulfate ($FeSO_4$), and other such elements to the soil. It is true that foliar fertilization with EDTA of $FeSO_4$ and $ZnSO_4$ is more successful than other methods (Shivay et al., 2016); some of the important studies on this strategy are summarized in Table 3. There have only been a few efforts made to research the function that soil and foliar applications of micronutrients have in raising the level of Fe and Zn in the shoots and grains of food legumes. The grain Fe concentration of field pea was successfully increased by the use of Fe foliar sprays under Fe-deficient conditions. Treating plants with EDDHA [ethylenediamine-*N*, *N′*-bis (2-hydroxyphenylacetic acid)] at the beginning of blooming, followed by $FeSO_4$ (73.7 mg/l Fe), proved to be the most efficient method of resolving this deficit. In addition to this, the concentration of Fe in the grains showed a considerable rise with the use of any foliar spray during grain filling (Kabir et al., 2016). In a similar vein, Poblaciones and Rengel (2016) applied the foliar Zn treatments, either on their own or in conjunction with the soil Zn applications, which increased the amount of Zn in the field pea grain. Unfortunately, the majority of these investigations are still in the research phase at the moment.

Agronomic biofortification is a time-consuming and inefficient approach due to the necessity for timely chemical application (López-Morales et al., 2020). Whereas it is possible to cut down on expenses and labor costs by combining the application of foliar Zn and pesticide sprays, this strategy is not recommended (Poblaciones & Rengel, 2016; Pal et al., 2019a, b). Despite this, there are growing concerns over the farmers' careless use of pesticides, which can lead to the leaking of these chemicals into the groundwater. Furthermore, the soil quality factors that influence mineral mobility and bioavailability, as well as crop physiology that influences phloem mineral mobility and seed accumulation, all contribute to the effectiveness of agronomic biofortification (Shivay et al., 2016; López-Morales et al., 2020; Joshi-Saha et al., 2022; Dwivedi et al., 2023; Lagoriya et al., 2023). A variety of different agronomic practices, such as intercropping and crop rotations, as well as an increase in the activity of soil microorganisms, can be utilized to better facilitate the processes of solubilization and mobilization of soil micronutrients (Shivay et al., 2016;

Pal et al., 2019a, 2021; Silva et al., 2021). Due to the harmful effects of chemicals on the environment and the fact that these chemicals can interact with one another to alter the nutritional status of crops, dosage recommendations are of the utmost importance. In order to have a better understanding of these interactions, a lot of research has been done on cereals. In plants, the absorption of Zn and phosphorus (P) is influenced by each other. Zn fertilization has been found to increase the bioavailability of Zn because it decreases P absorption and phytate synthesis in seeds (Duhan et al., 2002; Islam et al., 2018; López-Morales et al., 2020; Barman et al., 2023). On the flip side, large levels of P, either applied or present, can cause Zn to precipitate, increasing the plant's need for the element. It is also well known that Zn has a wide range of interactions with other minerals, including Ca, sulfur (S), and boron (B) (Prasad et al., 2014; Shivay et al., 2016; Islam et al., 2018; Lagoriya et al., 2023). Soil pH, composition, and characteristics should all be taken into account when making agronomic biofortification recommendations, as well as the crop (Poblaciones & Rengel, 2016; Hidoto et al., 2017; Pal et al., 2021; López-Morales et al., 2020; Lagoriya et al., 2023). In the future, it will be necessary to make coordinated efforts for the evaluation of different biofortification procedures in order to increase the concentrations of micronutrients in the grains of food and legumes under field circumstances. Other important aspects that can have an effect on the nutrient enrichment or density of grain include cultivars, environmental factors, type of soil, the fertility status of the soil, crop species, and so on. All of these aspects should be taken into consideration before making any recommendations of this kind.

4.2 Genetic Biofortification in Legumes

Genetic biofortification is the most cost-effective, practicable, and sustainable way now available to increase the mineral micronutrient content of food (Ludwig & Slamet-Loedin, 2019; Dhanasekar et al., 2021; Tererai & Mcebisi, 2023). Plants' ability to absorb, translocate, and distribute nutrients to edible portions of the plant can be improved by introducing genetic modifications through traditional breeding, mutant breeding, or transgenic techniques (Saltzman et al., 2017; Garg et al., 2018; Tan et al., 2018; Malik & Maqbool, 2020; Gupta et al., 2021; Dhanasekar et al., 2021). When breeders have access to sufficient genotypic variety for the required nutritional qualities, they can increase mineral concentrations in high-yielding genotypes using the traditional breeding program's biofortification strategy. If there is a significant variation not present for the desired characteristics, breeders have to resort to alternative techniques, such as remote hybridization and mutagenesis (Gaikwad et al., 2020; Sen Gupta et al., 2021; Pandey et al., 2022). The "Health Grain Project (2005–2010)" and the "HarvestPlus program" are two examples of programs run by multiple international organizations that involve a large number of partners from a wide range of nations (HarvestPlus, 2014; Garg et al., 2018; HarvestPlus, 2021; Parihar et al., 2021). The goals of these programs are to increase the nutritional value of a variety of crops through breeding in order to provide nutritional security to undernourished populations (Gupta et al., 2021). As traditional crop breeding is more widely accepted than other methods, it has been used to biofortify a wide variety of crops. While some recent attempts have been made to biofortify food legumes like lentils, cowpeas, and common beans, the topic had not received nearly as much attention previously as it deserved. In cooperation with the HarvestPlus initiative, the International Center for Agricultural Research in Dry Areas (ICARDA) has introduced a number of lentil varieties that contain high levels of Fe and Zn. They include lentil varieties in Bangladesh (five), Nepal (seven), India (two), Ethiopia (one), and Syria (two). Furthermore, four varieties of cowpea have been produced by Govind Ballabh Pant University of Agriculture and Technology, Pantnagar, India, in collaboration with HarvestPlus that have been biofortified for Fe content by traditional breeding. As part of the HarvestPlus effort, biofortified high-Fe common beans have been distributed in six African countries and eight Latin American and Caribbean nations, and experiments for their distribution in many more countries are currently under way (HarvestPlus, 2014; Bouis & Saltzman, 2017; Garg et al., 2018; Gupta et al., 2021; HarvestPlus, 2021; Parihar et al., 2021).

It was shown that the level of Fe and Zn in various cereals and food legumes, including chickpeas, had a positive association, which suggests that these characteristics might be co-selected. These cereals and food legumes include wheat, rice, sorghum, and millet (Vandemark et al., 2018; Ludwig & Slamet-Loedin, 2019; Misra et al., 2020; Naik et al., 2020; Dhaliwal et al., 2021; Pal et al., 2021; De Steur et al., 2022). Nonetheless, there is a significant need for more research into the relationship between mineral and micronutrient concentrations and yield-contributing characteristics. For example, a number of research studies on chickpeas and other crops have found that a negative association exists between the amount of Zn and the yield. So, the first step to genetic biofortification is to screen various accessions, such as cultivated varieties, accessible germplasm, and wild relatives, for existing genetic diversity in the mineral micronutrient content (Diapari et al., 2014; Phuke et al., 2017; Vandemark et al., 2018; Misra et al., 2020; Naik et al., 2020). This is useful for two reasons: determining the current state of the trait's diversity and establishing a starting point for the targeted increase in micronutrient content. In the recent past, evaluations of this kind have been carried out for a variety of food legumes, and the results have shown a substantial variation (Table 4). Food legumes, on average, have a greater initial concentration of Fe and Zn than cereals (Blair, 2013; Dhaliwal et al., 2021; Gupta et al., 2021; Affrifah et al., 2023). HarvestPlus reports demonstrate that the starting concentrations of Zn are found to be 16 and 25 ppm in rice and wheat, respectively. Consequently, 28 ppm and 37 ppm were chosen as the breeding goals. Pearl millet had a Fe level of 47 ppm at the beginning of the breeding process, and a goal of 77 ppm was set (HarvestPlus, 2014; Bouis & Saltzman, 2017; Gupta et al., 2021; HarvestPlus, 2021). Table 4 shows that such proportions are already present in a variety of food legume germplasms. Food staples per capita consumption is a major component in setting the breeding objectives for crops, and cereals tend to be preferred above food legumes. But we must give

Table 4 Research on biofortifying key food legumes with iron and zinc through breeding on a global scale

Food legume crop		Location	Number of samples	Biofortifying nutrients (mg/kg or ppm)		Determination approach	References
Common name	Scientific name			Iron (Fe)	Zinc (Zn)		
Chickpea	*C. arietinum*	India	19	64.9–89.9	32.5–43.2	AAS	Misra et al. (2020)
		India	83	25.5–96.3	25.8–81.5	AAS	Saha et al. (2017)
		Canada	8	48.6–55.6	21.1–28.3	AAS	Ray et al. (2014)
		Pakistan	4	24–41	35–60	AAS	Zia-Ul-Haq et al. (2007)
Common bean	*Ph. vulgaris*	Canada	10	48–56	27.6–34	AAS	Ray et al. (2014)
		Colombia	2 genotypes and RILs of the cross between the two	Genotype 1: 80 Genotype 2: 46 RILs: 52–66	Genotype 1: 34 Genotype 2: 25 RILs: 23–39	AAS and ICP-OES	Blair et al. (2010)
		Portugal	155	32–88	11.5–45	AAS	Pinheiro et al. (2010)
		Colombia	8	48–74	17–28	ICP–ES	Ariza-Nieto et al. (2007)
Cowpea	*V. unguiculata*	India	13	81.7–93.2	28.4–34.6	AAS	Raina et al. (2020)
		Brazil	5	51–64	35–45	ICP–AES	Pereira et al. (2014)
		Brazil	87	36.5–137	36–58	AAS	Santos and Boiteux (2013)
		Brazil	30	61–81	27–44	AES	Carvalho et al. (2012)
		Africa	NR	27.6–28.9	32.6–31.5	AAS	Mamiro et al. (2011)
		South Africa	10	10–350	15–30	ICP–OES	Madodé et al. (2011)

(continued)

Table 4 (continued)

Food legume crop				Biofortifying nutrients (mg/kg or ppm)			
Common name	Scientific name	Location	Number of samples	Iron (Fe)	Zinc (Zn)	Determination approach	References
Lentil	*L. culinaris* Medikus subsp. *culinaris*	Syria	234	40.54–85.61	22.36–60.40	AAS	Sarker et al. (2018)
Mung bean	*V. radiata*	United States	95, Nashville TN	44.69 ± 17.66	71.75 ± 41.69	ICP–OES	Wu et al. (2020)
		Pakistan	4	144.04	28	AAS	Samreen et al. (2017)
		India	10	35–87	21–62	ICP–EMS	Nair et al. (2015)
		India	16	30–60	20–40	NR	Nair et al. (2013)
		India (northern part)	30	17.5–65.8	15.3–38	AAS	Singh et al. (2013)
		India	16	44.5–106.1	23.21–40.46	AAS	Taunk et al. (2012)
Pigeonpea	*C. cajan*	India	1	99.1	NR	AAS	Duhan et al. (2002)
		India	9	32.3–45	20.7–24.1	AAS	Singh et al. (1984)
Urdbean	*V. mungo*	India (two locations)	83, Indo-Gangetic Plains	117 (19–235) 91 (16–255)	44 (5–134) 29 (12–59)	AAS	Gupta et al. (2020)
		India	NR	104.7 ± 17.5 (seed coat)	11.5 ± 0.4 (seed coat)	AAS	Arockianathan et al. (2019)
		India	26	89 (71–100)	39.6 (19–60.5)	AAS	Singh et al. (2017a, b)

Abbreviations: AAS Atomic Absorption Spectroscopy, *AES* Atomic Emission Spectroscopy, *ICP-AES* Inductively Coupled Plasma-Atomic Emission Spectroscopy, *ICP-EMS* Inductively Coupled Plasma-Emission Spectroscopy, *ICP-ES* Inductively Coupled Plasma-Emission Spectrometry, *ICP-OES* Inductively Coupled Plasma-Optical Emission Spectrometry, *NR* Not Reported, *RILs* Recombinant Inbred Line food legumes more consideration than we do staple foods. As in the preceding sections, food legumes have been addressed in relation to the significance they hold in human diets and health.

To properly establish the standards, it is necessary to conduct a complete study that takes into account more than one growing season and several different types of environments (Phuke et al., 2017; Saha et al., 2017; Gupta et al., 2020; Naik et al., 2020; Misra et al., 2020). These kinds of research are presently being concentrated mostly on chickpeas. In a study of 19 chickpea genotypes, it was shown that the location (environment) impact contributed the most to the observed variability, followed by the genotype effect and then the genotype X location effect (Misra et al., 2020). The same kind of findings was found in 17 different genotypes of another food legume crop called groundnut, which were tested in over 150 different places in Thailand (Putto et al., 2008). It is interesting to note that the amplitude of genotype × location interactions for the Fe level was greater than that for the Zn level. Moreover, the variation due to genotype × location was nearer to the variance due to genotypes, which indicates that the environment has a greater impact on Fe level (Gaikwad et al., 2020; Naik et al., 2020; Misra et al., 2020; Gupta et al., 2021).

4.3 Biotechnological and Molecular Approaches for Legume Biofortification

The vast majority of transgenic studies involving food legumes have concentrated on soybeans, and a great number of transgenic soybean cultivars with increased levels of oleic acid and stearidonic acid have been commercially marketed in the United States of America, Canada, Japan, New Zealand, and Australia (Garg et al., 2018; Tan et al., 2018; Malik & Maqbool, 2020; Shelke et al., 2023; Soni et al., 2023). As far as we are aware, there have been a few reports of a food legume crop that has been biofortified through the use of transgenic technology (Garg et al., 2018; Tan et al., 2018; Shelke et al., 2023). In addition, the dissemination of such cultivars is made more difficult by the restrictions imposed by regulatory, legal, and socio-political frameworks. Gaining better knowledge of the physiological mechanisms involved in mineral micronutrient absorption, transport, and translocation to the seeds, as well as the repercussions of doing so, has been the primary contribution of transgenic research. Transgenic methods have been used for several genes in an effort to boost Fe concentration (Connorton & Balk, 2019; Majeed et al., 2020; Malik & Maqbool, 2020; Shelke et al., 2023). According to the findings of genome-wide association studies, these genes also have a significant role in the biofortification of Fe. In recent years, the concept of "gene stacking" has been offered as a method for concurrently enhancing numerous nutrients in staple foods by employing a combination of traditional breeding techniques and metabolic engineering (Van Der Straeten et al., 2020; Gupta et al., 2021; Chakraborty & Roychoudhury, 2023). Fe and/or Zn concentration or their bioavailability can be increased by knocking down genes such as the vacuolar Fe transporter (VIT) and genes that govern phytic acid accumulation. Next-generation genome editing technologies, such as CRISPR-Cas9, are now available for use to accomplish this goal (Dhanasekar et al., 2021; Joshi-Saha et al., 2022; Huertas et al., 2023). Where there is no genetic variation for the desired feature in the target crop species, the transgenic method of biofortification becomes the most effective. However, past studies have shown that the majority of food legumes have a significant variation in the content of both Fe and Zn (Table 4), which is something that may be addressed by conventional breeding methods.

During the course of the past 4–5 years, researchers and scientists have been making use of molecular markers for the purpose of identifying and mapping genes of interest. Their use in molecular breeding to transfer characteristics to elite backgrounds has been shown to be an effective method for producing trait-specific cultivars of a wide variety of agricultural plants, including cereals and food legumes (Ojiewo et al., 2019; Gupta et al., 2021; Tererai & Mcebisi, 2023). The process of establishing acceptable mapping populations for producing molecular markers connected with high levels of micronutrient loci involves large-scale analyses of existing germplasm sets of diverse food legumes (Sab et al., 2020; Wu et al., 2020; Gupta et al., 2021). This process is quite similar to conventional breeding. The mapping of genetic variation has been done using these populations (Prathet et al., 2012; Diapari et al., 2014; Ates et al., 2018). As a consequence of the introduction of next-generation sequencing technology, the process of molecular breeding has developed to integrate genomics-assisted breeding as an integral part of its methodology (Egan & Vatanparast, 2019; Ojiewo et al., 2019; Gaikwad et al., 2020; Ha & Lee, 2020; Pandey et al., 2022). It combines the use of numerous "omics" methodologies with high-throughput phenotyping, with the goal of predicting phenotypes based on genotypes and providing assistance to breeding operations (Aldemir et al., 2017; Ojiewo et al., 2019; Ha & Lee, 2020; Varshney et al., 2020; Gupta et al., 2021; Pandey et al., 2022).

At the moment, the genomes of a great number of food legumes have been sequenced, although the sequences themselves are of diverse quality, ranging from reference to draft to unassembled (Aldemir et al., 2017; Egan & Vatanparast, 2019; Ha & Lee, 2020; Sab et al., 2020; Wu et al., 2020). Having access to such genomic data would be of tremendous assistance in the process of developing molecular markers. These types of molecular markers include simple sequence repeats (SSRs) and single nucleotide polymorphisms (SNPs), both of which may be used in high-throughput systems to detect linkages and marker characteristics, respectively (Khazaei et al., 2017; Egan & Vatanparast, 2019; Pandey et al., 2022; De Souza Rodrigues Marinho et al., 2023; Ghosh et al., 2023). With the use of linkages and marker characteristics that are related to the mineral micronutrient concentration, a high mineral concentration may be preferentially introgressed into a background that already produces a high yield. In addition to the already-utilized marker-assisted selection (MAS), marker-assisted backcross breeding (MABC), and marker-assisted recurrent selection (MARS) in breeding programs, whole genome sequence data may be employed in genome-wide association studies (GWAS) to distinguish haplotypes for different traits for haplotype-based breeding (HBB) (Bevan et al., 2017; Khazaei et al., 2017; Singh et al., 2017b; Katuuramu et al., 2018; Sab et al., 2020; Wu et al., 2020; Gupta et al., 2021; Ghosh et al., 2023). Because the lines that are going to be crossed in genomic selection (GS) are chosen because they have high genome-estimated breeding values (GEBVs) evaluated on genome-wide marker information, there is no need for information about marker-trait associations (Khazaei et al., 2017; Singh et al., 2017b; Gupta et al., 2021; De Souza Rodrigues Marinho et al., 2023). Initiatives aimed at the biofortification of food legumes have only started putting these techniques and technology into practice. At the present time, molecular markers have only been linked or associated with quantitative trait loci (QTLs) for Fe and/or Zn in a limited range of food legumes (Table 5).

5 Biofortification of Food Legumes: Where We Are Now and What We Have Accomplished

In recognition of his efforts to introduce high-yielding wheat and rice varieties to India in the 1960s, when the country was threatened by widespread famine, Professor M.S. Swaminathan, an Indian agronomist, agricultural scientist, plant geneticist, administrator, and humanitarian, was named the first recipient of the World Food Prize in 1987 (World Food Prize, 2023). The country became food secure as wheat output doubled in a short period of time, sparing millions of people from starvation. Professor Swaminathan, widely recognized as the "Father of the Green Revolution in India" or the "Father of the Indian Green Revolution," is a pioneer in agricultural science and a man of vision who has long advocated within the scientific community for a greater emphasis on nutritional security instead of food security. Professor Swaminathan stated that, in addition to calories and protein, all important micronutrients should be included in everyone's diet. Only by combining efforts in the three pillars of agriculture, health, and nutrition can this be possible (Yadava et al., 2018; Bamji et al., 2021; Jaiswal et al., 2022; Dwivedi et al., 2023; Ramachandra & Vidya, 2023). As a result, the idea of biofortification justifies the requirement, given the current state of affairs, to provide a tool for combating malnutrition. Of the several methods of biofortification, breeding-led biofortified crops have shown the greatest potential for improving the overall nutritional profile of crops over the course of time (Saltzman et al., 2017; Garg et al., 2018; Dhanasekar et al., 2021; Gupta et al., 2021). In review reports published by different groups of researchers and scientists in 2018 and 2020, they discussed many significant targeted crops that had been biofortified through the use of breeding, agronomy, and transgenic methods (Garg et al., 2018; Malik & Maqbool, 2020). This study provided evidence that biofortified crops played a significant part in the overall

Table 5 Significant contributions on mineral biofortification in food legumes by utilizing molecular markers and quantitative trait loci

Common names	Scientific name	Nutrient	Plant part	Identified Markers and QTLs		Accessions/ cultivars/ lines	References
				Markers	QTLs		
Chickpea	*C. arietinum*	Fe	Seed	–	11 QTLs	MNK-1 and Annigeri 1	Sab et al. (2020)
		Zn		–	8 QTLs		
		Fe	Seed	10 SNPs	6 QTL	ICC 4958 and ICC8261	Upadhyaya et al. (2016)
		Zn		10 SNPs	5 QTLs,		
		Fe	Seed	4SNPs	–	94 accessions	Diapari et al. (2014)
		Zn		5 SNPs	–		
Common beans	*P. vulgaris*	Ca	Seed	5 SNPs	–	206 accessions	Katuuramu et al. (2018)
		Fe		5 SNPs	–		
		Zn		6 SNPs	–		
		Fe	Seed	–	8 QTLs	G14519 and G4825	Blair et al. (2010)
		Zn		–	9 QTLs		
		Fe	Seed	–	3 QTLs	DOR 364 and G19833	Cichy et al. (2009)
		Zn		–	3 QTLs		
		Phosphate ($PO4^{3-}$)	Root	–	26 QTLs	G19833 and DOR 364	Beebe et al. (2006)
Cowpea	*V. unguiculata*	Fe	Seed	9 SNPs		353 accessions	Ravelombola et al. (2018)
Lentil	*L. culinaris*	Mn	Seed	–	6 QTLs	CDC Redberry and ILL7502	Ates et al. (2018); Ates et al. (2016)
		Se		–	4 QTLs		
		Fe	Seed	–	21 QTLs	ILL 8006 X CDC Milestone	Aldemir et al. (2017)
		Fe	Seed	9 SNPs	–	138 accessions	Khazaei et al. (2017)
		Zn		12 SNP	–		

Mung bean	*V. radiata*	Fe	Seed	5 SNPs,	–	95 genotypes	Wu et al. (2020)
		Mn		5 SNPs	–		
		P		8 SNPs	–		
		K		9 SNPs	–		
		S		9 SNPs	–		
		Zn		7 SNPs	–		
		Fe	Seed	–	2 QTLs	ML776 X Sattya	Singh et al. (2017a, b)
		Zn		–	15 QTLs		
		Fe	Leaf (Fe deficiency chlorosis)	2 AFLP and 4 SSR	2 QTLs	Kamphaeng Saen 2 and NM10–12	Prathet et al. (2012)

Abbreviations: *AFLP* Amplified Fragment Length Polymorphism, *Ca* Calcium, *Fe* Iron, *K* Potassium, *Mn* Manganese, *P* Phosphorus, *QTLs* Quantitative Trait Loci, *S* Sulphur, *Se* Selenium, *SNP* Single Nucleotide Polymorphisms, *SSR* Simple Sequence Repeat, *Zn* Zinc improvement of the quality of life for millions of people all over the world. Biofortified crops account for all cereals, all legumes, all vegetables, and all fruits (58.1%, 13.2%, 19.8%, and 9%, respectively) that have been developed using the breeding strategy (Table 6 and Fig. 2). In continuing this, an agronomic strategy was utilized to share cereals in the portion of 69.4%, legumes in the portion of 14.5%, oilseeds in the portion of 3.2%, vegetables in the portion of 11.3%, and fruits in the portion of 1.6% (Table 6 and Fig. 2). There were 44% biofortified cereals, 10.3% biofortified legumes, 12.6% biofortified oilseeds, 17.7% biofortified vegetables, 9% biofortified fruits, and 2.3% biofortified fodder developed using the transgenic strategy (Table 6 and Fig. 2). Up till the date, around 290 biofortified cultivars of 12 different food crops have been registered or are in the process of being tested in over 60 countries throughout the globe (Garg et al., 2018). At the current time, there are more than 30 million people all over the world who cultivate at least one of the 12 biofortified crops and consume it in some form. These crops have been biofortified with the micronutrients Fe and Zn, as well as the provitamin vitamin A, both of which are essential for the human body to be able to carry out certain biological and other functions. By the year 2030, the objective is to provide biofortified crops and their varieties to more than one billion people so that they can gain the associated nutritional and agronomic advantages. The common bean (Fe), lentil (Fe and Zn), and cowpea (Fe and Zn) are three of the 12 biofortified crops that are most often sold and consumed as part of food legumes (HarvestPlus, 2021). There are now 17 biofortified crop varieties available for cultivation in India. They include 2 biofortified rice varieties, 2 biofortified wheat varieties, 2 biofortified pearl millet varieties, 4 biofortified maize varieties, 1 biofortified cauliflower variety, 1 biofortified pomegranate variety, and 1 biofortified sweet potato variety. One lentil variety, "Pusa Ageti Masoor," high in Fe content, was developed among the eight biofortified crops (Yadava et al., 2017). On October 16, 2020, the Prime Minister of India delivered this variety to the Food and Agriculture Organization (FAO), which was celebrating its 75th anniversary at the time. In food legumes, mutation breeding has led to the development of about improved 466 varieties (FAO/IAEA, 2021; Gupta et al., 2021).

Table 6 A comparison of the percentage of biofortification in selected crops utilizing different strategies

	Different biofortification strategies		
Targeted biofortified crops	Agronomic	Breeding	Transgenic
Cereals	69.4	58.1	44
Legumes	14.5	13.2	10.3
Vegetables	11.3	19.8	17.7
Fruits	1.6	9	13.1
Oilseeds	3.2	–	12.6
Fodder	–	–	2.3

Sources: Garg et al. (2018)

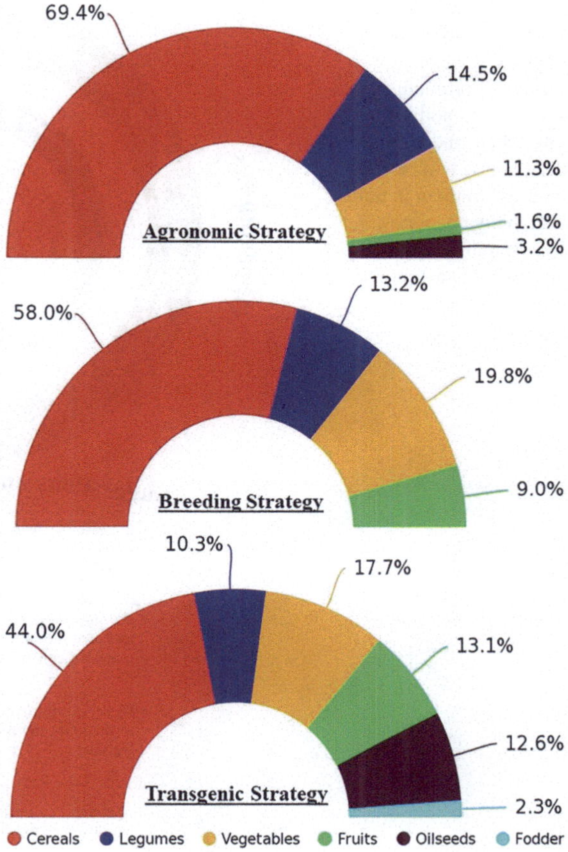

Fig. 2 A comparative presentation on the utilization of different approaches for biofortification of the crops that have been specifically targeted. (Garg et al., 2018)

6 Challenges, Opportunities, and Suggestions in Biofortification of Food Legumes

Fortification of food legumes presents a number of obstacles as well as possibilities, and ideas for researchers and scientists have been depicted in Fig. 3 that might greatly improve their grasp of the topic and the quality of their experiments. The following challenges and potential solutions have been encountered when working toward the goal to develop food legume crops more nutritionally dense through the use of biofortification (Table 7A). The policies of the nation determine whether or not the needy and hungry have a reliable source of food available to them. Regardless of this, the focus has switched from food security to nutritional security due to the fact that the latter better matches the healthy lifestyle paradigm of the people and has been proven to be an effective weapon in the struggle against malnutrition. As a result, the strategies that are detailed below can be implemented in the future to solve this problem (Table 7B).

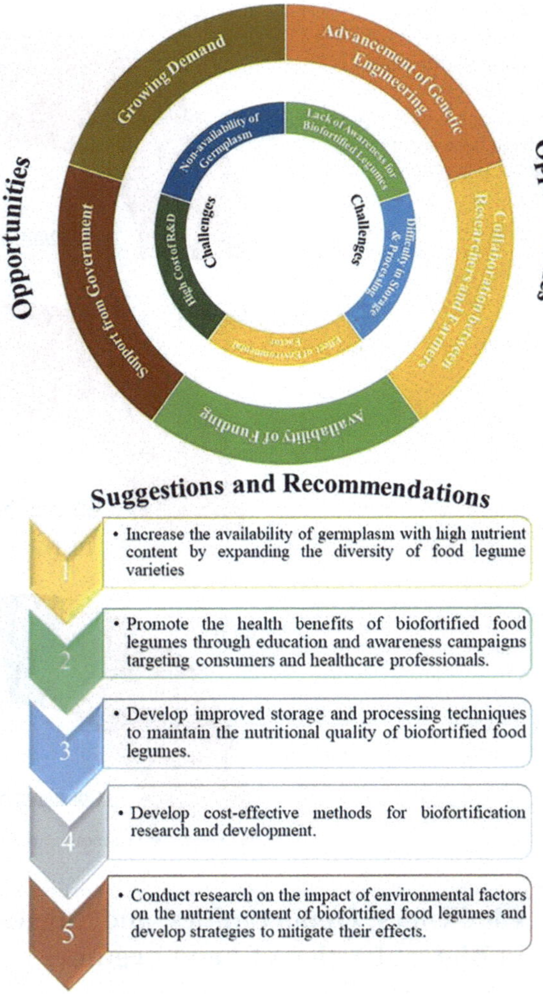

Fig. 3 Concerning the fortification of food legumes, there are a number of problems, possibilities, and ideas for researchers and scientists, all of which will be of great use to them in terms of gaining a better understanding and successfully carrying out their studies and experiments

7 Concluding Remarks

The majority of people who live in underdeveloped nations are suffering as a result of an inadequate supply of food grains and low consumption of a balanced diet. This results in a lack of essential micronutrients, which in turn leads to poor health and social discontent in both the community and the country as a whole. These are the fundamental issues. Considering this, the most critical chain-link activities that confer food, as well as nutritional security, are those that provide accessibility, reachability, and equal distribution of nutritionally rich food grains among individuals. Food legumes are a better alternative than other crops for the purpose of biofortification since they can be used for a variety of purposes and are often consumed worldwide as an important source of minerals and proteins in diets. There is

Table 7 List of a number of challenges, as well as opportunities and suggestions that have been presented by biofortification of food legumes

Remarks	References
A. *Difficulties and solutions for developing more nutritionally dense legume crops*	
The most significant barrier to the proper biofortification of food legumes is represented by antinutritional components. The effect that biofortified food legume crops have on the health of people is currently receiving a significant amount of attention.	Garg et al. (2018); Margier et al. (2018); López-Morales et al. (2020); Joshi-Saha et al. (2022)
The widespread ignorance about the causes of nutritional insufficiency among the inhabitants of poor and emerging nations in general on the nutritional profile of food legume crops is the most fundamental impediment to resolving the issue of nutritional insufficiency.	Singh and Pratap (2016); Venkidasamy et al. (2019); Yadava et al. (2018); Bamji et al. (2021); Affrifah et al. (2023); Pratap and Revanasidda (2023)
It is challenging to provide those who are economically disadvantaged with biofortified foods and legumes, both of which are essential components of a balanced diet.	Garg et al. (2018); Semba et al. (2021); De Steur et al. (2022); Joshi-Saha et al. (2022); Odesina et al. (2022); Affrifah et al. (2023)
The limited bioavailability of minerals (Fe and Zn) presents a significant challenge to the scientific community when it comes to the development of free phytic acid varieties.	Singh et al. (2017a); Yadava et al. (2017); Katuuramu et al. (2018); Dwivedi et al. (2023)
The low-cost, long-term, and ecologically friendly solutions to nutritional inadequacies offered by biofortification, such as crop breeding aimed at increasing nutrient content, are provided by biofortification. However, the lessons learned from Golden Rice suggest that genetically modified crops might be the most effective solution to the problems caused by a lack of micronutrients.	Gaikwad et al. (2020); Ojiewo et al. (2019); Roriz et al. (2020); Dhanasekar et al. (2021); Gupta et al. (2021); De Steur et al. (2022); Jaiswal et al. (2022); Barman et al. (2023)
In a similar vein, obtaining approval for genetically modified (transgenic) crops from the regulatory agencies that oversee food crops is not a simple task, and some countries are hesitant to do so. Because of this, the widespread production and commercialization of nutrient-dense legumes that have been biofortified are being prevented and discouraged.	Smart et al. (2017); Smyth (2017); Garg et al. (2018); Tan et al. (2018); Egan and Vatanparast (2019); Ha and Lee (2020); Malik and Maqbool (2020); Joshi-Saha et al. (2022); Amirhusin (2023)
The testing of several genotypes in order to uncover characteristics that are rich in micronutrients requires additional effort and money.	Phuke et al. (2017); Gupta et al. (2020); Misra et al. (2020)
Farmers that cultivate the biofortified food legume crop and variety have the ability to sell their yield at the MSP (minimum support price), but this is the only benefit that they receive from the government. Because of this, the government of an emerging nation should have to pay particular focus for purchasing biofortified food crops at the MSP.	Yadava et al. (2017); Yadava et al. (2018); Sheoran et al. (2022)
B. *Strategies implementation to address challenges in future*	
Any new solutions to the widespread Fe and Zn deficits that afflict the majority of the world's population need to be cost-effective and offer long-term advantages in order to be considered viable options.	Jha and Warkentin (2020); Monika et al. (2022); Tererai and Mcebisi (2023)

(continued)

Table 7 (continued)

Remarks	References
If landraces, wild cultivars, or primitive cultivars that are high in micronutrients could be identified, it would be of great assistance to the breeder who is working on developing novel varieties.	Singh et al. (2017b); Yadava et al. (2017); Wu et al. (2020); Gaikwad et al. (2020); Gupta et al. (2021); Pandey et al. (2022)
Importance and attention should be given to putting in place and bolstering a fast-track breeding program to develop nutrient-dense cultivars in response to local agro-climatic circumstances and dietary preferences.	Gaikwad et al. (2020); Dhanasekar et al. (2021); Gupta et al. (2021); Sen Gupta et al. (2021)
The breeding of novel plant varieties should be with enhanced capacity for micronutrient uptake, transport, and accumulation in consumable plant tissues. Moreover, those genes and processes should be investigated that improve the efficiency with which food legumes convert micronutrients into usable forms.	Ates et al. (2016); Poblaciones and Rengel (2016); Yadava et al. (2017); Singh et al. (2017b); Gupta et al. (2021)
The possibility of developing trait-specific (biofortified) varieties of food legume crops should be explored by using genome editing technologies, as well as molecular-based nontransgenic reverse technology.	Khazaei et al. (2017); Yadava et al. (2017); Katuuramu et al. (2018); Ha and Lee (2020); Sab et al. (2020); Wu et al. (2020); Dhanasekar et al. (2021)
Such newly produced varieties should be disseminated to the people who use them most—the consumers/farmers—so that they can commercialize them.	Yadava et al. (2017); Ojiewo et al. (2019); HarvestPlus (2021); Affrifah et al. (2023)
Consideration ought to be given to conducting more studies on the bioavailability of several micronutrients rather than phytic acids that are found in food legumes, most notably Fe and Zn.	Poblaciones and Rengel (2016); Singh et al. (2017a); Katuuramu et al. (2018); Dwivedi et al. (2023)
Governments ought to institute MSPs that are distinct for biofortified food crops. If farmers understood that they could make more money by growing food crops that have been biofortified, it should be encouraged for them to cultivate more of these types of crops.	Yadava et al. (2018); Sheoran et al. (2022)
To enhance the variety of biofortified food crops that are used, attention should be paid to establishing a market chain that extends from farmers to consumers, as well as other supporting infrastructure.	Yadava et al. (2017); Garg et al. (2018); Ojiewo et al. (2019); HarvestPlus (2021); Joshi-Saha et al. (2022); Affrifah et al. (2023)

currently a wide variety of choices accessible; however, only certain of these solutions are both cost-efficient and capable of sustainably supplying necessary micronutrients and provitamin A (vitamin A). Biofortification serves as a critical part of the process of endowing the edible sections of the plant with a greater density of certain characteristics. The conventional and marker-assisted breeding procedures are both extremely satisfying and have been shown to detect (QTLs and SNPs) and probe such desirable characteristics in targeted food crops. Additionally, both approaches are long-term ways that raise the amounts of vital minerals in edible sections and are cost-effective. The agronomic strategy is appropriate for use in the short term for the purpose of enriching the crops with the necessary micronutrients using the exogenous application of fertilizers (both soil and foliar). In addition, mineralization and higher accumulations in the edible section of plants are thought to be a faster and cheaper strategy to increase the nutritional profile of food legume crops. The presence of phytic acid in seeds can have an effect on the amount of Fe, Zn, and other minerals that are bioavailable in food legume crops. However, the accessibility of micronutrients, which are crucial to human health and effective in the battle against malnutrition, is significantly reduced by this antinutritional factor. As a consequence of this, researchers and medical experts are becoming more and more interested in finding answers to the topic of how to boost the bioavailability of minerals of such types.

References

Affrifah, N. S., Uebersax, M. A., & Amin, S. (2023). Nutritional significance, value-added applications, and consumer perceptions of food legumes: A review. *Legume Science, e192.*

Aldemir, S., Ateş, D., Temel, H. Y., et al. (2017). QTLs for iron concentration in seeds of the cultivated lentil (Lens culinaris Medic.) via genotyping by sequencing. *Turkish Journal of Agriculture and Forestry, 41*(4), 243–255.

Amirhusin, B. (2023). Genetically modified organism status, regulation, approval, labeling, and consumer perception in ASEAN. In *GMOs and political stance* (pp. 129–150). Academic Press.

Ariza-Nieto, M., Blair, M. W., Welch, R. M., & Glahn, R. P. (2007). Screening of iron bioavailability patterns in eight bean (Phaseolus vulgaris L.) genotypes using the Caco-2 cell in vitro model. *Journal of Agricultural and Food Chemistry, 55*(19), 7950–7956.

Arockianathan, P. M., Rajalakshmi, K., & Nagappan, P. (2019). Proximate composition, phytochemicals, minerals and antioxidant activities of *Vigna mungo* L. seed coat. *Bioinformation, 15*(8), 579.

Ates, D., Aldemir, S., Yagmur, B., et al. (2018). QTL mapping of genome regions controlling manganese uptake in lentil seed. *G3: Genes, Genomes, Genetics, 8*(5), 1409–1416.

Ates, D., Sever, T., Aldemir, S., et al. (2016). Identification QTLs controlling genes for Se uptake in lentil seeds. *PLoS One, 11*(3), e0149210.

Athar, T., Khan, M. K., Pandey, A., et al. (2020). Biofortification and the involved modern approaches. *Journal of Elementology, 25*(2).

Bamji, M. S., Murty, P. V. V. S., & Sudhir, P. D. (2021). Nutritionally sensitive agriculture—An approach to reducing hidden hunger. *European Journal of Clinical Nutrition, 75*(7), 1001–1009.

Bar-El Dadon, S., Abbo, S., & Reifen, R. (2017). Leveraging traditional crops for better nutrition and health—The case of chickpea. *Trends in Food Science & Technology, 64*, 39–47.

Barman, A., Bera, A., Saha, P., et al. (2023). Agronomic Zn biofortification of cereal crops a sustainable way to ensuring nutritional security: A review. *International Journal of Environment and Climate Change, 13*(3), 151–168.

Bazzano, L. A., He, J., Ogden, L. G., et al. (2001). Legume consumption and risk of coronary heart disease in US men and women: NHANES I Epidemiologic Follow-up Study. *Archives of Internal Medicine, 161*(21), 2573–2578.

Beebe, S. E., Rojas-Pierce, M., Yan, X., et al. (2006). Quantitative trait loci for root architecture traits correlated with phosphorus acquisition in common bean. *Crop Science, 46*(1), 413–423.

Bevan, M. W., Uauy, C., Wulff, B. B., et al. (2017). Genomic innovation for crop improvement. *Nature, 543*(7645), 346–354.

Blair, M. W. (2013). Mineral biofortification strategies for food staples: The example of common bean. *Journal of Agricultural and Food Chemistry, 61*(35), 8287–8294.

Blair, M. W., Medina, J. I., Astudillo, C., et al. (2010). QTL for seed iron and zinc concentration and content in a Mesoamerican common bean (Phaseolus vulgaris L.) population. *Theoretical and Applied Genetics, 121*(6), 1059–1070.

Bouis, H. E., & Saltzman, A. (2017). Improving nutrition through biofortification: A review of evidence from HarvestPlus, 2003 through 2016. *Global Food Security, 12*, 49–58.

Carvalho, A. F. U., de Sousa, N. M., Farias, D. F., et al. (2012). Nutritional ranking of 30 Brazilian genotypes of cowpeas including determination of antioxidant capacity and vitamins. *Journal of Food Composition and Analysis, 26*(1–2), 81–88.

Chakraborty, S., & Roychoudhury, A. (2023). Metabolic engineering of essential micronutrients in plants to ensure food security. In S. K. Upadhyay & S. P. Singh (Eds.), *Plants as bioreactors for industrial molecules* (pp. 255–272. ISBN: 9781119875116). Wiley. https://doi.org/10.1002/9781119875116.ch10

Chan, Y. S., Zhang, Y., & Ng, T. B. (2013). Brown kidney bean Bowman–Birk trypsin inhibitor is heat and pH stable and exhibits anti-proliferative activity. *Applied Biochemistry and Biotechnology, 169*, 1306–1314.

Cichy, K. A., Caldas, G. V., Snapp, S. S., & Blair, M. W. (2009). QTL analysis of seed iron, zinc, and phosphorus levels in an Andean bean population. *Crop Science, 49*(5), 1742–1750.

Connorton, J. M., & Balk, J. (2019). Iron biofortification of staple crops: Lessons and challenges in plant genetics. *Plant and Cell Physiology, 60*(7), 1447–1456.

D'Amato, R., Regni, L., Falcinelli, B., et al. (2020). Current knowledge on selenium biofortification to improve the nutraceutical profile of food: A comprehensive review. *Journal of Agricultural and Food Chemistry, 68*(14), 4075–4097.

De Souza Rodrigues Marinho, J., Valdisser, P. A. M. R., Brondani, C., et al. (2023). Molecular markers for assessing the inter-and intra-racial genetic diversity and structure of common bean. *Genetic Resources and Crop Evolution, 70*(1), 263–279.

De Steur, H., Stein, A. J., & Demont, M. (2022). From Golden Rice to Golden Diets: How to turn its recent approval into practice. *Global Food Security, 32*, 100596.

Dhaliwal, S. S., Sharma, V., Shukla, A. K., et al. (2021). Comparative efficiency of mineral, chelated and nano forms of zinc and iron for improvement of zinc and iron in chickpea (Cicer arietinum L.) through biofortification. *Agronomy, 11*(12), 2436.

Dhanasekar, P., Souframanien, J., & Suprasanna, P. (2021). Breeding cowpea for quality traits: A genetic biofortification perspective. In D. S. Gupta, S. Gupta, & J. Kumar (Eds.), *Breeding for enhanced nutrition and bio-active compounds in food legumes*. Springer. https://doi.org/10.1007/978-3-030-59215-8_7

Diapari, M., Sindhu, A., Bett, K., et al. (2014). Genetic diversity and association mapping of iron and zinc concentrations in chickpea (Cicer arietinum L.). *Genome, 57*(8), 459–468.

Duhan, A., Khetarpaul, N., & Bishnoi, S. (2002). Changes in phytates and HCl extractability of calcium, phosphorus, and iron of soaked, dehulled, cooked, and sprouted pigeon pea cultivar (UPAS-120). *Plant Foods for Human Nutrition, 57*(3), 275–284.

Dwivedi, S., Garcia Oliveira, A. L., Govindaraj, M., & Ortiz, R. (2023). Biofortification to avoid malnutrition in humans in a changing climate: Enhancing micronutrient bioavailability in seed, tuber, and storage roots. *Frontiers in Plant Science, 14*. https://doi.org/10.3389/fpls.2023.1119148

Egan, A. N., & Vatanparast, M. (2019). Advances in legume research in the genomics era. *Australian Systematic Botany, 32*(6), 459–483.

FAO. (2023). *India at a glance*. Accessed May 10, 2023. https://www.fao.org/india/fao-in-india/india-at-a-glance/en/

FAO/IAEA. (2021). *Mutant variety database*. Joint Food & Agriculture Organization and International Atomic Energy Agency (FAO/IAEA) programmme. Accessed January 20, 2023. https://mvd.iaea.org/

FAOSTAT. (2018). *Compare data*. Accessed January 20, 2023. http://www.fao.org/faostat/en/#compare

FAOSTAT. (2019). *Crops and livestock products*. Food and Agriculture Organization-statistical database (FAO-STAT) of the United Nations Food and Agriculture Organization (FAO) statistical division. Accessed January 20, 2023. http://www.fao.org/faostat/en/#data/QC

Feregrino-Pérez, A. A., Berumen, L. C., García-Alcocer, G., et al. (2008). Composition and chemopreventive effect of polysaccharides from common beans (Phaseolus vulgaris L.) on azoxymethane-induced colon cancer. *Journal of Agricultural and Food Chemistry, 56*(18), 8737–8744.

Gaikwad, K. B., Rani, S., Kumar, M., et al. (2020). Enhancing the nutritional quality of major food crops through conventional and genomics-assisted breeding. *Frontiers in Nutrition, 7*, 533453.

Garg, M., Sharma, N., Sharma, S., et al. (2018). Biofortified crops generated by breeding, agronomy, and transgenic approaches are improving lives of millions of people around the world. *Frontiers in Nutrition, 12*.

Ghosh, R., Tarafdar, A., Rao, K., et al. (2023). Pigeonpea crop improvement: Genomics and post-genomics. In *Smart plant breeding for field crops in post-genomics era* (pp. 271–292). Springer.

Glover, D., & Poole, N. (2019). Principles of innovation to build nutrition-sensitive food systems in South Asia. *Food Policy, 82*, 63–73.

GNR (Global Nutrition Report). (2020). *Country nutrition profiles: India*. Accessed on January 20, 2023. https://globalnutritionreport.org/resources/nutrition-profiles/asia/southern-asia/india/

Gödecke, T., Stein, A. J., & Qaim, M. (2018). The global burden of chronic and hidden hunger: Trends and determinants. *Global Food Security, 17*, 21–29.

Gupta, D. S., Gupta, S., & Kumar, J. (Eds.). (2021). *Breeding for enhanced nutrition and bioactive compounds in food legumes*. Springer.

Gupta, D. S., Singh, U., Kumar, J., et al. (2020). Estimation and multi-variate analysis of iron and zinc concentration in a diverse panel of urdbean (*Vigna mungo* L. Hepper) genotypes grown under differing soil conditions. *Journal of Food Composition and Analysis, 93*, 103605.

Ha, J., & Lee, S. H. (2020). Updates on legume genome sequencing. In *Legume genomics* (pp. 1–18). Humana.

Haider, M. U., Farooq, M., Nawaz, A., & Hussain, M. (2018). Foliage applied zinc ensures better growth, yield and grain biofortification of mungbean. *International Journal of Agriculture and Biology, 20*(12), 2817–2822.

Haider, M. U., Hussain, M., Farooq, M., Ul-Allah, S., et al. (2021). Zinc biofortification potential of diverse mungbean [Vigna radiata (L.) Wilczek] genotypes under field conditions. *PLoS One, 16*(6), e0253085.

HarvestPlus. (2014). *Biofortification progress briefs: iron and zinc lentils*. Accessed on January 25, 2023. https://www.harvestplus.org/sites/default/files/publications/Biofortification_Progress_Briefs_August2014_WEB_2_0.pdf

HarvestPlus. (2021). *Biofortification's growing global reach*. Accessed on January 25, 2023. https://www.harvestplus.org/knowledge-market/in-the-news/biofortification%E2%80%99s-growing-global-reach

Hidoto, L., Worku, W., Mohammed, H., & Bunyamin, T. (2017). Effects of zinc application strategy on zinc content and productivity of chickpea grown under zinc deficient soils. *Journal of Soil Science and Plant Nutrition, 17*(1), 112–126.

Huertas, R., Karpinska, B., Ngala, S., et al. (2023). Biofortification of common bean (*Phaseolus vulgaris* L.) with iron and zinc: Achievements and challenges. *Food and Energy Security, 12*(2), e406.

Islam, M. M., Karim, M. R., Oliver, M. M. H., et al. (2018). Impacts of trace element addition on lentil (Lens culinaris L.) agronomy. *Agronomy, 8*(7), 100.

Jaiswal, D. K., Krishna, R., Chouhan, G. K., et al. (2022). Bio-fortification of minerals in crops: Current scenario and future prospects for sustainable agriculture and human health. *Plant Growth Regulation, 98*(1), 5–22.

Jha, A. B., & Warkentin, T. D. (2020). Biofortification of pulse crops: Status and future perspectives. *Plants, 9*(1), 73.

Jimenez-Cruz, A., Bacardi-Gascon, M., Turnbull, W. H., et al. (2003). A flexible, low-glycemic index Mexican-style diet in overweight and obese subjects with type 2 diabetes improves metabolic parameters during a 6-week treatment period. *Diabetes Care, 26*(7), 1967–1970.

Jimenez-Lopez, J. C., Singh, K. B., Clemente, A., et al. (2020). Legumes for global food security. *Frontiers in Plant Science, 11*, 926.

Joshi-Saha, A., Sethy, S. K., Misra, G., Dixit, G. P., et al. (2022). Biofortified legumes: Present scenario, possibilities and challenges. *Field Crops Research, 279*, 108467.

Kabir, A. H., Paltridge, N., & Stangoulis, J. (2016). Chlorosis correction and agronomic biofortification in field peas through foliar application of iron fertilizers under Fe deficiency. *Journal of Plant Interactions, 11*(1), 1–4.

Katuuramu, D. N., Hart, J. P., Porch, T. G., et al. (2018). Genome-wide association analysis of nutritional composition-related traits and iron bioavailability in cooked dry beans (*Phaseolus vulgaris* L.). *Molecular Breeding, 38*(4), 1–18.

Khazaei, H., Podder, R., Caron, C. T., et al. (2017). Marker–trait association analysis of iron and zinc concentration in lentil (Lens culinaris Medik.) seeds. *The Plant Genome, 10*(2).

King, J. C., Brown, K. H., Gibson, R. S., et al. (2015). Biomarkers of nutrition for development (BOND)—Zinc review. *The Journal of Nutrition, 146*(4), 858S–885S.

Kumar, A. (2022). Agriculture for combating global starvation. In *Augmenting crop productivity in stress environment* (pp. 345–365). Springer.

Lagoriya, D. S., Harishma, S. J., & Singh, S. K. (2023). Agronomical approaches for biofortification of cereal crops. In *Biofortification in cereals: Progress and prospects* (pp. 1–19). Springer.

López-Morales, D., De La Cruz-Lazaro, E., Sánchez-Chávez, E., et al. (2020). Impact of agronomic biofortification with zinc on the nutrient content, bioactive compounds, and antioxidant capacity of cowpea bean (*Vigna unguiculata* L. Walpers). *Agronomy, 10*(10), 1460.

Ludwig, Y., & Slamet-Loedin, I. H. (2019). Genetic biofortification to enrich rice and wheat grain iron: From genes to product. *Frontiers in Plant Science, 10*, 833.

Madodé, Y. E., Houssou, P. A., Linnemann, A. R., et al. (2011). Preparation, consumption, and nutritional composition of West African cowpea dishes. *Ecology of Food and Nutrition, 50*(2), 115–136.

Majeed, A., Minhas, W. A., Mehboob, N., et al. (2020). Iron application improves yield, economic returns and grain-Fe concentration of mungbean. *PLoS One, 15*(3), e0230720.

Malik, K. A., & Maqbool, A. (2020). Transgenic crops for biofortification. *Frontiers in Sustainable Food Systems, 4*, 571402.

Mamiro, P. S., Mbwaga, A. M., Mamiro, D. P., et al. (2011). Nutritional quality and utilization of local and improved cowpea varieties in some regions in Tanzania. *African Journal of Food, Agriculture, Nutrition and Development, 11*(1).

Mandal, M., Ghosh, D., Karmakar, M., et al. (2023). Rural child health in India: The persistent nature of deprivation, undernutrition and the 2030 Agenda. In *Environment, Development and Sustainability* (pp. 1–24) 26.

Margier, M., Georgé, S., Hafnaoui, N., et al. (2018). Nutritional composition and bioactive content of legumes: Characterization of pulses frequently consumed in France and effect of the cooking method. *Nutrients, 10*(11), 1668.

Maruyama, C., Araki, R., Kawamura, M., et al. (2008). Azuki bean juice lowers serum triglyceride concentrations in healthy young women. *Journal of Clinical Biochemistry and Nutrition, 43*(1), 19–25.

Mishra, G. P., Dikshit, H. K., Kukreja, B., et al. (2022). Historical overview of biofortification in crop plants and its implications. In *Biofortification of staple crops* (pp. 31–61). Springer.

Misra, G., Joshi-Saha, A., Salaskar, D., Reddy, K. S., et al. (2020). Baseline status and effect of genotype, environment and genotype × environment interactions on iron and zinc content in Indian chickpeas (Cicer arietinum L.). *Euphytica, 216*(9), 1–16.

Mollard, R. C., Luhovyy, B. L., Panahi, S., et al. (2012). Regular consumption of pulses for 8 weeks reduces metabolic syndrome risk factors in overweight and obese adults. *British Journal of Nutrition, 108*(S1), S111–S122.

Monika, G., Kim, S. R. M., Kumar, P. S., et al. (2022). Biofortification: A long-term solution to improve global health—A review. *Chemosphere, 137713*.

Naik, S. M., Raman, A. K., Nagamallika, M., et al. (2020). Genotype × environment interactions for grain iron and zinc content in rice. *Journal of the Science of Food and Agriculture, 100*(11), 4150–4164.

Nair, R. M., Thavarajah, P., Giri, R. R., et al. (2015). Mineral and phenolic concentrations of mungbean [Vigna radiata (L.) R. Wilczek var. radiata] grown in semi-arid tropical India. *Journal of Food Composition and Analysis, 39*, 23–32.

Nair, R. M., Yang, R. Y., Easdown, W. J., et al. (2013). Biofortification of mungbean (Vigna radiata) as a whole food to enhance human health. *Journal of the Science of Food and Agriculture, 93*(8), 1805–1813.

Nkhata, S. G., Chilungo, S., Memba, A., & Mponela, P. (2020). Biofortification of maize and sweetpotatoes with provitamin A carotenoids and implication on eradicating vitamin A deficiency in developing countries. *Journal of Agriculture and Food Research, 2*, 100068.

Obaid, H., Shrestha, R. K., Liu, D., et al. (2022). Biofortification of maize with zinc and its effect on human health. *Journal of Soil Science and Plant Nutrition, 22*(2), 1792–1804.

Odesina, I., Gonzuk, N., Daodu, E., & Akintunde, S. (2022). Legume genetic resource security as main requirement for future challenges. *Legumes Research, 1*, 119.

Ojiewo, C., Monyo, E., Desmae, H., et al. (2019). Genomics, genetics and breeding of tropical legumes for better livelihoods of smallholder farmers. *Plant Breeding, 138*(4), 487–499.

Pal, V., Singh, G., & Dhaliwal, S. S. (2019a). Agronomic biofortification of chickpea with zinc and iron through application of zinc and urea. *Communications in Soil Science and Plant Analysis, 50*(15), 1864–1877.

Pal, V., Singh, G., & Dhaliwal, S. S. (2019b). Yield enhancement and biofortification of chickpea (*Cicer arietinum* L.) grain with iron and zinc through foliar application of ferrous sulfate and urea. *Journal of Plant Nutrition, 42*(15), 1789–1802.

Pal, V., Singh, G., & Dhaliwal, S. S. (2021). A new approach in agronomic biofortification for improving zinc and iron content in chickpea (*Cicer arietinum* L.) grain with simultaneous foliar application of zinc sulphate, ferrous sulphate and urea. *Journal of Soil Science and Plant Nutrition, 21*(2), 883–896.

Pandey, S., Singh, A., Parida, S. K., & Prasad, M. (2022). Combining speed breeding with traditional and genomics-assisted breeding for crop improvement. *Plant Breeding, 141*(3), 301–313.

Parihar, A. K., Dixit, G. P., Singh, U., et al. (2021). Potential of field pea as a nutritionally rich food legume crop. In D. S. Gupta, S. Gupta, & J. Kumar (Eds.), *Breeding for enhanced nutrition and bio-active compounds in food legumes* (pp. 47–82). Springer. https://doi.org/10.1007/978-3-030-59215-8_3

Pereira, E. J., Carvalho, L. M., Dellamora-Ortiz, G. M., et al. (2014). Effects of cooking methods on the iron and zinc contents in cowpea (Vigna unguiculata) to combat nutritional deficiencies in Brazil. *Food & Nutrition Research, 58*(1), 20694.

Pérez-Escamilla, R. (2017). Food security and the 2015–2030 sustainable development goals: From human to planetary health: Perspectives and opinions. *Current Developments in Nutrition, 1*(7), e000513.

Phuke, R. M., Anuradha, K., Radhika, K., et al. (2017). Genetic variability, genotype × environment interaction, correlation, and GGE biplot analysis for grain iron and zinc concentration and other agronomic traits in RIL population of sorghum (*Sorghum bicolor* L. Moench). *Frontiers in Plant Science, 8*, 712.

Pinheiro, C., Baeta, J. P., Pereira, A. M., et al. (2010). Diversity of seed mineral composition of Phaseolus vulgaris L. germplasm. *Journal of Food Composition and Analysis, 23*(4), 319–325.

Poblaciones, M. J., & Rengel, Z. (2016). Soil and foliar zinc biofortification in field pea (*Pisum sativum* L.): Grain accumulation and bioavailability in raw and cooked grains. *Food Chemistry, 212*, 427–433.

Prasad, R., Shivay, Y. S., & Kumar, D. (2014). Agronomic biofortification of cereal grains with iron and zinc. *Advances in Agronomy, 125*, 55–91.

Pratap, A., & Revanasidda, A. (2023). Nutritional and health benefits of power-packed pulses. *Pulse India, IX*, 37–39.

Prathet, P., Somta, P., & Srinives, P. (2012). Mapping QTL conferring resistance to iron deficiency chlorosis in mungbean [Vigna radiata (L.) Wilczek]. *Field Crops Research, 137*, 230–236.

PULSEPOD. (2020). *A global perspective on pulse production*. Accessed on January 11, 2023. https://pulsepod.globalpulses.com/pod-feed/post/a-global-perspective-on-pulse-production

Putto, W., Patanothai, A., Jogloy, S., & Hoogenboom, G. (2008). Determination of mega-environments for peanut breeding using the CSM-CROPGRO-peanut model. *Crop Science, 48*(3), 973–982.

Raina, A., Laskar, R. A., Tantray, Y. R., et al. (2020). Characterization of induced high yielding cowpea mutant lines using physiological, biochemical and molecular markers. *Scientific Reports, 10*(1), 3687.

Ramachandra, S., & Vidya, B. (2023). Agricultural revolutions which moved India from food deficit to food surplus since independence. *Delta National Journal of Multidisciplinary Research, 10*(Spl), 110–112.

Ravelombola, W., Qin, J., Shi, A., et al. (2018). Population structure analysis and association mapping for iron deficiency chlorosis in worldwide cowpea (*Vigna unguiculata* (L.) Walp) germplasm. *Euphytica, 214*, 1–15.

Ray, H., Bett, K., Tar'an, B., Vandenberg, A., et al. (2014). Mineral micronutrient content of cultivars of field pea, chickpea, common bean, and lentil grown in Saskatchewan, Canada. *Crop Science, 54*(4), 1698–1708.

Reboul, E. (2019). Mechanisms of carotenoid intestinal absorption: Where do we stand? *Nutrients, 11*(4), 838.

Rehman, H. M., Cooper, J. W., Lam, H. M., & Yang, S. H. (2019). Legume biofortification is an underexploited strategy for combatting hidden hunger. *Plant, Cell & Environment, 42*(1), 52–70.

Roriz, M., Carvalho, S. M., Castro, P. M., & Vasconcelos, M. W. (2020). Legume biofortification and the role of plant growth-promoting bacteria in a sustainable agricultural era. *Agronomy, 10*(3), 435.

Sab, S., Lokesha, R., Mannur, D. M., et al. (2020). Genome-wide SNP discovery and mapping QTLs for seed iron and zinc concentrations in chickpea (*Cicer arietinum* L.). *Frontiers in Nutrition, 7*, 559120.

Saha, A. J., Mehzabin, P., Reddy, K. S., & Ramachandran, V. (2017). Effect of seasonal variation on micronutrient content in chickpea (*Cicer arietinum* L.) and identification of accessions having high iron and zinc. *Journal of Food Legumes, 31*(1), 1–4.

Saltzman, A., Birol, E., Oparinde, A., et al. (2017). Availability, production, and consumption of crops biofortified by plant breeding: Current evidence and future potential. *Annals of the New York Academy of Sciences, 1390*(1), 104–114.

Samreen, T., Shah, H. U., Ullah, S., & Javid, M. (2017). Zinc effect on growth rate, chlorophyll, protein and mineral contents of hydroponically grown mungbeans plant (Vigna radiata). *Arabian Journal of Chemistry, 10*, S1802–S1807.

Santos, C. A. F., & Boiteux, L. S. (2013). Breeding biofortified cowpea lines for semi-arid tropical areas by combining higher seed protein and mineral levels. *Genetics and Molecular Research, 12*, 6782.

Sarker, A., Rizvi, A. H., & Singh, M. (2018). Genetic variability for nutritional quality in lentil (Lens culinaris Medikus subsp. culinaris). *Legume Research-An International Journal, 41*(3), 363–368.

Sayre, R. T. (2022). Biofortification of cassava: Recent progress and challenges facing the future. In S. Kumar, H. K. Dikshit, G. P. Mishra, & A. Singh (Eds.), *Biofortification of staple crops*. Springer. https://doi.org/10.1007/978-981-16-3280-8_16

Semba, R. D., Ramsing, R., Rahman, N., et al. (2021). Legumes as a sustainable source of protein in human diets. *Global Food Security, 28*, 100520.

Sen Gupta, D., Barpete, S., Kumar, J., & Kumar, S. (2021). Breeding for better grain quality in Lathyrus. In D. S. Gupta, S. Gupta, & J. Kumar (Eds.), *Breeding for enhanced nutrition and bio-active compounds in food legumes* (pp. 131–156). Springer.

Sharma, P., Dwivedi, S., & Singh, D. (2016). Global poverty, hunger, and malnutrition: A situational analysis. In *Biofortification of food crops* (pp. 19–30). Springer.

Shelke, D. B., Chambhare, M. R., Nikalje, G. C., & Nikam, T. D. (2023). Improvement of soybean crop for yield, stress tolerance, and value-added products using a transgenic approach. *Advances in Agriculture, 2023*.

Sheoran, S., Kumar, S., Ramtekey, V., et al. (2022). Current status and potential of biofortification to enhance crop nutritional quality: An overview. *Sustainability, 14*(6), 3301.

Shivay, Y. S., Singh, U., Prasad, R., & Kaur, R. (2016). Agronomic interventions for micronutrient biofortification of pulses. *Indian Journal of Agronomy, 61*(4th IAC Special Issue), S161–S172.

Silva, V. M., Nardeli, A. J., de Carvalho Mendes, et al. (2021). Agronomic biofortification of cowpea with zinc: Variation in primary metabolism responses and grain nutritional quality among 29 diverse genotypes. *Plant Physiology and Biochemistry, 162*, 378–387.

Singh, J., Kanaujia, R., Srivastava, A. K., et al. (2017a). Genetic variability for iron and zinc as well as antinutrients affecting bioavailability in black gram (*Vigna mungo* (L.) Hepper). *Journal of Food Science and Technology, 54*(4), 1035–1042.

Singh, N. P., & Pratap, A. (2016). Food legumes for nutritional security and health benefits. In *Biofortification of food crops* (pp. 41–50). Springer.

Singh, S. P., Vogel-Mikuš, K., Arčon, I., et al. (2013). Pattern of iron distribution in maternal and filial tissues in wheat grains with contrasting levels of iron. *Journal of Experimental Botany, 64*(11), 3249–3260.

Singh, U., Jain, K. C., Jambunathan, R., & Faris, D. G. (1984). Nutritional quality of vegetable pigeonpeas [Cajanus cajan (L.) Mill sp.]: Mineral and trace elements. *Journal of Food Science, 49*(2), 645–646.

Singh, V., Yadav, R. K., Yadav, N. R., et al. (2017b). Identification of genomic Regions/genes for high iron and zinc content and cross transferability of SSR markers in mungbean (*Vigna radiata* L.). *Legume Research, 40*(6), 1004–1011.

Smart, R. D., Blum, M., & Wesseler, J. (2017). Trends in approval times for genetically engineered crops in the United States and the European Union. *Journal of Agricultural Economics, 68*(1), 182–198.

Smyth, S. J. (2017). Genetically modified crops, regulatory delays, and international trade. *Food and Energy Security, 6*(2), 78–86.

Soni, K., Frew, R., & Kebede, B. (2023). A review of conventional and rapid analytical techniques coupled with multivariate analysis for origin traceability of soybean. *Critical Reviews in Food Science and Nutrition*, 1–20.

Tan, G. Z., Das Bhowmik, S. S., Hoang, T. M., et al. (2018). Investigation of baseline iron levels in Australian chickpea and evaluation of a transgenic biofortification approach. *Frontiers in Plant Science, 9*, 788.

Taunk, J., Yadav, N. R., Yadav, R. C., & Kumar, R. (2012). Genetic diversity among greengram [*Vigna radiata* (L.) Wilczek] genotypes varying in micronutrient (Fe and Zn) content using RAPD markers. *Indian Journal of Biotechnology, 11*, 48–53.

Tererai, T., & Mcebisi, M. (2023). Genetic bio-fortification of cereals from a plant breeding perspective. *Journal of Plant Breeding and Crop Science, 15*(1), 42–56.

UN (United Nations). (2019). *World population prospects*. Accessed on January 15, 2023. https://population.un.org/wpp/Download/Standard/CSV/

UNICEF/WHO/WB. (2021). *UNICEF-WHO-The World Bank: Joint child malnutrition estimates—levels and trends—2021 edition*. Accessed on May 10, 2023. https://data.unicef.org/resources/jme-report-2021/

Upadhyaya, H. D., Bajaj, D., Das, S., et al. (2016). Genetic dissection of seed-iron and zinc concentrations in chickpea. *Scientific Reports, 6*(1), 1–12.

Van Der Straeten, D., Bhullar, N. K., De Steur, H., et al. (2020). Multiplying the efficiency and impact of biofortification through metabolic engineering. *Nature Communications, 11*(1), 1–10.
Vandemark, G. J., Grusak, M. A., & McGee, R. J. (2018). Mineral concentrations of chickpea and lentil cultivars and breeding lines grown in the US Pacific Northwest. *The Crop Journal, 6*(3), 253–262.
Varshney, R. K., Sinha, P., Singh, V. K., et al. (2020). 5Gs for crop genetic improvement. *Current Opinion in Plant Biology, 56*, 190–196.
Venkidasamy, B., Selvaraj, D., Nile, A. S., et al. (2019). Indian pulses: A review on nutritional, functional and biochemical properties with future perspectives. *Trends in Food Science & Technology, 88*, 228–242.
Verma, D. K., Thakur, M., Singh, S., et al. (2021). The emphasis of effect of cooking and processing methods on anti-nutritional phytochemical of legumes and their significance in human health. In D. K. Verma (Ed.), *Phytochemicals in food and health: Perspectives for research and technological development*. Apple Academic Press.
Wang, X. (2022). Poverty criteria and world poverty. In *Multidimensional poverty measurement: Theory and methodology* (pp. 23–52). Springer.
Winham, D. M., & Hutchins, A. M. (2007). Baked bean consumption reduces serum cholesterol in hypercholesterolemic adults. *Nutrition Research, 27*(7), 380–386.
World food prize. (2023). *1987: Swaminathan*. The World Food Prize Foundation. Accessed on January 10, 2023. https://www.worldfoodprize.org/en/laureates/19871999_laureates/1987_swaminathan/
Wu, X., Islam, A. F., Limpot, N., et al. (2020). Genome-wide Snp identification and association mapping for seed mineral concentration in mung bean (*Vigna Radiata* L.). *Frontiers in Genetics, 11*, 656.
Yadava, D. K., Choudhury, P. R., Hossain, F., & Kumar, D. (2017). *Biofortified varieties: Sustainable way to alleviate malnutrition*. Indian Council of Agricultural Research.
Yadava, D. K., Hossain, F., & Mohapatra, T. (2018). Nutritional security through crop biofortification in India: Status & future prospects. *The Indian Journal of Medical Research, 148*(5), 621.
Zia-Ul-Haq, M., Iqbal, S., Ahmad, S., et al. (2007). Nutritional and compositional study of desi chickpea (Cicer arietinum L.) cultivars grown in Punjab, Pakistan. *Food Chemistry, 105*(4), 1357–1363.

Secondary Metabolites of Pigeonpea [*Cajanus cajan* (L.) Millsp.] with Their Biological and Pharmacological Activities

Pawan Prabhakar, Soubhagya Tripathy, Deepak Kumar Verma, Mamoni Banerjee, and Prem Prakash Srivastav

Abstract Pigeonpea [*Cajanus cajan* (L.) Millsp.] is mostly grown in tropical and subtropical regions of the world. This member of the Fabaceae family is a protein-rich legume that, besides its regular use as a *dal*, has been used for a very long time in traditional medicine to cure a broad variety of diseases, including diabetes, food poisoning, constipation, pain, and inflammation. These therapeutic characteristics are associated with a wide variety of classes of secondary metabolites, the majority of which are categorized as phenolic compounds and include apigenin, cajanin, naringenin, quercetin, vitexin, and isovitexin. As a result of the various biological and pharmacological actions that have been recorded in published research over the course of the past several years, it is currently thought of as a possible herbal therapeutic intervention for a wide variety of ailments. This chapter presents an overview of the secondary metabolites of *C. cajan*, along with trends and technical breakthroughs in the work that is now being done on the biological and pharmacological activities of *C. cajan* for human food, health, and medicine.

Keywords Secondary metabolites · Flavonoids and phenolic compounds · β-carotene-linoleic acid test · Phosphor-molybdenum assay · Climate resilience · Oxidative stress

P. Prabhakar · M. Banerjee (✉)
Bio-Research Laboratory, Rajendra Mishra School of Engineering Entrepreneurship, Indian Institute of Technology Kharagpur, Kharagpur, West Bengal, India
e-mail: mamoni@see.iitkgp.ac.in

S. Tripathy · D. K. Verma · P. P. Srivastav
Agricultural and Food Engineering Department, Indian Institute of Technology Kharagpur, Kharagpur, West Bengal, India

A. Pratap et al. (eds.), *Integrated Improvement of Food Legumes*,
https://doi.org/10.1007/978-3-031-82629-0_2

1 Introduction

Pigeonpea [*Cajanus cajan* (L.) Millsp.], a member of the Fabaceae family, is considered to be one of the most important pulse crops, often grown in tropical regions (Tungmunnithum & Hano, 2020; Verma et al., 2021; Prabhakar et al., 2023). Currently, around five million hectares (ha) are dedicated to the production of pigeonpea, making it the sixth most important legume food crop in the world. More than one billion people in the less developed countries depend on this crop as their major source of protein, and millions of farmers with limited resources in Asia, Africa, South America, Central America, and the Caribbean depend on it as a cash crop for their survival (Varshney et al., 2012; Vaz Patto et al., 2015). This perennial plant is now acclimatized to several tropical countries and is being cultivated in Africa, Asia, and Egypt since prehistoric times, eventually being introduced to the American continent. Pigeonpea is resistant to drought, as well as resilient and quick to adapt to new environments. The capacity of this crop to endure drought makes it almost absolutely necessary for maintaining food security in regions where rainfall is erratic and droughts are likely to occur (Vaz Patto et al., 2015; Agesa et al., 2019). In a great number of indigenous groups and communities of ethnic origin, this plant is utilized as a form of traditional medicine to cure a wide range of illnesses. The "Garo" tribe of Bangladesh and the "Kabui Naga tribe" of Manipur, India, both employ the leaves of this plant as an ethnomedicinal antidiabetic and anti-measles treatment (Wu et al., 2009; Pal et al., 2011). According to Agbafor et al. (2017), the leaves of *C. cajan* are utilized in the treatment of constipation, colic, and food poisoning in Trinidad and Tobago. According to Anadebe et al. (2017), the leaves of the pigeonpea plant are utilized in traditional Chinese medicine to staunch bleeding, alleviate pain, and eradicate parasites. In certain parts of Tamil Nadu, India, the leaves, the seeds, and the young stems are utilized in the treatment of gingivitis and stomatitis and are used as a toothbrush. People in the eastern part of Rajasthan use fresh juice or the leaf extract after boiling as a laxative and to counteract the symptoms of drunkenness when they give it to someone orally (Pal et al., 2011). Leaf paste and sometimes leaf tea are used to heal oral ulcers and inflammations. *C. cajan* seeds are used in the traditional medicine of Omani people for the treatment of a variety of persistent ailments. It is important to point out that in the past, humans utilized the juice extracted from pigeonpea leaves to cure a range of skin conditions, including those that occurred within the mouth (Nanna et al., 2013). This could be owing to the presence of several key secondary metabolites in the plant that have substantial biological and pharmacological properties (Orni et al., 2018). According to research conducted on the chemical components of *C. cajan* leaves, these contain a significant amount of flavonoids as well as stilbenes. They also included saponins, a considerable quantity of tannins, and only trace levels of reducing sugars, resins, and terpenoids in their composition (Yang et al., 2020). Because of its high protein content and its fascinating flavonoid and phenolic component profiles (Gargi et al., 2022), pigeonpea is now the subject of research, with new findings being published. Flavonoids and other phenolic compounds of *C. cajan* are presently attracting increasing attention, which indicates that there may be additional biological activities for these secondary metabolites that have not yet been thoroughly investigated (Verma & Thakur, 2021; Prabhakar et al., 2023). As a result, there is a growing interest in the study of the flavonoids and other phenolics that are extracted from this leguminous plant. This leads one to believe that other possible biological activities derived from these secondary metabolites are just waiting to be investigated (Verma & Thakur, 2021). Therefore, the purpose of this chapter is to present an overview of the secondary metabolites of *C. cajan*, along with trends and technical breakthroughs in the research work that is now being done on its biological and pharmacological activities for human food, health, and medicine. Specifically, this chapter will focus on the pharmacological applications of these secondary metabolites.

2 Secondary Metabolites

The composition of the active compounds found in plant matrices and the concentration of those active compounds have a significant impact on the bioactive effects of those compounds (Banwo et al., 2021; Kaur et al., 2021; Silva et al., 2021; Verma & Thakur, 2021; Bautista-Hernández et al., 2022; Tripathy et al., 2022). While numerous compounds have been reported in *C. cajan*, the main bioactive compounds found can be broadly classified into the groups of stilbenes, phenolics, and flavonoids. Cajanol, longistylins A and C, genistein, and biochanin A were identified to be among the phenolic compounds that were present, according to the screening of secondary metabolites that was based on the published literature (Nix et al., 2015; Sarkar et al., 2020). The primary components of pigeonpea roots are a pair of isomeric compounds that go by the names genistein and apigenin. These compounds are responsible for a wide variety of pharmacological effects (Orni et al., 2018). An investigation using high-performance liquid chromatography (HPLC) was carried out in order to ascertain the constituent chemicals of an ethanolic extract of *C. cajan* leaf (Rinthong & Maneechai, 2018). In the course of this investigation, a total of seven flavonoids, namely, pinostrobin, orientin, naringenin, apigenin, apigenin-6, 8-di-C-L-arabinopyranoside, and pinostrobin chalcone, were identified. Additionally, two stilbenes, namely cajaninstilbene acid and longistyline C, were found. Zhang et al. (2014) revealed the structure of naringenin-3′-isoprenyl-7-methyl ether 1, a new prenylated flavanone that was obtained from *C. cajan*. Another investigation reported that *C. cajan* contains reducing sugars, tannins, resins, saponins, glycosides, and acidic compounds (Oke, 2014). In a separate study conducted by Gui-Yun et al. (2019), the leaves of *C. cajan* were used to isolate six novel prenylated stilbenes, which were given the designations cajanusins A–C, a new natural product, cajanusin D, and six known derivatives such as 5-hydroxy-6,7-(2″, 2″-dimethylchromene) flavanone, longistytline A, longistytline C, cajanolactone A, pi-nostrobin, and cajanonic acid A.

The leaves of the *C. cajan* plant contain the highest concentration of flavonoids as compared to any other part of the plant. These include six flavones, two isoflavones, two flavonols, two flavanones, one isoflavanone, and one chalcone. As has been seen in a variety of other plant species, *C. cajan* exclusively possesses flavones in its leaves (Nix et al., 2015). This is likely associated with their function in photoprotection against UV radiation, which has been seen in several other plant species. These secondary metabolites are prominently present in each part of the plant, such as pinostrobin chalcone, luteolin, apigenin, orientin, vitexin, isovitexin, quercetin, cajanol, etc., in the leaves of *C. cajan*, while roots and stems are rich in isoflavone, daidzein, genistein, quercetin, and formonentin, and seeds and pods are rich in quercetin and its derivatives, cajanol, and cajanin (Fig. 1) (Gargi et al., 2022). Among a wide variety of phenolic compounds found in *C. cajan*, each of these compounds has one or more aromatic rings in addition to one or more hydroxyl groups (Prabhakar et al., 2023). According to Tungmunnithum et al. (2018), there are presently more than 8000 unique phenolic structures that have been discovered. These structures can range from very simple compounds like phenolic acids to highly polymerized substances like tannins. They are the most common kind of secondary metabolites found in plants and may be found all across the plant kingdom. The quantity of hydroxyl groups present in the molecular structure of phenolic compounds is directly related to the functionality of these compounds. According to Wink (2015), the distinct structural makeup of each subgroup is contributed to by the degree and pattern of hydroxylation, methoxylation, prenylation, or glycosylation that it possesses. Fig. 2 provides an explanation of the structural makeup of the molecules that make up distinct groups of phenolic phytocompounds.

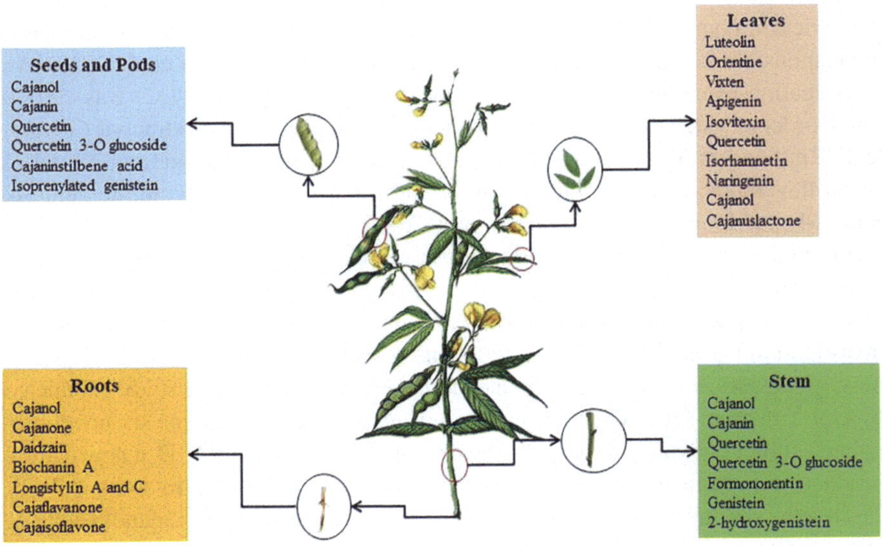

Fig. 1 Phenolic compounds found in various parts of *C. cajan* plants (Gargi et al., 2022)

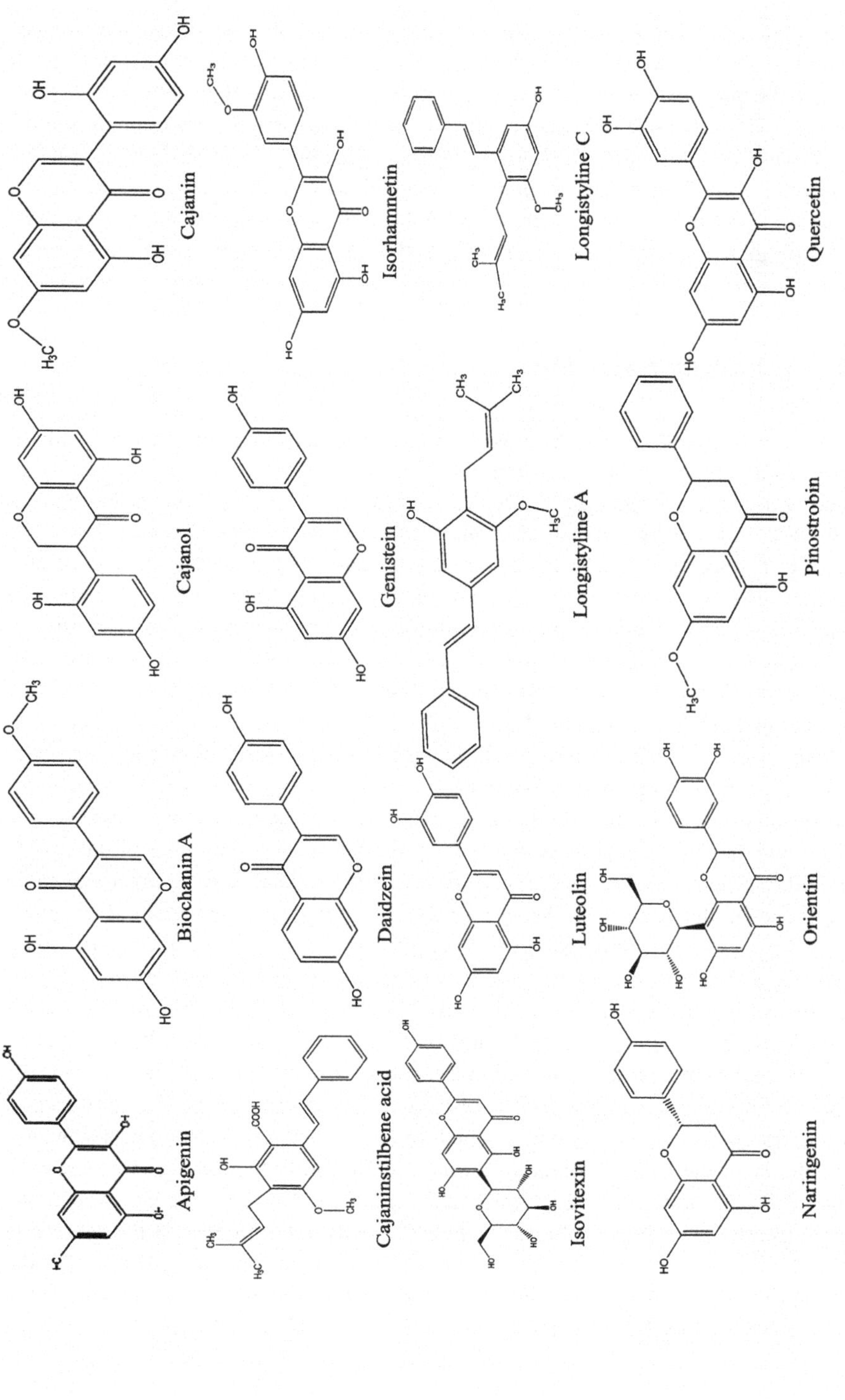

Fig. 2 Chemical structures of major secondary metabolites found in *C. cajan* plants

3 Biological and Pharmacological Activities

C. cajan demonstrated a wide variety of biological and pharmacological actions as a result of the existence of numerous classes of secondary metabolites in it, most notably the preponderance of phenolic compounds. Antioxidant, anti-plasmodial, anti-malarial, anticancer, hypoglycemic, hypocholesterolemic, insecticidal, neuroprotective, anti-inflammatory, hepatoprotective, nephroprotective, and so on are only some of the biological and pharmacological actions that are carried out by the compound. In the following sections, we will go through some of the most significant biological and pharmacological effects that these secondary metabolites have.

3.1 Antioxidant Activities

Because of its potential to reduce oxidative damage brought about by free radicals, antioxidants are one of the most popular types of additives used in food (Alvarez-Pérez et al., 2020; Banwo et al., 2021; Tripathy et al., 2021; Bautista-Hernández et al., 2022). According to Poljsak et al. (2013), reactive oxygen species (ROS) are naturally formed in the mitochondria as a byproduct of the reaction that generates energy in the cell. An excessive amount of ROS can be produced as a result of a variety of situations, such as illness, respiratory burst, ischemia, and severe exercise, to name just a few of these potential triggers. Consuming foods rich in antioxidant components is one easy approach to maintaining the equilibrium between ROS and the human body's antioxidant level (Zeb, 2020; Banwo et al., 2021; Bautista-Hernández et al., 2022). As a result of the fact that leguminous plants are one of the most important sources of antioxidants, *C. cajan* is a leguminous food with a significant potential for antioxidant activity (Fig. 3). This plant is a significant source of antioxidant phytocompounds due to the presence of a huge number of flavonoids and stilbenes in a variety of its parts. Using a DPPH radical-scavenging assay and a β-carotene-linoleic acid test, Wu et al. (2009) reported that four main compounds isolated from the ethanolic extract of *C. cajan* leaf were tested for their antioxidant properties. These compounds were (1) cajaninstilbene acid (3-hydroxy-4-prenylmethoxystilbene-2-carboxylic acid), (2) pinostrobin, (3) vitexin, and (4) orientin. With IC_{50} values of 242.01 and 404.91 μg/mL, the ethanolic extracts of *C. cajan* leaf demonstrated much higher levels of antioxidant activity in the DPPH system than the aqueous extracts did, respectively. The value of the IC_{50} for the ethyl acetate fraction was 194.98 μg/mL, which indicated that it had the highest scavenging activity of the four fractions. In terms of their ability to scavenge free radicals, cajaninstilbene acid (302.12 μg/mL) and orientin (316.21 μg/mL) proved to be more effective than pinostrobin and vitexin. It was determined that the inhibition ratio (%) of the ethyl acetate fraction in the β-carotene-linoleic acid test was the highest (94.13 ± 3.41%), and it was almost as high as the inhibitory capacity of the positive control butylated hydroxytoluene (93.89 ± 1.45%) at 4 mg/mL. In contrast

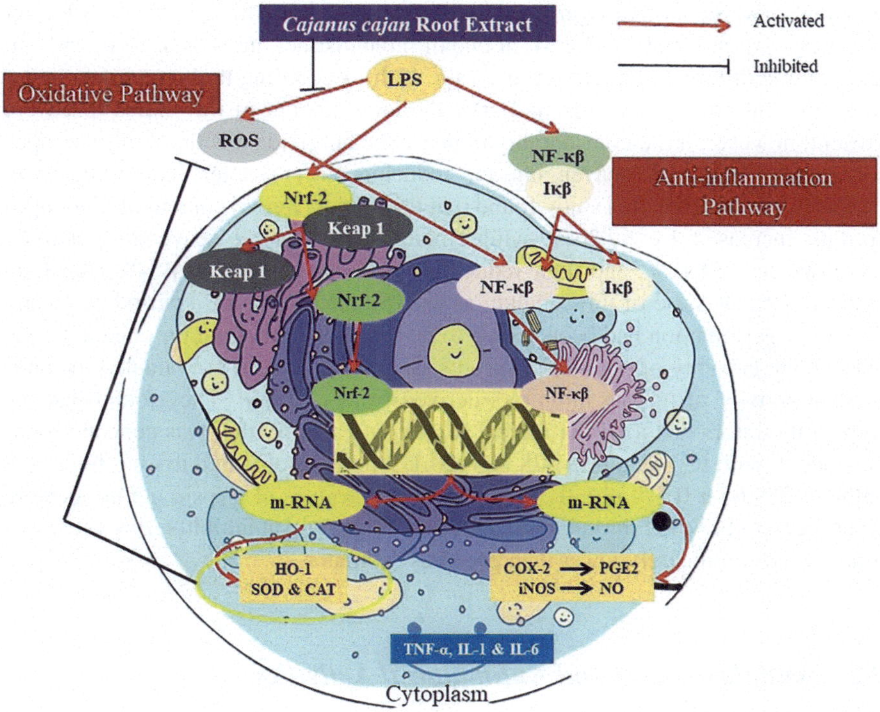

Fig. 3 Antioxidant and anti-inflammatory pathway of *C. cajan* root extract (Vo et al., 2020, 2022)

to cajaninstilbene (321.53 μg/mL) and orientin (444.61 μg/mL), the antioxidant activity of pinostrobin (>500 μg/mL) and vitexin (>500 μg/mL) was insignificant (Wu et al., 2009). In another experiment conducted by Nahar et al. (2014), the antioxidant study and quantification of the phenolic compound in the methanolic extract of *C. cajan* root revealed that they have a high antioxidant capacity as the DPPH was strongly scavenged by the methyl extract as the IC_{50} value was found to be 17.44 μg/mL and the total antioxidant capacities were found to be 826.87 ± 10.27 mg/g equivalent of ascorbic acid. According to Mahitha et al. (2015), the IC_{50} values for the DPPH assay for the aqueous extract of *C. cajan* leaf were 0.69 mg/mL, while the IC_{50} value for the ethanol extract was 0.79 mg/mL. Leaf extracts in both water and ethanol showed an increase in reducing power that was proportional to the concentration of the extract. The phosphor-molybdenum assay showed that the aqueous extract had an effective antioxidant capacity of up to 55.97 nM GAE/g, and the maximal hydroxyl radical scavenging activity reached up to 83.67% in the aqueous extract compared to 78.75% in the ethanol extract. The aqueous extract was shown to have the maximum antioxidant activity across all of the different tests that were carried out. This was found to be the case because of the presence of polyphenols, which are responsible for the antioxidant characteristics of leaf extracts (Mahitha et al., 2015). In a separate study, Kristiani et al. (2019) examined the antioxidant potential of sprouts of *C. cajan* that had been germinated for a total of 48 h after being evoked with NaCl (50 mM) at one of three distinct times (8 h, 12 h, or 16 h) and at three different elicitation times (8 h, 12 h). According to the reports, elicitation time did not significantly influence the total amount of flavonoids that were present; however, it did considerably improve the antioxidant activity of pigeonpea sprouts (DPPH radical scavenging and reducing power). After being exposed to NaCl 50 mM for 16 h, this study found that the total flavonoid content of pigeonpea sprouts increased by 69.36%, while their DPPH radical scavenging activity increased by 134.18%, and their reducing power increased by 24.54% (Kristiani et al., 2019). Because of this, elicitation with 50 mM NaCl for a period of 16 and 48 h after germination had been shown to be an efficient method for enhancing the antioxidant capacity of pigeonpea sprouts. Shamsi et al. (2018) evaluated the antioxidant activity of trypsin inhibitors derived from *C. cajan*. They found that the trypsin inhibitors had a DPPH radical scavenging activity that was concentration-dependent, with IC_{50} values of 478.285 ± 0.123 μg/mL compared to ascorbic acid's value of 259.65 ± 0.172 μg/mL. While the FRAP values of trypsin inhibitors were determined to be 0.002 mM at 1000 μg/mL of cajan trypsin inhibitor, this value was found to be significantly lower than previous estimates.

3.2 *Antiplasmodial and Antimalarial Activities*

C. cajan extracts from the roots and leaves were fractionated according to their bioactivity, and the eight compounds that were produced as a consequence were as follows: (1) betulinic acid, (2) biochanin A, (3) cajanol, (4) genistein, (5) 2′-hydroxygenistein, (6) longistylin A, (7) longistylin C, and (8) pinostrobin. The stilbenes longistylin A and C, as well as betulinic acid, were observed to exhibit a level of activity that was quite robust against a chloroquine-resistant strain of *Plasmodium falciparum* strain 3D7 (Duker-Eshun et al., 2004). In Nigerian ethnobotany, *C. cajan* has been reported as having characteristics that are effective against malaria. The ethyl acetate fraction of *C. cajan* leaves was used in the isolation of cajachalcone, which is also known as 2′,6′-dihydroxy-4-methoxy chalcone. In a parasite lactate dehydrogenase experiment using the multiresistant strain of *P. falciparum (K1)*, it was recommended to be an antimalarial drug since it showed an IC_{50} value of 2.0 μg/mL (7.4 μM) compared to its methanol fraction, which had an IC_{50} value of merely 53.5 μg/mL (Ajaiyeoba et al., 2013).

3.3 *Anticancer Activities*

Cancerous cells have proven to be particularly susceptible to the cytotoxic effects of the secondary metabolites that were identified from *C. cajan*. Roots of *C. cajan* have cajanol, an isoflavanone that is an important phytoalexin. The anticancer activities of this compound were explored and evaluated using MCF-7 human breast cancer cells by researchers (Luo et al., 2010; Fu et al., 2015; Schuster et al., 2016; Rao et al., 2018; Prabhakar et al., 2023). Other factors, such as DNA fragmentation assay, cell cycle distribution and morphological assessment of nuclear change, disruption of mitochondrial membrane potential, generation of ROS, expression of caspase-3 and caspase-9, Bcl-2, PARP, and cytochrome-C, were measured in order to determine the mechanism by which cajanol inhibits the growth of cells (Fig. 4). Cajanol has shown that it can decrease MCF-7 cell growth in a time- and dose-dependent manner, leading to apoptosis via a mitochondria-dependent approach and arresting the cell cycle in the G2/M phase (Luo et al., 2010). Cajanol has also proven that it can prevent the cell cycle from continuing past the G2/M phase. Another interesting secondary metabolite called cajaninstilbene acid was identified from the *C. cajan* plant. It has structural similarities with estrogen and exhibits anti-estrogenic and cytotoxic activities on estrogen receptor-α (Fu et al., 2015). This indicates that it may be useful in the treatment of breast cancer cells. Compounds such as orientin, pinostrobin, cajaninstilbene acid, pinosylvin, and vitexin were shown to have considerable anti-inflammatory action when tested on lipopolysaccharide-stimulated macrophages. These compounds were found to inhibit the release of pro-inflammatory cytokines, interleukin-6 (IL-6) (Schuster et al., 2016). In addition to this, these compounds demonstrated potent cytotoxic activity against three distinct cancer cell lines, namely (1) the human cervical adenocarcinoma cell line (HeLa), (2) the human colorectal adenocarcinoma cell line

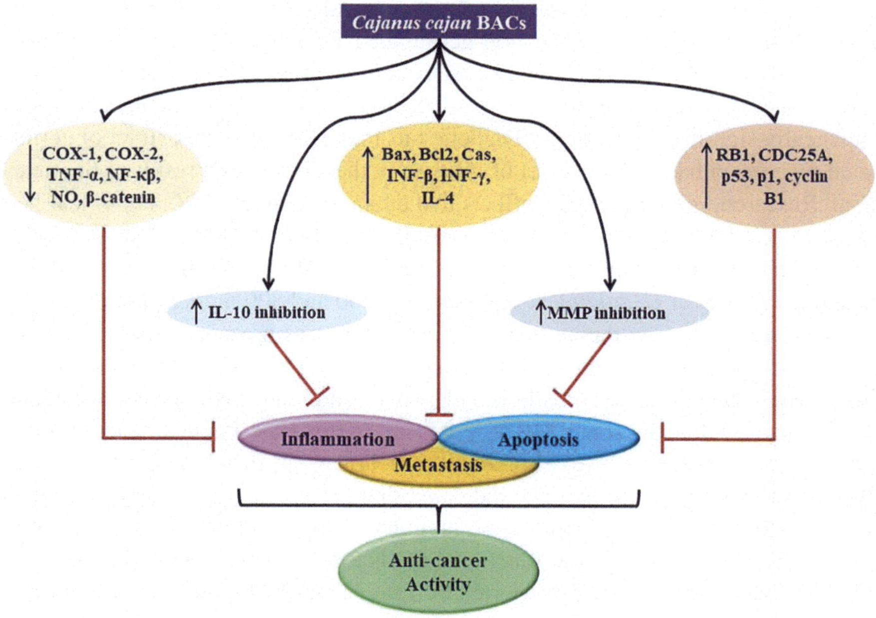

Fig. 4 Anticancer mechanism of *C. cajan* BACs (Rao et al., 2018)

(CaCo-2), and (3) the human breast adenocarcinoma cell line (MCF-7). These significant cytotoxic effects provided evidence that PPARγ activation results in anti-inflammatory actions. In a separate experiment, the ethanol extract of *C. cajan* dramatically decreased cell proliferation at a variety of doses and in a dose-dependent way. This resulted in a reduction in the release of matrix metalloproteinase-2 and vascular endothelial growth factor-2. The iNOS/COX-2/NF-B signaling pathways were the ones that were downregulated as part of the process. In addition, EECRp70 lowered the migration and invasion of SCC25 cells when it was paired with a cancer chemotherapeutic agent (cisplatin or taxol). It did this by inhibiting the generation of ROS within the cell, blocking cell proliferation, and arresting the cell cycle in the subG1 phase. EECRp70 was shown to reduce proliferation and metastatic activity, which may be related to its relatively high genistein absorption (2.71%) in SCC25 cells over the course of a 12-h period (Vo et al., 2020, 2022). The findings suggest that EECRp70 suppresses proliferation and metastatic activity.

3.4 Antidiabetic Activities

Diabetes mellitus is a metabolic illness that is often associated with irregularities in the metabolism of a number of substrates, which ultimately leads to increased levels of glucose in the blood. The most common risk factor for developing diabetes mellitus is a family history of the disease (Banwo et al., 2021; Prabhakar et al., 2023). According to the researchers, the inhibitory potential of the germinated pigeonpea extract against α-glucosidase was found to be considerably greater after the germination process compared to the non-germinated pigeonpea extract (Uchegbu & Ishiwu, 2016; Prabhakar et al., 2023). After being exposed to alloxan, the level of glucose that was present in the rats' fasting blood increased in a manner that was statistically significant ($p < 0.05$). Diabetic rats that were given an extract of germinated pigeonpea had a lower level of glucose in their blood as a result of the treatment. Researchers examined the effects that ethanolic extracts of *C. cajan* leaf had on the mice's blood glucose levels to determine whether or not they had an antihyperglycemic impact (Manzo & Vitor II, 2017). They found that mice that were fed ethanolic extracts of *C. cajan* leaf at dosages of 400 and 800 mg/kg of body weight exhibited a reduction in the amount of glucose that was present in their blood. To predict the dipeptidyl peptidase-4 inhibitory activity of papain hydrolysates, another experiment was conducted in silico and in vitro utilizing eight physicochemical characteristics (Boachie et al., 2019). As a result of this experiment, a random forest model with a reasonable level of prediction accuracy was developed. According to Boachie et al. (2019), the findings indicated that proteins derived from pigeonpea are precursors of dipeptidyl peptidase-4 inhibitors. These proteins have the potential to be used in the formulation of functional foods for the management of type 2 diabetes (Boachie et al., 2019). As reported by Yang et al. (2020), an ethanol extract of *C. cajan* was shown to inhibit carbohydrate-digesting enzymes (α-glucosidase and α-amylase), and the IC_{50} values for these enzymes were found to be 39 and 120 μg/

mL. These were much superior to those of acarbose (IC_{50} = 303 and 970 μg/mL), and the compound additionally demonstrated anti-glycation effects that were substantially equal to those of aminoguanidine (220 μg/mL). The ethanolic extract of *C. cajan* was able to effectively inhibit the synthesis of α-glucosidase, α-amylase, and advanced glycation end products (AGEPs) (IC_{50} values for each of these enzymes were 0.12, 0.32, and 0.50 mg/mL, respectively). The induction of methylglyoxal (MGO) results in a significant increase in serum levels of blood glucose, glycosylated hemoglobin, insulin resistance homeostasis model assessment, AGEPs, lipid biochemical values, and atherogenic index, whereas *C. cajan* ethanolic extract reduces these levels in a dose-dependent manner. In insulin tolerance tests, the ethanol extract of *C. cajan* can also function as an insulin sensitizer, significantly bringing down blood glucose levels after an intraperitoneal injection of insulin (by 47%, which is statistically significant at $p < 0.05$) (Yang et al., 2020). The hypoglycemic and antihyperlipidemic mechanisms of *C. cajan* ethanolic extract are likely to occur via a variety of different mechanisms (Yang et al., 2022). These mechanisms include the inhibition of carbohydrate-hydrolyzing enzymes (α-glucosidase and α-amylase) as well as the enhancement of MGO trapping effects on the inhibition of AGEPs formation (Fig. 5).

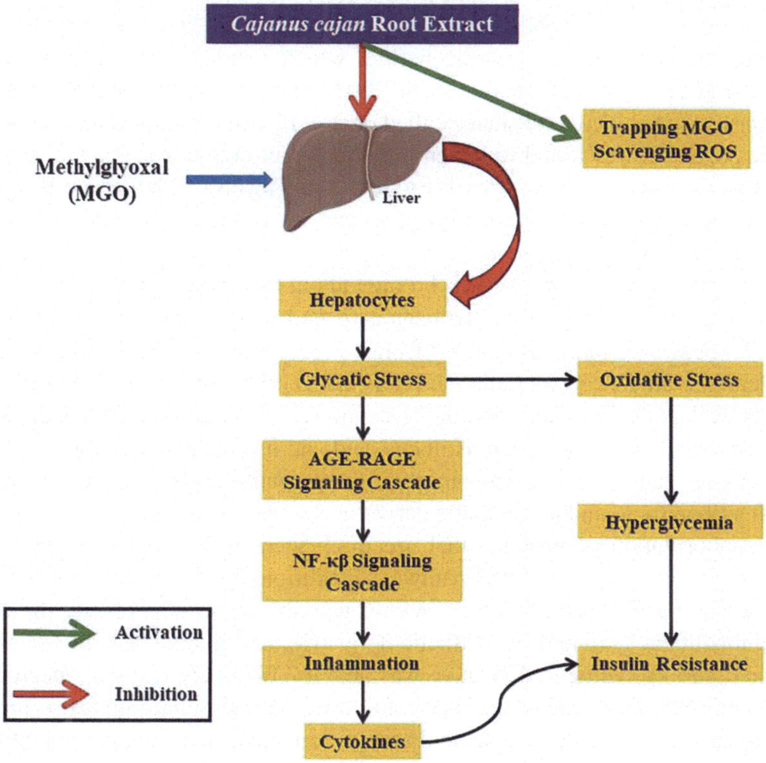

Fig. 5 Antidiabetic pathway of *C. cajan* root extract (Yang et al., 2022)

Methylglyoxal (MGO) is the most significant precursor of advanced glycation end products (AGEPs) and has received the greatest attention from researchers. The accumulation of AGEPs in vivo and the activation of various signaling pathways, including the mitogen-activated protein kinase (MAPK), nuclear factor kappa β (NF-κβ), and signal transducer and transcriptional activator pathways, are closely related to the development of a number of different chronic metabolic diseases such as diabetes. It has been found that an accumulation of AGEPs can contribute to inflammation and oxidative stress through an increase in ROS production. This is one of the primary causes of insulin resistance (IR), which is recognized as a medical condition. The activation of RAGE signaling by extracellular AGEPs is another factor that contributes to oxidative stress, inflammation, and insulin resistance. It is important to consider the inhibition of AGE production as a potentially effective treatment technique for the reduction of glycative stress-induced IR.

3.5 *Anti-inflammatory Activities*

It was shown that the cyanidin-3-monoglucoside that was extracted from the *C. cajan* plant was able to reduce the production of inflammatory cytokines, such as tumor necrosis factor-α (TNF-α), IL-1β, and IL-6, in RAW264.7 macrophages that had been treated with lipopolysaccharide, which resulted in anti-inflammatory actions (Lai et al., 2012). Hassan et al. (2016) found that the hexane extract of *C. cajan* seeds included a substance called phytol. Phytol is composed of sterols and fatty acids, and it was found to lessen the inflammation that was generated by carrageenan. The hexane extract was able to reduce the levels of TNF-α and IL-6 in the serum of rats that had inflammation produced by carrageenan. Additionally, the hexane extract of *C. cajan* seeds induced a considerable decrease in the levels of IgG serum in inflammation that was brought about by carrageenan.

In murine macrophages that had been activated with lipopolysaccharide (LPS), cajaninstilbene acid that was isolated from *C. cajan* seeds and its synthetic derivatives demonstrated a considerable inhibitory effect on the production of nitric oxide (NO) as well as the inflammatory markers tumor TNF-α and IL-6. When applied to wounded zebrafish larvae, cajaninstilbene acid and its synthetic derivatives led to a considerable reduction in the migration of neutrophils and macrophages. Cajaninstilbene acid and its synthetic derivatives were able to dramatically suppress the phosphorylation of proteins that are implicated in the NF-κβ and mitogen-activated protein kinase (MAPK) pathways. Additionally, the proliferator-activated receptor gamma (PPARγ) inhibitor GW9662 was able to partially reverse the effects of cajaninstilbene acid and its synthetic derivative, and treatment with cajaninstilbene acid and its synthetic derivative was able to effectively resist the decrease in PPARγ, mRNA, and protein that was caused by LPS stimulation. These findings demonstrated that cajaninstilbene acid and its derivatives have the potential to possess anti-inflammatory characteristics and may be helpful anti-inflammatory lead compounds (Huang et al., 2016). On the basis of a study conducted on Wistar rats, it was observed that the rat paw oedema caused by carrageenan was greatly decreased by the methanolic extract of *C. cajan* (Ganga Raju & Kumara Swamy, 2018). Anti-inflammatory properties have been found in the polyphenol content, flavonoids, and saponins of the methanolic extract of *C. cajan*, which makes sense given that inflammation has been shown in several studies to be associated with the denaturation of proteins. These compounds have the ability to form strong bonds with cations, which stops the membrane of proteins from becoming denatured. Therefore, protein membranes were significantly protected against denaturation by the presence of the extract. The mechanism of action of the anti-inflammatory activity of the ethanolic extract of *C. cajan* was described by Vo et al. (2020, 2022). This extract blocked the NF-κβ signaling pathway, which resulted in the extract having anti-inflammatory properties. As reported by Nwaechefu et al. (2022), *C. cajan* leaf extract was able to elicit a hepatoprotective action on CCl_4-induced liver injury in Wistar rats. This was accomplished by inhibiting the opening of the mitochondrial permeability transition (mPT) pore, preventing CCl_4-induced hepatic oxidative stress, and suppressing the inflammatory response in the liver. Stilbene was evaluated for its in vitro anti-inflammatory properties by Tan et al. (2022). The researchers found that stilbene had a somewhat inhibitory impact on the production of NO in lipopolysaccharide (LPS)-induced RAW264.7 cells, but it did not cause substantial cytotoxicity. It has been demonstrated in Fig. 3 that the bioactive compounds found in *C. cajan* have an anti-inflammatory effect.

3.6 Neuroprotective Activities

Epilepsy, Alzheimer's disease, Amyotrophic Lateral Sclerosis, Parkinson's disease, and major depressive disorder are all examples of neurodegenerative disorders that are often characterized by neuronal loss in the hippocampus and cerebral cortex (Lamptey et al., 2022). Other neurodegenerative diseases include Huntington's disease, Friedreich ataxia, Lewy body disease, and spinal muscular atrophy. In light of this, protecting nerve cells has emerged as a key therapeutic method for the treatment of neurodegenerative diseases (Lamptey et al., 2022). Researchers have been investigating a wide variety of naturally occurring bioactive compounds as part of their mission to find effective treatments for disorders that cause neurodegeneration. The utilization of plant extracts and the bioactive components of such plants is emerging as one of the most promising approaches to the treatment of neurological illnesses. As a result, the researchers are looking into whether or not *C. cajan* has any neuroprotective properties. Damage models produced by corticosterone (Cort) or glutamate in the rat pheochromocytoma cell line PC12 cells were used to evaluate the neuroprotective effects of the four stilbenes derived from *C. cajan* leaves (Liu et al., 2015). The results showed that the stilbenes have neuroprotective properties. Again, Liu and co-workers found that pre-treatment with longistyline C at a concentration of 8 mol/L was able to greatly diminish the overexpression of NMDAR/NR2B and Ca^{2+}/calmodulin-dependent protein kinase II (CaMKII), as well as raise the production of *p*-ERK and *p*-CREB and decrease ER stress (Liu et al., 2017). Longistyline C most likely exerts its neuroprotective impact against glutamate-induced damage in PC12 cells via altering the NMDAR/NR2B-ERK1/2-linked pathway and restoring endoplasmic reticulum function. This was an inference that can be drawn from the research of Liu et al. (2017). Further, the hexane extract of *C. cajan* stem bark, which had the lowest flavonoid content, showed the strongest anti-AChE and anti-BChE activity (Sinan et al., 2021). In contrast, the ethyl acetate extract, which had the highest flavonoid content, had the least efficacy in reducing the AChE enzyme activity (1.25 ± 0.14 mg GALAE/g) associated with Alzheimer's disease. As a consequence of this, it might contend that the level of cholinesterase activity in extracts does not always correlate with the quantity of flavonoid that is present in such extracts (Sinan et al., 2021). Research has also been done on the flavonoid vitexin, which is present in *C. cajan* extracts (Amedu & Obu, 2022). This flavonoid has proven to have neuroprotective characteristics. According to the findings of Amedu and Obu (2022), giving Wistar rats vitexin and *C. cajan* greatly reduced the amount of oxidative stress, brain damage, and microglial activation that would have been induced by the toxicity of lead. In addition, both vitexin and *C. cajan* helped rats keep a healthy working memory, which was demonstrated by the rats' improved performance. In spite of the fact that both vitexin and *C. cajan* exhibited neuroprotective characteristics, the findings obtained with vitexin were significantly more significant (Amedu & Obu, 2022).

3.7 Antimicrobial Activities

The antibacterial activity of the extract of *C. cajan* against a wide variety of bacterial and fungal species demonstrates that this extract of *C. cajan* has a broad range of potential applications. The antibacterial activity of plant extracts showed that *K. pneumoniae* and *P. mirabilis* were resistant to all of the extracts, but the organic solvent extracts of *C. cajan* were able to inhibit *E. coli*, *S. aureus*, and *S. typhi*. The aqueous extract of *C. cajan* was only able to inhibit *E. coli* and *S. aureus*. The IC_{50} values of the extract varied between 0.125 and 0.25 mg/mL for *E. coli*, 0.125 mg/mL for *S. aureus*, and 0.0325–0.0625 mg/mL for *S. typhi*, respectively. This was the case for all three strains. The cidal or static effect was affected by factors such as the concentration of the leaf extracts, the organisms, and the kind of solvent used to extract the leaf material (Okigbo & Omodamiro, 2007). According to the research that was carried out, it has been found that both methanolic and ethanolic extracts of *C. cajan* leaves have substantial antibacterial activity against gram-positive and gram-negative bacterial species, respectively. However, the ethanolic extracts of *C. cajan* leaf revealed the highest amount of inhibition against *E. coli*, with varied zones of inhibition. The methanolic extracts of *C. cajan* leaf showed the highest level of action against *S. aureus*. Table 1 provides information on the antibacterial activity of several solvent extracts against a number of bacterial and fungal species.

Table 1 The minimum inhibitory concentration of some solvent extracts of *C. cajan* demonstrating their antibacterial and antifungal properties

Plant part used	Used solvent	Bacterial/fungal species	Minimum inhibitory concentration (μg/mL)	References
Leaves	Methanol	*C. albicans*	0.25×10^3	Dinore and Farooqui (2022)
		E. coli	0.05×10^3	
Leaves	Solvent free	*B. subtilis*	1.06×10^3	Qi et al. (2014)
		P. acnes	2.12×10^3	
Leaves	Ethanol	*C. albicans*	0.256×10^3	Brito et al. (2012)
		C. krusei	0.512×10^3	
		C. tropicalis	0.256×10^3	
Roots	Ethanol	*C. albicans*	0.512×10^3	
		C. krusei	0.512×10^3	
		C. tropicalis	0.512×10^3	
Leaves	Methanol	*B. subtilis*	200×10^3	Nwachukwu and Uzoeto (2010)
		E. coli	6.25×10^3	
		S. aureus	50×10^3	
Leaves	Ethanol	*B. subtilis*	400×10^3	
		E. coli	3.13×10^3	
		S. aureus	100×10^3	
		S. typhi	12.5×10^3	
Leaves	Acetone	*B. subtilis*	400×10^3	
		E. coli	12.5×10^3	
		S. aureus	50×10^3	
		S. typhi	3.13×10^3	
Leaves	Hot water	*B. subtilis*	800×10^3	
		S. typhi	200×10^3	
		S. typhi	200×10^3	
Leaves	Petroleum ether	*B. subtilis*	6.25×10^3	Pratima and Mathad (2011)
		E. coli	12.5×10^3	
		K. pneumoniae	65×10^3	
		S. aureus	60×10^3	
		S. typhi	70×10^3	
Leaves	Chloroform	*B. subtilis*	6.25×10^3	
		S. aureus	75×10^3	
Leaves	Methanol	*B. subtilis*	6.25×10^3	
		E. coli	12.5×10^3	
		K. pneumoniae	3.125×10^3	
		S. aureus	6.25×10^3	
		S. typhi	6.25×10^3	
Leaves	Methanol	*B. subtilis*	6.25×10^3	
		E. coli	6.25×10^3	
		K. pneumoniae	70×10^3	
		P. aeruginosa	50×10^3	
		S. aureus	3.125×10^3	
		S. typhi	6.25×10^3	

(continued)

Table 1 (continued)

Plant part used	Used solvent	Bacterial/fungal species	Minimum inhibitory concentration (μg/mL)	References
Leaves	Chloroform/ methanol	*E. coli*	0.25 × 10^3	Okigbo and Omodamiro (2007)
		S. aureus	0.125 × 10^3	
		S. typhi	0.0625 × 10^3	
Leaves	Ethanol	*E. coli*	0.25 × 10^3	
		S. aureus	0125 × 10^3	
		S. typhi	0.0325 × 10^3	
Leaves	Petroleum ether	*E. coli*	0.125 × 10^3	
		S. aureus	0.125 × 10^3	
		S. typhi	0.0625 × 10^3	
Leaves	Water	*E. coli*	0.25 × 10^3	
		S. typhi	0.125 × 10^3	
Seeds	Acetone	*S. aureus*	0.032–0.064 × 10^3	Balida et al. (2022)
Seeds	Ethanol	*C. albicans*	6.25 × 10^3	Chidebelu et al. (2019)
		E. coli	25 × 10^3	
		L. ivanovii	25 × 10^3	
		S. fumigatus	12.5 × 10^3	
Seeds	Acetate	*K. pneumonia*	12.5 × 10^3	Siyabonga et al. (2016)
		S. aureus	6.25 × 10^3	
		S. typhimurium	12.5 × 10^3	

Note: Some values have been converted from mg/ml to μg/ml by the conversion factor of 1 mg/ml = 10^3 μg/ml

3.8 Insecticidal Activities

Ammonium sulfate precipitations followed by ion exchange, affinity, and gel filtration chromatography were utilized in the process of isolating a proteinase inhibitor (RgPI) from *C. cajan* (Prasad et al., 2010b). The activity of RgPI against bovine pancreatic trypsin and chymotrypsin had respective inhibition constants of 292 and 2265 nM when tested against each of these enzymes. It was effective throughout a broad pH range, ranging from 2 all the way up to 12, and it remained unaffected by temperatures as high as 80 °C. Once the compound was decreased with either dithiothreitol or 2-mercaptoethanol, the inhibitory action against trypsin and chymotrypsin was no longer present. Additionally, as a consequence of this factor, the level of activity of trypsin-like proteinases found in the larval midgut of the tobacco hornworm (*Manduca sexta*) was decreased. It was observed to inhibit the growth and development of these larvae when it was administered in the food, which provided further proof of its insecticidal function (Prasad et al., 2010b). In a different investigation, it was reported that pure Bowman-Birk proteinase inhibitors from *C. cajan* were evaluated for their capacity to eliminate lepidopteran insect pests (Prasad et al., 2010b). *C. cajan* demonstrated an impressive level of inhibitory action against the trypsin-like proteinases that had been identified in the midgut of *Achaea janata*. Proteinase inhibitors that were produced by *C. cajan* had the ability to inhibit the midgut trypsin-like proteinases of *A. janata* and *H. armigera*, as well as the midgut trypsin- and chymotrypsin-like proteinases of *S. litura*. In feeding experiments, a dose-dependent drop in body weight and survival rate of these larvae was observed, and this decrease happened with an increase in the concentration of inhibitors that were added to the diet. There was a substantial decrease in both the body weight and the survival rate of *A. janata* larvae when they were exposed to proteinase inhibitors derived from *C. cajan* (Prasad et al., 2010a). In another study, Balasubramanian et al. (2013) found that the insecticidal action of *C. cajan* urease reduced the percentage of *C. chinensis* larvae that survived. It is essential to keep in mind that the insecticidal action of plant urease can only be noticed in insects whose digestive systems are made up of enzymes that are similar to cathepsin. In contrast to the control, a 0.025% (w/w) *C. cajan* urease solution that contained artificial seed meal caused *C. chinensis* to die off entirely after 5 days. Furthermore, researchers investigated the effectiveness of mature *C. cajan* leaf extracts as larvicides against *Culex vishnui*, the mosquito that transmits Japanese encephalitis (Rawani, 2021). The crude extract of *C. cajan* leaves, as well as three separate solvent extracts of *C. cajan* leaves, were employed in a larvicidal bioassay against all instars of *Cx. vishnui* larvae. The mortality rate of the larvae reached its peak at a concentration of 0.5% crude extract following a 72-h period of exposure. One of the three solvent extracts that were examined, ethyl acetate was shown to have the most efficient larvicidal activity. The other two solvent extracts that were studied were *n*-hexane and a chloroform:methanol mixture (1:1 volume/volume). After being exposed for 72 h, 100 ppm of ethyl acetate extract caused first instar larvae to die off completely, i.e., 100% mortality was recorded. According to the chemical analysis, the key component was a flavonoid, which may be the source of the toxicity to larvae.

4 Concluding Remarks

Pigeonpea is not only beneficial to one's health from a nutritional standpoint, but it is also abundant in secondary metabolites, which research has shown to participate in a wide variety of biological and pharmacological processes. Flavonoids, which are abundant in pigeonpea and are well-known for their antioxidant properties, are one of the most significant secondary metabolites identified in pigeonpea. Flavonoids such as quercetin, kaempferol, and rutin have been found in pigeonpea in quite high concentrations in many studies. It has been shown that these chemicals can scavenge free radicals and protect against oxidative stress. Oxidative stress is a crucial factor in the development of a variety of chronic illnesses, including cancer, diabetes, and cardiovascular disorders. Alkaloids, which are present in pigeonpea, have been found to possess a variety of biological and pharmacological actions, including anti-diabetic, anti-inflammatory, and antibacterial qualities. Pigeonpea alkaloids, such as cajaninstilbene acid and cajanin, have been the subject of research and have been found to have the ability to reduce blood glucose levels and enhance insulin sensitivity. As a result, these alkaloids might be a possible therapeutic agent for the treatment of diabetes. Pigeonpea is also an excellent source of a wide variety of secondary metabolites, including flavonoids, alkaloids, tannins, and phenolic acids, in addition to flavonoids and alkaloids. It has been shown that these chemicals participate in a wide range of biological processes, including antiviral, anticancer, and anti-inflammatory activities. Pigeonpea extract has been demonstrated to limit the development of cancer cells and decrease inflammation throughout the body. Therefore, in a nutshell, pigeonpea is not only a rich source of protein and other vital nutrients, but it is also a remarkable source of secondary metabolites with a variety of biological and pharmacological actions. Pigeonpea has been shown to have a wide range of bioactive compounds. Therefore, in order to properly investigate the potential benefits of pigeonpea and the secondary metabolites and deploy them for therapeutic use, human health, and well-being, elaborate investigations are still required.

Acknowledgments The first author thanks the Ministry of Education (formerly the Ministry of Human Resource Development), Government of India, for an Institute Research Assistantship, and also acknowledges Rajendra Mishra School of Engineering Entrepreneurship, Indian Institute of Technology, Kharagpur, for their assistance in this study.

References

Agbafor, K. N., Nwaka, A. C., Dasofunjo, K., Asuk, A. A., & Ugwu, M. N. (2017). Creatinine, urea and uric acid levels in albino rats treated with leaf extract of *Cajanus cajan* (Pigeonpea). *IDOSR Journal of Applied Sciences, 2*, 1–9.

Agesa, B. L., Onyango, C. M., Kathumo, V. M., Onwonga, R. N., & Karuku, G. N. (2019). Climate change effects on crop production in Yatta sub-county: Farmer perceptions and adaptation strategies. *African Journal of Food, Agriculture, Nutrition and Development, 19*(1), 14010–14042.

Ajaiyeoba, E. O., Ogbole, O. O., Abiodun, O. O., Ashidi, J. S., Houghton, P. J., & Wright, C. W. (2013). Cajachalcone: An antimalarial compound from *Cajanus cajan* leaf extract. *Journal of Parasitology Research, 2013*, 1.

Alvarez-Pérez, O. B., Ventura-Sobrevilla, J. M., Ascacio-Valdés, J. A., Rojas, R., Verma, D. K., & Aguilar, C. N. (2020). Valorization of *Flourensia cernua* DC as source of antioxidants and antifungal bioactives. *Industrial Crops and Products, 152*, 17. https://doi.org/10.1016/j.indcrop.2020.112422

Amedu, N. O., & Obu, M. O. (2022). Neuroprotective effects of vitexin and *Cajanus cajan* extract against Pb-induced neurotoxicity in Wistar rats. *Pharmaceutical and Biomedical Research, 8*(4), 291–300.

Anadebe, V. C., Okafor, N. A., Ezeugo, J. O., Amanjide, I. J., & Ogide, B. C. (2017). GC-MS analysis of phytochemical compounds in *Cajanus cajan* leaf. *Journal of Chemical and Pharmaceutical Research, 9*(5), 360–363.

Balasubramanian, A., Durairajpandian, V., Elumalai, S., Mathivanan, N., Munirajan, A. K., & Ponnuraj, K. (2013). Structural and functional studies on urease from pigeonpea (*Cajanus cajan*). *International Journal of Biological Macromolecules, 58*, 301–309.

Balida, L. A. P., Regalado, J. T. A., Teodosio, J. J. R., Dizon, K. A. H., Sun, Z., Zhan, Z. Q., et al. (2022). Antibiotic isoflavonoids, anthraquinones, and pterocarpanoids from pigeon pea (*Cajanus cajan* L.) seeds against multidrug-resistant Staphylococcus aureus. *Metabolites, 12*(4), 279.

Banwo, K., Olojede, A. O., Adesulu-Dahunsi, A. T., Verma, D. K., Thakur, M., Tripathy, S., Singh, S., Patel, A. R., Gupta, A. K., Aguilar, C. N., & Utama, G. L. (2021). Functional importance of bioactive compounds of foods with Potential Health Benefits: A review on recent trends. *Food Bioscience, 43*, 101320. https://doi.org/10.1016/j.fbio.2021.101320

Bautista-Hernández, I., Aguilar, C. N., Martínez-Ávila, G. C., Ilina, A., Torres-León, C., Verma, D. K., & Chávez-González, M. L. (2022). Phenolic compounds and antioxidant activity of *Lippia graveolens* Kunth residual leaves fermented by two filamentous fungal strains in solid-state process. *Food and Bioproducts Processing, 136*, 24–35. https://doi.org/10.1016/j.fbp.2022.09.001

Boachie, R. T., Okoro, F. L., Imai, K., Sun, L., Elom, S. O., Nwankwo, J. O., et al. (2019). Enzymatic release of dipeptidyl peptidase-4 inhibitors (gliptins) from pigeonpea (*Cajanus cajan*) nutrient reservoir proteins: In silico and in vitro assessments. *Journal of Food Biochemistry, 43*(12), e13071.

Brito, S. A., Rodrigues, F. F., Campos, A. R., & Da Costa, J. G. (2012). Evaluation of the antifungal activity and modulation between *Cajanus cajan* (L.) Millsp. leaves and roots ethanolic extracts and conventional antifungals. *Pharmacognosy Magazine, 8*(30), 103.

Chidebelu, P., Onovo, I., & Nweze, E. I. (2019). Antimicrobial activities of *Cajanus cajan* L., Phaseolus vulgaris L. and Vigna unguiculata L. against some bacterial and fungal isolates. *Journal of Basic Pharmacology and Toxicology, 3*(1), 29–33.

Dinore, J. M., & Farooqui, M. (2022). GC-MS and LC-MS: An integrated approach towards the phytochemical evaluation of methanolic extract of Pigeonpea [*Cajanus cajan* (L.) Millsp] leaves. *Natural Product Research, 36*(8), 2177–2181.

Duker-Eshun, G., Jaroszewski, J. W., Asomaning, W. A., Oppong-Boachie, F., & Brøgger Christensen, S. (2004). Antiplasmodial constituents of *Cajanus cajan. Phytotherapy Research: An International Journal Devoted to Pharmacological and Toxicological Evaluation of Natural Product Derivatives, 18*(2), 128–130.

Fu, Y., Kadioglu, O., Wiench, B., Wei, Z., Wang, W., Luo, M., et al. (2015). Activity of the antiestrogenic cajanin stilbene acid towards breast cancer. *The Journal of Nutritional Biochemistry, 26*(11), 1273–1282.

Ganga Raju, M., & Kumara Swamy, K. (2018). Anti-inflammatory and antiradical potential of methanolic extract of *Cajanus cajan. Asian Journal of Pharmacy and Pharmacology, 4*(6), 860–864.

Gargi, B., Semwal, P., Jameel Pasha, S. B., Singh, P., Painuli, S., Thapliyal, A., & Cruz-Martins, N. (2022). Revisiting the nutritional, chemical and biological potential of *Cajanus cajan* (L.) Millsp. *Molecules, 27*(20), 6877.

Gui-Yun, W. U., Zhang, X., Xue-Ying, G. U. O., Lu-Qiong, H. U. O., Hong-Xin, L. I. U., Xiao-Ling, S. H. E. N., & Hai-Bo, T. A. N. (2019). Prenylated stilbenes and flavonoids from the leaves of *Cajanus cajan. Chinese Journal of Natural Medicines, 17*(5), 381–386.

Hassan, E. M., Matloub, A. A., Aboutabl, M. E., Ibrahim, N. A., & Mohamed, S. M. (2016). Assessment of anti-inflammatory, antinociceptive, immunomodulatory, and antioxidant activities of *Cajanus cajan* L. seeds cultivated in Egypt and its phytochemical composition. *Pharmaceutical Biology, 54*(8), 1380–1391.

Huang, M. Y., Lin, J., Lu, K., Xu, H. G., Geng, Z. Z., Sun, P. H., & Chen, W. M. (2016). Anti-inflammatory effects of cajaninstilbene acid and its derivatives. *Journal of Agricultural and Food Chemistry, 64*(14), 2893–2900.

Kaur, K., Asthir, B., & Verma, D. K. (2021). Biosynthesis, bioavailability, and metabolism of plant polyphenols: Biological activities and their potential benefits in human health. In D. K. Verma (Ed.), *Phytochemicals in food and health: Perspectives for research and technological development*. Apple Academic Press.

Kristiani, L. S., Rajendra, F. M., & Ariviani, S. (2019). Antioxidant capacity of pigeonpea (*Cajanus cajan* L.) sprouts elicited using NaCl with various elicitation time. In *IOP conference series: Materials science and engineering* (Vol. 633(1), p. 012039). IOP Publishing.

Lai, Y. S., Hsu, W. H., Huang, J. J., & Wu, S. C. (2012). Antioxidant and anti-inflammatory effects of pigeonpea (*Cajanus cajan* L.) extracts on hydrogen peroxide-and lipopolysaccharide-treated RAW264. 7 macrophages. *Food & Function, 3*(12), 1294–1301.

Lamptey, R. N., Chaulagain, B., Trivedi, R., Gothwal, A., Layek, B., & Singh, J. (2022). A review of the common neurodegenerative disorders: Current therapeutic approaches and the potential role of nanotherapeutics. *International Journal of Molecular Sciences, 23*(3), 1851.

Liu, Y. M., Shen, S. N., Xia, F. B., Chang, Q., & Liu, X. M. (2015). Neuroprotection of stilbenes from leaves of *Cajanus cajan* against oxidative damage induced by corticosterone and glutamate in differentiated PC12 cells. *Chinese Herbal Medicines, 7*(3), 238–246.

Liu, Y., Zhao, N., Li, C., Chang, Q., Liu, X., Liao, Y., & Pan, R. (2017). Longistyline C acts antidepressant in vivo and neuroprotection in vitro against glutamate-induced cytotoxicity by regulating NMDAR/NR2B-ERK pathway in PC12 cells. *PLoS One, 12*(9), e0183702.

Luo, M., Liu, X., Zu, Y., Fu, Y., Zhang, S., Yao, L., & Efferth, T. (2010). Cajanol, a novel anticancer agent from Pigeonpea [*Cajanus cajan* (L.) Millsp.] roots, induces apoptosis in human breast cancer cells through a ROS-mediated mitochondrial pathway. *Chemico-Biological Interactions, 188*(1), 151–160.

Mahitha, B., Archana, P., Ebrahimzadeh, M. H., Srikanth, K., Rajinikanth, M., & Ramaswamy, N. (2015). In vitro antioxidant and pharmacognostic studies of leaf extracts of *Cajanus cajan* (l.) millsp. *Indian Journal of Pharmaceutical Sciences, 77*(2), 170.

Manzo, J. A. M., & Vitor, R. J. S., II. (2017). Antihyperglycemic effects of *Cajanus cajan* L.(pigeonpea) ethanolic extract on the blood glucose levels of ICR mice (Mus musculus l.). *National Journal of Physiology, Pharmacy and Pharmacology, 7*(8), 860.

Nahar, L., Nasrin, F., Zahan, R., Haque, A., Haque, E., & Mosaddik, A. (2014). Comparative study of antidiabetic activity of *Cajanus cajan* and *Tamarindus indica* in alloxan-induced diabetic mice with a reference to in vitro antioxidant activity. *Pharmacognosy Research, 6*(2), 180.

Nanna, R. S., Banala, M., Pamulaparthi, A., Kurra, A., & Kagithoju, S. (2013). Evaluation of phytochemicals and fluorescent analysis of seed and leaf extracts of *Cajanus cajan* L. *International Journal of Pharmaceutical Sciences Review and Research, 22*(1), 11–18.

Nix, A., Paull, C. A., & Colgrave, M. (2015). The flavonoid profile of pigeonpea, *Cajanus cajan*: A review. *Springerplus, 4*, 1–6.

Nwachukwu, E., & Uzoeto, H. O. (2010). Antimicrobial activities of leaf of Vitex doniana and *Cajanus cajan* on some bacteria. *Research, 2*(3), 37–47.

Nwaechefu, O. O., Olaolu, T. D., Akinwunmi, I. R., Ojezele, O. O., & Olorunsogo, O. O. (2022). *Cajanus cajan* ameliorated CCl4-induced oxidative stress in Wistar rats via the combined mechanisms of anti-inflammation and mitochondrial-membrane transition pore inhibition. *Journal of Ethnopharmacology, 289*, 114920.

Oke, D. G. (2014). Proximate and phytochemical analysis of *Cajanus cajan* (Pigeonpea) leaves. *Chemical Science Transactions, 3*(3), 1172–1178.

Okigbo, R. N., & Omodamiro, O. D. (2007). Antimicrobial effect of leaf extracts of pigeonpea (*Cajanus cajan* (L.) Millsp.) on some human pathogens. *Journal of Herbs, Spices & Medicinal Plants, 12*(1–2), 117–127.

Orni, P. R., Ahmed, S. Z., Monefa, M., Khan, T., & Dash, P. R. (2018). Pharmacological and phytochemical properties of *Cajanus cajan* (L.) Huth.(Fabaceae): A review. *International Journal of Pharmaceutical Sciences and Research, 3*(2), 27–37.

Pal, D., Mishra, P., Sachan, N., & Ghosh, A. K. (2011). Biological activities and medicinal properties of *Cajanus cajan* (L) Millsp. *Journal of Advanced Pharmaceutical Technology & Research, 2*(4), 207.

Poljsak, B., Šuput, D., & Milisav, I. (2013). Achieving the balance between ROS and antioxidants: When to use the synthetic antioxidants. *Oxidative Medicine and Cellular Longevity, 2013*.

Prabhakar, P., Mukherjee, S., Kumar, A., Kumar, S., Verma, D. K., Dhara, S., et al. (2023). Optimization of microwave-assisted extraction (MAE) of key phenolic compounds from pigeon pea (*Cajanus cajan* L.), their characterization, and measurement of their anti-diabetic and cytotoxic potential. *Journal of Food Measurement and Characterization*, 1–24.

Prasad, E. R., Dutta-Gupta, A., & Padmasree, K. (2010a). Insecticidal potential of Bowman–Birk proteinase inhibitors from red gram (*Cajanus cajan*) and black gram (Vigna mungo) against lepidopteran insect pests. *Pesticide Biochemistry and Physiology, 98*(1), 80–88.

Prasad, E. R., Merzendorfer, H., Madhurarekha, C., Dutta-Gupta, A., & Padmasree, K. (2010b). Bowman−Birk Proteinase inhibitor from *Cajanus cajan* seeds: Purification, characterization, and insecticidal properties. *Journal of Agricultural and Food Chemistry, 58*(5), 2838–2847.

Pratima, H., & Mathad, P. (2011). Antibacterial activity of various leaf extract of *Cajanus cajan* L. *Bioscan, 6*(1), 111–114.

Qi, X. L., Li, T. T., Wei, Z. F., Guo, N., Luo, M., Wang, W., et al. (2014). Solvent-free microwave extraction of essential oil from pigeonpea leaves [*Cajanus cajan* (L.) Millsp.] and evaluation of its antimicrobial activity. *Industrial Crops and Products, 58*, 322–328.

Rao, S., Chinkwo, K. A., Santhakumar, A. B., & Blanchard, C. L. (2018). Inhibitory effects of pulse bioactive compounds on cancer development pathways. *Diseases, 6*(3), 72.

Rawani, A. (2021). Larvicidal efficacy of the mature leaf extract of *Cajanus cajan* against the vector of Japanese encephalitis. *International Journal of Tropical Insect Science, 41*, 1155–1161.

Rinthong, P. O., & Maneechai, S. (2018). Total phenolic content and tyrosinase inhibitory potential of extracts from *Cajanus cajan* (L.) Millsp. *Pharmacognosy Journal, 10*(6s).

Sarkar, S., Panda, S., Yadav, K. K., & Kandasamy, P. (2020). Pigeonpea (*Cajanus cajan*) an important food legume in Indian scenario—A review. *Legume Research-An International Journal, 43*(5), 601–610.

Schuster, R., Holzer, W., Doerfler, H., Weckwerth, W., Viernstein, H., Okonogi, S., & Mueller, M. (2016). *Cajanus cajan*–a source of PPARγ activators leading to anti-inflammatory and cytotoxic effects. *Food & Function, 7*(9), 3798–3806.

Shamsi, T. N., Parveen, R., Afreen, S., Azam, M., Sen, P., Sharma, Y., et al. (2018). Trypsin inhibitors from *Cajanus cajan* and *Phaseolus limensis* possess antioxidant, anti-inflammatory, and antibacterial activity. *Journal of Dietary Supplements, 15*(6), 939–950.

Silva, M. J. D., Périco, L. L., Da Silva, M. A., Rinaldo, D., Gambero, A., Santos, L. C. D., Hiruma-Lima, C. A., Vilegas, W., Verma, D. K., & Thakur, M. (2021). Phytomedicinal study on Brazilian medicinal plants for their potential in chronic diseases treatment. In D. K. Verma (Ed.), *Phytochemicals in food and health: Perspectives for research and technological development*. Apple Academic Press.

Sinan, K. I., Mahomoodally, M. F., Eyupoglu, O. E., Etienne, O. K., Sadeer, N. B., Ak, G., et al. (2021). HPLC-FRAP methodology and biological activities of different stem bark extracts of *Cajanus cajan* (L.) Millsp. *Journal of Pharmaceutical and Biomedical Analysis, 192*, 113678.

Siyabonga, S. J., Ayanda, M., Sydney, M. T., & Oluwole, S. F. (2016). Comparative evaluation of antibacterial activity of induced and non-induced *Cajanus cajan* seed extract against selected gastrointestinal tract bacteria. *African Journal of Microbiology Research, 10*(10), 319–323.

Tan, Y., Zhang, X., & Cheang, W. S. (2022). Isoflavones daidzin and daidzein inhibit lipopolysaccharide-induced inflammation in RAW264. 7 macrophages. *Chinese Medicine, 17*(1), 1–10.

Tripathy, S., Verma, D. K., Thakur, M., Chakravorty, N., Singh, S., & Srivastav, P. P. (2022). Recent trends in extraction, identification and quantification methods of *Centella asiatica* phytochemicals with potential applications in food industry and therapeutic relevance: A review. *Food Bioscience, 101864*. https://doi.org/10.1016/j.fbio.2022.101864

Tripathy, S., Verma, D. K., Thakur, M., Patel, A. R., Srivastav, P. P., Singh, S., Chávez-González, M. L., & Aguilar, C. N. (2021). Encapsulated food products as a strategy to strengthen immunity against COVID-19. *Frontiers in Nutrition, 8*, 245. https://doi.org/10.3389/fnut.2021.673174

Tungmunnithum, D., & Hano, C. (2020). Cosmetic potential of *Cajanus cajan* (L.) millsp: Botanical data, traditional uses, phytochemistry and biological activities. *Cosmetics, 7*(4), 84.

Tungmunnithum, D., Thongboonyou, A., Pholboon, A., & Yangsabai, A. (2018). Flavonoids and other phenolic compounds from medicinal plants for pharmaceutical and medical aspects: An overview. *Medicine, 5*(3), 93.

Uchegbu, N. N., & Ishiwu, C. N. (2016). Germinated Pigeonpea (*Cajanus cajan*): A novel diet for lowering oxidative stress and hyperglycemia. *Food Science & Nutrition, 4*(5), 772–777.

Varshney, R. K., Chen, W., Li, Y., Bharti, A. K., Saxena, R. K., Schlueter, J. A., et al. (2012). Draft genome sequence of pigeonpea (*Cajanus cajan*), an orphan legume crop of resource-poor farmers. *Nature Biotechnology, 30*(1), 83.

Vaz Patto, M. C., Amarowicz, R., Aryee, A. N., Boye, J. I., Chung, H. J., Martin-Cabrejas, M. A., & Domoney, C. (2015). Achievements and challenges in improving the nutritional quality of food legumes. *Critical Reviews in Plant Sciences, 34*(1–3), 105–143.

Verma, D. K., & Thakur, M. (2021). *Phytochemicals in food and health: Perspectives for research and technological development*. Apple Academic Press, . ISBN: 9781771889124. https://doi.org/10.1201/9781003082125

Verma, D. K., Thakur, M., Singh, S., Sandhu, K. S., Kaur, M., & Srivastav, P. P. (2021). The emphasis of effect of cooking and processing methods on anti-nutritional phytochemical of legumes and their significance in human health. In D. K. Verma (Ed.), *Phytochemicals in food and health: Perspectives for research and technological development*. Apple Academic Press.

Vo, T. L. T., Yang, N. C., Yang, S. E., Chen, C. L., Wu, C. H., & Song, T. Y. (2020). Effects of *Cajanus cajan* (L.) millsp. roots extracts on the antioxidant and anti-inflammatory activities. *Chinese Journal of Physiology, 63*(3), 137.

Vo, T. L. T., Yang, S. E., Huang, L. G., Li, P. H., Chen, C. L., & Song, T. Y. (2022). Anti-proliferative and anti-metastatic effects of ethanol extract from *Cajanus cajan* (L.) Millsp. roots and its sub-fractions in oral squamous cell carcinoma. *Agriculture, 12*(12), 1995.

Wink, M. (2015). Modes of action of herbal medicines and plant secondary metabolites. *Medicine, 2*(3), 251–286.

Wu, N., Fu, K., Fu, Y. J., Zu, Y. G., Chang, F. R., Chen, Y. H., et al. (2009). Antioxidant activities of extracts and main components of pigeonpea [*Cajanus cajan* (L.) Millsp.] leaves. *Molecules, 14*(3), 1032–1043.

Yang, S. E., Lin, Y. F., Liao, J. W., Chen, J. T., Chen, C. L., Chen, C. I., et al. (2022). Insulin sensitizer and antihyperlipidemic effects of *Cajanus cajan* (L.) millsp. root in methylglyoxal-induced diabetic rats. *Chinese Journal of Physiology, 65*(3), 125.

Yang, S. E., Vo, T. L. T., Chen, C. L., Yang, N. C., Chen, C. I., & Song, T. Y. (2020). Nutritional composition, bioactive compounds and functional evaluation of various parts of *Cajanus cajan* (L.) Millsp. *Agriculture, 10*(11), 558.

Zeb, A. (2020). Concept, mechanism, and applications of phenolic antioxidants in foods. *Journal of Food Biochemistry, 44*(9), e13394.

Zhang, N., Huang, R., Zhu, Y., Fu, M., Cai, J., Yang, J., et al. (2014). A new isoprenylated flavanone from *Cajanus cajan*. *Chemistry of Natural Compounds, 50*, 438–439.

Molecular Breeding: A Precise and Faster Option for Accelerated Breeding in Food Legumes

Vishal Chugh and Vigya Mishra

Abstract Legumes, a diverse family of plants encompassing important crops like beans, peas, lentils, and chickpeas, play a crucial role in global agriculture due to their nutritional value, environmental benefits, and ability to fix atmospheric nitrogen. Enhancing legume crops for yield, resilience, and nutritional quality is imperative to meet the growing demand for plant-based protein sources. In the context of improving legumes, this chapter explores the developments and applications of marker-assisted breeding (MAB). The significance of legumes in sustainable agriculture and the challenges faced by breeders in developing improved varieties are well known. Traditional breeding methods, although effective, are time-consuming and labor-intensive. Marker-assisted breeding emerges as a powerful tool that accelerates the breeding process by integrating molecular markers associated with desirable traits, enabling breeders to make informed selection decisions at early stages of plant development. The chapter further discusses the fundamental principles of marker-assisted breeding, elucidating the types of molecular markers used, such as Single Nucleotide Polymorphisms (SNPs), Simple Sequence Repeats (SSRs), and Insertion/Deletion (InDel) markers. Case studies have been illustrated involving the exploitation of MAB to enhance disease resistance, abiotic stress tolerance, and nutritional content. However, the full realization of MAB's potential hinges on accessible genomic information, marker validation, and interdisciplinary collaboration among geneticists, molecular biologists, and breeders. Marker-assisted breeding stands as a promising avenue for revolutionizing legume improvement, enabling breeders to develop high-yielding, resilient, and nutritious varieties tailored to meet the demands of a changing world.

V. Chugh (✉)
Department of Basic and Social Sciences, College of Horticulture, Banda University of Agriculture and Technology, Banda, Uttar Pradesh, India

V. Mishra
Department of Post Harvest Technology, College of Horticulture, Banda University of Agriculture and Technology, Banda, Uttar Pradesh, India

A. Pratap et al. (eds.), *Integrated Improvement of Food Legumes*,
https://doi.org/10.1007/978-3-031-82629-0_3

Keywords Next-generation sequencing (NGS) · Marker-assisted breeding (MAB) · Molecular markers · Breeding techniques · QTL mapping · Gene cloning · Stress-responsive genes

1 Introduction

Legumes are the third-largest family of flowering plants and offer significant amounts of food, fodder, oil, and fiber to humans (Varshney et al., 2019). The ability of the majority of species in the legume family to work with nitrogen-fixing bacteria, known as "rhizobia" to convert atmospheric nitrogen to ammonia, is its most notable characteristic. Legumes are essential parts of both natural and agricultural ecosystems because of the readily available reduced nitrogen that results, which also explains why they often have a high protein content and are therefore valuable as grain and fodder. Legumes are significant providers of secondary metabolites that promote health, as well as micro- and macronutrients from minerals. Interestingly, many of these compounds also defend plants from a variety of infections and pests (Grusak, 2002; Deavours & Dixon, 2005; He & Dixon, 2000). The highest protein content is found in soybean (33–45%) > common bean (21–39%) > wing bean (30–37%) > cowpea (21–35%) > groundnut (24–34%) > pea (21–33%) > moth bean and urd bean (21–31%) > lentil (20–31%) > grass pea (23–30%) > chickpea (15–30%) > horse gram (19–29%) > rice bean (18–27%) (Pandey et al., 2016). These crops are receiving more attention because of their increased protein and nutrient content in light of the rising percentage of malnourished people, especially in the developing globe (Bohra et al., 2015). Abiotic and biotic constraints prevent the intended level of legume production from being increased. Additionally, poor conditions will become worse in the upcoming years (due to climate change), which may put protein crops like legumes at risk. For the past 20 years, conventional breeding techniques have concentrated on the issues relating to the low yield of legumes. Programs to improve legume crops should focus on quality-related traits (such as protein; Gnanasambandam et al., 2012) and the absence of anti-nutritional elements (such as tannins; Mikic et al., 2009). However, there has been limited success in genetically enhancing these crops using traditional breeding techniques. The current food production methods are under immense pressure due to the growing problem of feeding the expanding global human population. To maintain the grain yields of these crops despite repeated biotic and abiotic stress, breeding efforts for legume crops must be more effective. Modern genomic methods and technologies could significantly help achieve this goal. These crops have access to extensive genomic resources, including genome assemblies, genome-wide molecular markers, and a range of genes and QTLs affecting key agricultural properties (Bohra et al., 2019; 2020; Varshney et al., 2019). Gene discovery and breeding techniques have undergone a paradigm shift as a result of the development of sequencing tools. The use of marker-assisted breeding endeavors in improving legume crops with regard to biotic and abiotic stressors and nutritional enrichment are highlighted in this chapter.

2 Evolution of Molecular Marker Technology and Marker-Assisted Breeding (MAB)

Assays used to identify specific genome sequence differences between two or more individuals of an organism at a particular locus are referred to as DNA-based markers, also known as molecular markers (Langridge & Chalmers, 2004). Agriculture has undergone a revolution, thanks to the very advanced tools and methods that molecular markers have made possible. From low-throughput restriction fragment length polymorphisms (RFLPs) in the 1980s to high-throughput array-based markers in the 2000s to ultrahigh-throughput next-generation sequencing-based marker systems in the 2010s, the development and application of molecular marker technology has undergone continuous evolution (Mir & Varshney, 2013; Varshney, 2016). The advantages and disadvantages of some of the most widely used molecular markers are represented diagrammatically in Fig. 1. Legumes are classified as an orphan crop since the development of marker technology in them is comparatively slower than in cereals (Gresta et al., 2016). First-generation markers such as restriction fragment length polymorphism (RFLP), amplified fragment length polymorphism (AFLP), and random amplification of polymorphic DNA (RAPD) are being developed gradually. These markers are primarily used in the study of diversity in winged bean, pigeon pea, cluster bean, cowpea, and dolichos bean (Chapman, 2015; Kumar et al., 2020). Sequence-oriented markers, specifically single nucleotide polymorphism (SNP) and simple sequence repeat (SSR) markers, and their alternation to increase reproducibility and reliability, have also been documented to work effectively in trait mapping, fine mapping, and linkage mapping studies (Choi, 2018). SSR markers, or simple sequence repeats, have been employed to measure genetic variation. While inter-species SSRs have been used in lablab for the

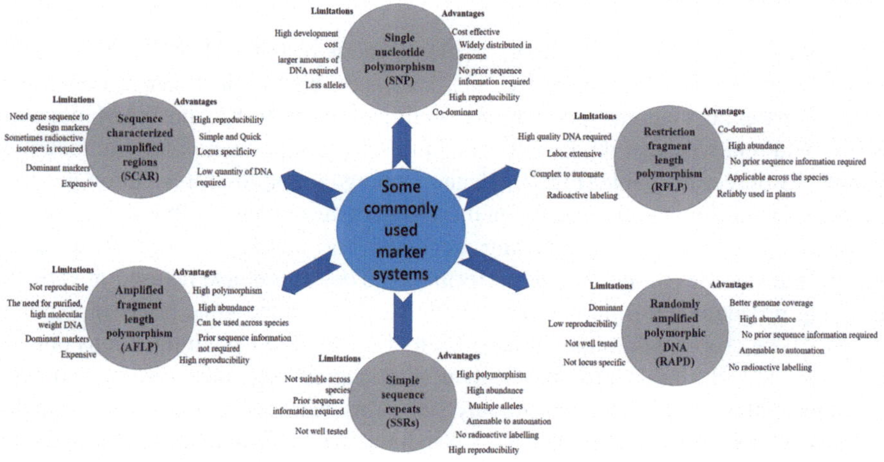

Fig. 1 Advantages and limitations of some of the commonly used molecular markers

assessment of diversity and other objectives, SSR markers have been used in cluster bean, pigeon pea, cowpea, and winged bean. Numerous research on dolichos beans have also evaluated the genetic diversity using SSR markers (Keerthi et al., 2018; Singh et al., 2021). Due to their widespread presence and abundance in the genetic material, SNPs have become a popular choice in markers. Cluster beans, cowpeas, winged beans, pigeon peas, and dolichos beans have all undergone genetic material-assisted breeding that has proven the efficacy of SNPs. As the desired markers are found, it gives us a mechanism to map problematic features for cloning as well as fine-tune the mapping (Bohra et al., 2019; Nawaz & Chung, 2020).

Contemporary marker systems are high/ultrahigh-throughput marker systems based on low-cost and high-throughput sequencing technologies, also known as next-generation sequencing (NGS) technologies, such as RAD-tag sequencing and genotyping-by-sequencing (GBS). Various applications of DNA markers are used in breeding. However, because of the method's exact qualities, as well as the fact that it is less time-consuming and gives maximal benefits (Afzal et al., 2020), the use of marker-assisted breeding for cultivar creation has grown in popularity. Furthermore, basic qualities are far more important when linked to complex or many genes. MAB can be used in conjunction with conventional field breeding to speed the selection of desirable traits and boost selection efficacy (Collard & Mackill, 2007; Ragot & Lee, 2007). MAB, based on a certain marker system's genotype, can be utilized to improve the efficiency and effectiveness of marker selection over traditional breeding (Collard & Mackill, 2007). These genes or alleles can be used as diagnostic tools by breeders to discover desirable genes or gene of interest.

3 Genetic Resources and Application of Marker-Assisted Breeding for Important Traits

To improve legume crops, a plethora of plant genetic resources has been developed and is available. Germplasm collections, landraces, core collections, donor genotypes, biparental mapping populations, multiparent, MAGIC and NAM populations, wild relatives, and wild x cultivated introgression populations are among the genetic resources available (Fig. 2). Gene mapping/tagging/high-resolution genetic mapping/gene cloning rely heavily on these genetic resources. Furthermore, genetic resources would be good breeding materials for developing improved next-generation varieties based on trait performance. They could also be used in various crossing procedures to improve the genetic base of farmed gene pools in legume crops. These crops' wild relatives have been discovered to have tremendous potential to produce desirable attributes such as abiotic stress tolerance, disease and pest resistance, nutritional traits, cleistogamy, photo-insensitivity, and cytoplasmic male sterility (CMS) (Pazhamala et al., 2015; Varshney, 2016). Trait mapping for marker-assisted breeding programs is one of the most significant goals in the development of genetic and genomic resources in legume crops. Information on crop genetic

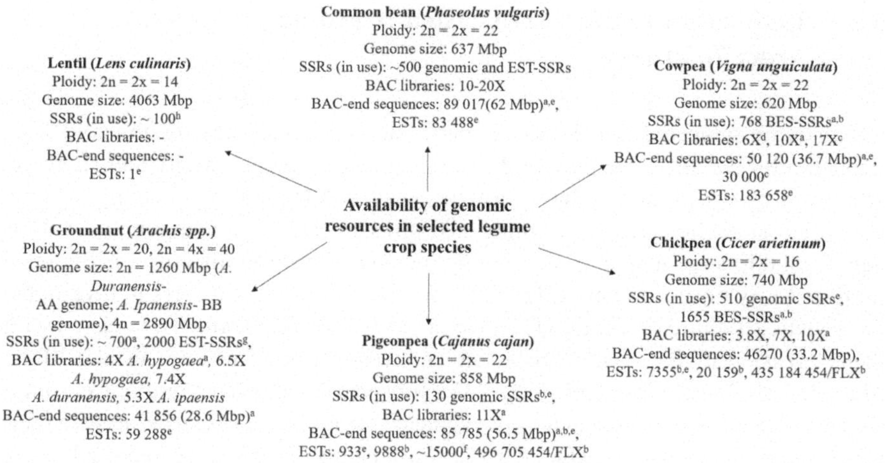

Fig. 2 Genomic resources in selected legume crop species. Annotations: [a] Source of information: UC Davis; [b] Source of information: ICRISAT; [c] Source of information: UC Riverside; [d] Source of information: University of Virginia; [e] Source of information: In public domain (e.g., NCBI); [f] Source of information: NRCPB, India; [g] Source of information: University of Georgia; [h] Source of information: USDA-ARS/Washington State University. (Varshney et al., 2019)

resources, which comprises pulse crops maintained in national and international gene banks, is accessible online at global portals such as Genesys (https://www.genesys-pgr.org/) and GRIN (https://www.ars-grin.gov/), as well as in a 1996 FAO publication titled "The second report on the state of the world's plant genetic resources for food and agriculture" (https://www.fao.org/3/i1500e/i1500e00.htm). In accordance to the FAO, 33,359 chickpea accessions, 35,891 common bean accessions, 13,289 pigeonpea accessions, 15,588 cowpea accessions, 10,864 lentil accessions, 6129 field pea accessions, 9186 faba bean accessions, and 3225 grass pea accessions have been deposited in Consortium of International Agricultural Research Centres.

Because of several constraints, such as the time-consuming and costly nature of traditional approaches to trait mapping, NGS-based approaches to trait mapping are currently recommended. Recent developments in NGS tools and technologies, as well as a greater decline in sequencing costs, enable trait mapping at the sequence level and plant selection at the nucleotide level. Recently, a large number of sequencing-based methodologies for trait mapping have been proposed and used, including trait mapping through sequencing a whole population and trait mapping through sequencing pooled samples (Pandey et al., 2017). Approaches based on sequenced trait mapping have also been begun for the identification of candidate genes/genomic areas for rust and LLS resistance in groundnuts (Pandey et al., 2017; Varshney, 2016). An exome sequencing technique was also used to sequence 250 groundnut lines in order to better understand the role of potential genes for targeted attributes (Pandey et al., 2017).

3.1 Application of Marker-Assisted Breeding for Abiotic Stress Tolerance

Legumes are subjected to a variety of abiotic stresses, including drought, cold, high temperatures, salinity, alkalinity, UV radiation, and mineral toxicity and shortage. Abiotic stressors frequently have an impact on crop legume productivity. Abiotic stress resistance heredity is a complicated process, and QTL mapping, as well as genetic and linkage mapping of genomic regions relevant to tolerance, is the most preferred way of discovering QTL. QTL analysis allows researchers to investigate the genetic structure of a trait. QTLs can identify genomic regions associated with the expression of the characteristic under investigation (Kushwah et al., 2021). Markers are effective in determining QTLs for specific traits from multiple chromosomal locations, identifying genotypes against various biotic stressors, and are multifunctional. Previously, QTLs in numerous legume crops were reported using linked markers (Luo et al., 2021; Zate, 2021). A genetic dissection was performed in the chickpea for complicated quantitative features such as drought/drought related traits and yield under drought (Varshney et al., 2014b). The efforts also resulted in the discovery of a QTL hotspot region on linkage group-4, which contains 12 significant QTLs for drought tolerance-related phenotypes and can explain up to 58.20% of phenotypic variation. A GBS technique was later used to fine-map the region (Jaganathan et al., 2015). Additionally fine mapping of this region was accomplished using two complementary approaches: a sliding window-based bin mapping approach and a genome-wide association study-based gene enrichment analysis of the RIL population's skim sequenced data (Kale et al., 2015). In the chickpea, whole genome scanning and candidate gene sequencing techniques were used to uncover QTLs/genes for drought- and heat-responsive characteristics (Thudi et al., 2014). Genes for determinacy have also been found using candidate gene sequencing and whole genome scanning (Mir et al., 2013). Stress-responsive genes have also been found that provide tolerance to salinity, drought, cold, and severe temperatures (Deeplanaik et al., 2013; Priyanka et al., 2010). Molecular mapping of genes/QTLs in groundnut has been performed effectively for a variety of characteristics. QTLs/genes for several abiotic and biotic stresses and quality attributes have been identified using different segregating biparental mapping populations/diverse germplasms (Varshney et al., 2013). Several QTLs (both main-effect and epistatic) for drought tolerance and related traits were discovered among the abiotic stresses (Ravi et al., 2010; Gautami et al., 2012). Table 1 highlights various QTL-based studies on abiotic stress resistance in legumes.

3.2 Application of Marker-Assisted Breeding for Biotic Stress Tolerance

A variety of biotic diseases, particularly those caused by fungus and nematodes, are threatening the legume crop. The fungal diseases such as charcoal rot (*Macrophomina phaseolina*), ascochyta blight (*Ascochyta rabiei*), alternaria blight (*Alternaria*

Table 1 QTLs identified against abiotic stresses for various legumes crops

Legume crop	Target abiotic stress	Linked markers	Population	Parents	References
Mung bean	Drought	SSRs	RILs	ZL × VC2917	Liu et al. (2017)
Faba bean	Cold	SNPs	RILs	Bean Gottingen Winter and Bean Pure Line 4628	Sallam et al. (2016)
`Soybean	Drought	SNPs	RILs	Magellan × PI 567,731, PI 567,690 × Pana	Ye et al. (2019)
Cowpea	Heat	SNPs	RILs	IT82E-18 × CB27	Lucas et al. (2013)
Cowpea	Drought	AFLPs	RILs	CB46 × IT93K503–1	Muchero et al. (2009)
Cowpea	Salinity	SSRs and RFLPs	$F_{2:5}$	Tokyo × S-100	Lee et al. (2004)
Chickpea	Drought	SSRs	RILs	ILC 3279 × ILC 588	Rehman et al. (2011)
Chickpea	Heat	SNPs	RILs	ICC 15,614 × ICC 4567	Paul et al. (2018)
Chickpea	Cold	SNPs	RILs	PI 489,777 and ICC 4958	Mugabe et al. (2019)
Chickpea	Drought	SSRs	RILs	*C. arietinum x C. reticulatum*	Kushwah et al., 2022
Lentil	Drought	SNPs and AFLPs	RILs	ILL 5888 × ILL 6002	Idrissi et al. (2016)
Bambara groundnut	Drought	DArTseq-based SNP markers	F_4	S19–3 × DodR	Gao et al. (2023)
Horse gram	Drought	SSRs	F_8 RILs	HPKM249 × HPK4	Katoch and Chahota (2021)
Soybean	Flooding	SSRs	RILs	Peking ×Tamahomare	Sayama et al. (2009)
Soybean	Low temperature	SSRs	BC_2F_3	Hongfeng × Harosoy	Zhang et al. (2012)
Common bean	Drought	AFLPs, RAPDs, SSRs	RILs	BAT 881 × G21212	Díaz et al. (2018)

alternate), powdery mildew (*Erysiphcpoiygoni*), cercospora leaf spot (*Cercospora canescens*), botrytis gray mold (*Botrytis cinerea*), collar rot (*Sclerotium rolfsii*), stemphylium blight (*Stemphylium* sp), phytophthora blight (*Phytophthora drechsleri*), fusarium wilt (*Fusarium oxysporum* f. sp. *ciceris*), vascular wilt (*F. oxysporum* f. sp. *lentis*), and wet root rot (*Rhizoctonia solani*) are contributed to major yield loss in legumes crops as compared to all others (Sampaio et al., 2020; Mahmoud, 2021; Pandey & Basandrai, 2020; Urva, 2021). Nematodes (Cyst and root-knot) are also a significant contributor to yield loss in legumes (Zwart et al., 2019). Using biparental mapping populations, gene mapping was also conducted for the two most important diseases in the chickpea, *Fusarium* wilt (FW) and *Ascochyta* blight (AB) (Sabbavarapu et al., 2013; Varshney, 2016). Similarly, QTLs for botrytis gray mold have been found (Anuradha et al., 2011). The two principal biotic stresses limiting output and productivity in the pigeonpea are FW and sterility mosaic disease (SMD). As a result, attempts have been made to identify genes/QTLs for these two critical disorders using various techniques and genetic resources (Pazhamala et al., 2015; Varshney, 2016). Transcriptomic approaches have also identified the gene for FW and SMD in the pigeonpea by differential gene expression (Raju et al., 2010; Dubey et al., 2011). QTLs for late leaf spot (LLS), leaf rust, early leaf spot, and the tomato spotted wilt virus have been found in groundnut using biparental mapping populations (Khedikar et al., 2010; Sujay et al., 2012; Wang et al., 2013). By introducing the resistance genes *Rps2* (*Phytophthora* rot resistance) and *Rmd-c* (powdery mildew resistance), together with a gene (*rj2*) associated with nodulation, Ramalingam et al. (2020) enhanced the host-plant resistance to *Phytophthora* rot and powdery mildew in soybean. Through the marker-assisted backcrossing (MABC) program, three soybean mosaic virus resistance genes—specifically, *RSC4, RSC8*, and *RSC14Q*—have been pyramided in soybean (Wang et al., 2017). Similarly, using the MABC technique, Obala et al. (2012) pyramided fusarium root rot resistance genes in the common bean variety. Through the use of MAS, Nzungize et al. (2011) introduced the pythium root rot resistance gene into 'Rwandan' susceptible common bean cultivars. Utilizing MABC, three superior groundnut types resistant to the rust disease have been created (Varshney et al., 2014a). Through a similar technique, Chu et al. (2011) pyramided *Rma, ahFAD2A*, and *ahFAD2B* genes for nematode resistance, together with high oleic acid, in a groundnut variety called "Tifguard." Three groundnut cultivars, TAG 24, ICGV 91114, and JL 24, have had foliar fungal disease resistance genes introduced by Janila et al. (2016). The introgressed cultivars showed disease resistance along with mean pod and haulm yield increases of 39–79%. Similar to this, using MABC techniques in which background selection was carried out using a groundnut SNP array, the resistance genes for foliar disease resistance and high oleic acid have been introgressed into three well-known Indian groundnut varieties, including GJG 9, GG 20, and GJGHPS 1 (Shasidhar et al., 2020). Table 2 enlists some of the marker-assisted applications for achieving biotic stress resistance in legumes.

3.3 Application of Marker-Assisted Breeding for Nutritional Improvement

The growing population continues to pose a challenge to the quantity and quality of food available globally. Parallel to this, a lack of micronutrients in the human diet causes malnutrition, which in turn makes a number of health issues known as "hidden hunger" more prevalent in developing nations all over the world. Because they are an inexpensive substitute for animal protein, grain legumes are widely known

Table 2 Marker-assisted applications for achieving biotic stress resistance in legumes

Legume crop	Target	QTL/gene	Markers employed	References
Chickpea	Fusarioum wilt race 2 resistance	*Foc 2*	SSR	Pratap et al., 2017
Chickpea	Fusarium wilt resistance	*foc-1, foc-3, foc-4, foc-5*	SSR	Bharadwaj et al., 2022
Black gram	Yellow mosaic virus resistance	Monogenic	STS-RGA	Basak et al. (2004)
Chickpea	Ascochyta blight resistance	QTL	RAPDs, SSR	Cobos et al. (2005); Iruela et al. (2007)
Chickpea	Fusarium wilt resistance	*foc-0, foc-3, foc-4, foc-5*	RAPDs, SSRs, STS	Cobos et al. (2005); Sharma et al. (2004); Cho et al. (2004); Iruela et al. (2007)
Common bean	Common bacterial blight	QTL	RAPD, SCAR, STS, SSR, RFLP	Jung et al. (1997); Yu et al. (2000); Tar'an et al. (2001); Kelly et al. (2003); Yu et al. (2004); Liu et al. (2008)
Common bean	Bean common mosaic virus	*bc-1, bc-2, bc-3*	RAPD, SCAR	Miklas et al. (2006)
Common bean	White mold resistance	QTL	AFLP, SCAR	Kolkman and Kelly (2003); Ender et al. (2007)
Cowpea	Aphid resistance	*Rac1*	RFLP	Myers et al. (1996)
Faba bean	Ascochyta blight resistance	*Af-1, Af-2*	RAPD	Román et al. (2003)
Faba bean	Rust resistance	*Uvf-1*	RAPD	Avila et al. (2003)
Black gram	Mung bean yellow mosaic India virus	QTL	KASP markers	Dhaliwal et al. (2022)
Black gram	Mung bean yellow mosaic virus	QTL	SSRs	Subramaniyan et al. (2022)
Mung bean	Cercospora leaf spot disease resistance	QTL	SSRs	Yundaeng et al. (2021)
Pigeonpea	Sterility mosaic disease	QTL	SSRs	Gnanesh et al. (2011)
Pigeonpea	Fusarium wilt resistance	QTL	SNPs	Saxena et al. (2017)
Mung bean	Powdery mildew resistance	QTL	AFLP, RFLP	Chaitieng et al. (2002); Humphry et al. (2003)
Lentil	Ascochyta blight resistance	QTL	RAPD	Ford et al. (1999)
Lentil	Stemphylium blight resistance	QTLs	SSRs, SRAPs, RAPDs	Saha et al. (2010a)
Lentil	Rust resistance	*R*	STS, SSRs, RFPLs, RAPDs, CAPS, dCAPS	Saha et al. (2010b)
Pea	Ascochyta blight resistance	QTL	SSR, RFLP	Dirlewanger et al. (1994); Tar'an et al. (2003)

for their high protein and nutrient content. Food legumes such as chickpea (*Cicer arietinum* L.), common bean (*Phaseolus vulgaris* L.), lentil (*Lens culinaris*), mung bean (*Vigna radiata* L.), pea (*Pisum sativum*), and pigeon pea (*Cajanus cajan* [L.] Millsp.) are rich reserves of dietary micronutrients and essential amino acids and thus might serve a pivotal role in confronting the rising severity of malnutrition-related problems. Optimizing nutrient bioavailability requires an understanding of the mineral acquisition, transport, and accumulation processes in legume seeds (Roorkiwal et al., 2021). The majority of the genes that control each of these activities are yet unknown. Numerous investigations have been made to identify the genes responsible for the uptake and mobilization of Zn (Palmgren et al., 2008; Urwat et al., 2021) and Fe (Palmer & Guerinot, 2009; Xiong et al., 2012; Sen Gupta et al., 2017) from various plant sections to the seeds.

One economically possible solution to the hunger issue is to increase grain micronutrient concentrations through plant breeding. Plant breeders, nutritionists, and agronomists all need to work together in order to biofortify crops. We could considerably benefit from doing a thorough analysis of the world's legume germplasm, encompassing both domesticated and wild species. Similarly, expanding the genetic diversity of legumes may help us identify the many QTLs and genomic regions that regulate the amount of micronutrients through biparental QTL mapping and genome-wide association studies (GWAS). The potential gene(s) underpinning the regulation of different micronutrient production and accumulation are now better understood thanks to the tremendous advances in functional genomics. Phytic acids (phytate or PA), oxalates, and tannins also chelate the minerals and reduce their bioavailability by limiting their absorption in our dietary system, which leads to a number of health issues even though grain legumes are a rich source of minerals.

According to reports, biofortification could be used to supplement grain legumes with micronutrients in an effort to reduce the problem of food and nutritional insecurity faced by the world's expanding population (Bohra et al., 2015). When consumed, antinutritional substances, including phytates, raffinose (RFO), oxalates, tannin, etc., have negative impacts on human health. Along with improving the bioavailability of micronutrients, reducing these factors or limiting their buildup provides an opportunity to increase the intake of legumes in the diet. For the purpose of creating biofortified grain legumes, integrated breeding techniques are frequently utilized to take advantage of the genetic variability for micronutrients that is currently available. These "omic"technologies include genomics, transcriptomics, ionomics, and metabolomics. The consumption of these by the human population will eventually result in a reduction in malnutrition because legumes, particularly pulses and oilseeds, are such an essential component of the human diet because of their nutritional composition. As a result, it is crucial to increase the quality of legumes since producing high-quality pulses and oilseeds will have a direct impact on people's health and way of life. Numerous quality characteristics, such as micronutrient content (iron, zinc, and calcium), protein content, bioactive chemicals, oil content, low anti-nutritional component content, etc., have not yet been well investigated. Target-specific genetic indicators for quality attributes have already been identified, and additional research into other genes is ongoing (Ojiewo et al., 2019).

In order to transmit the gene to the elite genotype and produce a biofortified crop variety, conventional breeding methods call for the screening of a genetically varied panel and the identification of possible donor parents fortified with the micronutrient(s) of our interest (Guindon et al., 2021). Although plant breeding techniques offer the most dependable, long-lasting, and economical methods for biofortification, the process of creating a bio-fortified product takes a lot of time. Utilizing molecular markers could speed up the process by rapidly identifying the superior donor who carries the desired gene. Successful biofortification requires genetic diversity for micronutrient concentration in the gene pool. The discovered potential genotypes can be used in biofortification projects for micronutrient breeding material development and mapping research. Marker-assisted selection can benefit from molecular markers associated with micronutrient content. Given that there are numerous quantitative trait loci (QTLs) controlling this attribute, genetic analysis of grain micronutrient content is still difficult. Unravelling the genetic control of significant features, such as grain micronutrient content, was made possible during the previous 10 years thanks to a paradigm change in the development of genomic resources in diverse legumes. Table 3 provides some instances of successful marker-assisted methods for improving the nutritional quality of legume crops.

Table 3 Marker-assisted applications for achieving nutritional improvement in legumes

Crops	Fortified elements/anti-nutritional compounds reduced	Marker employed	References
Chickpea	Fe and Zn	SNPs	Upadhyaya et al. (2016)
	Zn, Fe	SNPs	Diapari et al. (2014)
Common bean	Zinc and Fe content	SSR and RAPD Meta-QTL analysis	Blair et al. (2010); Izquierdo et al. (2018)
Soybean	Ca, Zn, P	QTL mapping	Ning et al. (2015)
	Tocopherol content	SNPs	Park et al. (2019)
Mung bean	Zn and Fe	SSR	Singh et al. (2017)
	S, P, K Mg	SNPs	Wu et al. (2020)
Faba bean	Tannin reduction, Vicine-covicine reduction	CAP	Gutierrez et al. (2006)
Pea	B, Ca, Mg, Mn, Mb, P, S	SNPs	Ma et al. (2017)
Lentil	Se,	SNPs	Ates et al. (2016)
	Mg	DArT SNPs	Ates et al. (2018)
	Zn, Fe	SNPs	Khazaei et al. (2017)

4 Future Prospects

Marker-assisted breeding holds immense promise for the future enhancement of legume crops, paving the way for substantial improvements in yield, quality, and resilience. Legumes, encompassing a diverse range of species such as soybean, chickpea, lentil, and peas are vital components of global agriculture due to their nutritional value, soil-enriching properties, and ability to fix atmospheric nitrogen. However, they often face challenges such as susceptibility to pests, diseases, and abiotic stresses, as well as limitations in yield and nutritional content. Marker-assisted breeding techniques offer innovative solutions to address these issues and unlock the full potential of legume crops. One of the key avenues for future progress in marker-assisted breeding is the utilization of advanced genomics tools. High-throughput sequencing technologies enable the rapid decoding of legume genomes, providing crucial insights into the genetic basis of desirable traits. This knowledge facilitates the identification of specific genes associated with traits like disease resistance, drought tolerance, and increased nutrient content. By employing techniques such as genome-wide association studies (GWAS) and quantitative trait locus (QTL) mapping, researchers can pinpoint these genes and accelerate the breeding process through marker-assisted selection (MAS). This not only expedites the development of improved legume varieties but also enables breeders to stack multiple beneficial traits into a single cultivar. Furthermore, the emergence of gene editing technologies, such as CRISPR-Cas9, revolutionizes the precision and speed of marker-assisted breeding. With CRISPR-Cas9, researchers can precisely modify or edit specific genes within the legume genome, thus generating crops with desired traits in a more targeted and efficient manner. This holds significant promise for introducing beneficial changes while minimizing unintended alterations. For instance, researchers can engineer legume crops to be more resistant to specific diseases, enhance nutrient uptake, or extend their adaptability to changing environmental conditions. The ability to harness such precise genetic modifications empowers breeders to tailor legume crops to diverse agro-climatic regions and evolving market demands. Moreover, the integration of omics technologies, including transcriptomics, proteomics, and metabolomics, enhances our understanding of complex biological processes in legumes. These techniques allow researchers to unravel the intricate molecular networks underlying traits of interest. By deciphering the interactions between genes, proteins, and metabolites, scientists can identify key regulatory elements and metabolic pathways that govern traits like seed yield, nutrient composition, and stress responses. This holistic approach enables a deeper exploration of the genetic potential of legumes and the development of comprehensive strategies for trait improvement. In the context of global challenges such as climate change and food security, marker-assisted breeding for legume improvement assumes even greater significance. As climates shift and resources become scarcer, the need for resilient and sustainable crops intensifies. Legumes, with their nitrogen-fixing abilities and potential for intercropping, can play a pivotal role in sustainable agricultural systems. Marker-assisted breeding equips these crops with the tools necessary to thrive in diverse and challenging environments, contributing to increased food production while reducing the environmental footprint of agriculture. Therefore, the future prospects for marker-assisted breeding in legume improvement are exceptionally bright. The combination of advanced genomics, gene editing, and omics technologies empowers researchers and breeders to expedite the development of legume varieties with enhanced traits, from disease resistance to nutritional content. These innovations not only address current challenges in agriculture but also align with the broader goals of sustainability, resilience, and food security. The growing pool of genetic resources will make it easier to eliminate detrimental loci from the variety and accumulate crucial alleles to develop new crop varieties. Furthermore, "breeder friendly markers" that are highly repeatable, simple to assay, and reasonably priced for genotyping should be used as molecular markers to hasten the use of molecular techniques in legume breeding programs. For instance, SNPs are well-known as the preferred molecular markers, and their expansion into cleaved amplified polymorphic sequences (CAPs), KASPar tests, and Illumina Veracode will allow for more use in programs for improving legume crops. Additionally, it is necessary to monitor and catalogue the allelic differences that exist between and even within various bean species' genome sequences (super-pangenome). The identification of superior haplotypes in leguminous crops, which can be used in legume breeding programs, will be aided in the future by haplotype-based breeding. As science continues to advance, the synergy between cutting-edge molecular techniques and traditional breeding methods promises a new era of innovation and progress for legume crops, fostering a more productive and sustainable agricultural landscape.

5 Conclusions

In conclusion, marker-assisted breeding (MAB) stands as a revolutionary tool in the realm of legume improvement, offering a potent synergy between cutting-edge molecular techniques and traditional breeding methods. By harnessing the power of molecular markers, MAB accelerates the selection of desired traits, ranging from disease resistance and drought tolerance to enhanced nutritional content. This approach not only expedites breeding cycles but also ensures the precision and effectiveness necessary to address the evolving challenges of global agriculture, including climate change and food security. However, successful implementation relies on the collaborative efforts of geneticists, molecular biologists, and breeders, along with the availability of comprehensive genomic data and rigorous marker validation. As MAB continues to evolve, ethical considerations such as genetically modified organisms and intellectual property rights must be navigated carefully. As we stand at the intersection of scientific advancement and agricultural innovation, marker-assisted breeding offers a promising path forward, fostering the development of legume varieties that are not only resilient and high-yielding but also align with the nutritional and environmental needs of a dynamic world.

References

Afzal, M., Alghamdi, S. S., Migdadi, H. H., et al. (2020). Legume genomics and transcriptomics: From classic breeding to modern technologies. *Saudi Journal of Biological Sciences, 27*, 543–555. https://doi.org/10.1016/j.sjbs.2019.11.018

Mikic, A., Perić, V., Đorđević, V., et al. (2009). Anti-nutritional factors in some grain legumes. *Biotechnology in Animal Husbandry, 25*(5-6-2), 1181–1188.

Anuradha, C., Gaur, P. M., Pande, S., et al. (2011). Mapping QTL for resistance to botrytis grey mould in chickpea. *Euphytica, 182*, 1–9. https://doi.org/10.1007/s10681-011-0394-1

Ates, D., Aldemir, S., Yagmur, B., et al. (2018). QTL mapping of genome regions controlling manganese uptake in lentil Seed. *G3: Genes, Genomes, Genetics, 8*, 1409–1416. https://doi.org/10.1534/g3.118.200259

Ates, D., Sever, T., Aldemir, S., et al. (2016). Identification QTLs controlling genes for se uptake in lentil seeds. *PLoS One, 15*, e0149210. https://doi.org/10.1371/journal.pone.0149210

Avila, C. M., Sillero, J. C., Rubiales, D., et al. (2003). Identification of RAPD markers linked to the Uvf-1 gene conferring hypersensitive resistance against rust (*Uromyces viciae-fabae*) in *Vicia faba* L. *Theoretical and Applied Genetics, 107*, 353–358. https://doi.org/10.1007/s00122-003-1254-8

Basak, J., Kundagrami, S., Ghose, T. K., & Pal, A. (2004). Development of Yellow Mosaic Virus (YMV) resistance linked DNA marker in *Vigna mungo* from populations segregating for YMV-reaction. *Molecular Breeding, 14*, 375–383. https://doi.org/10.1007/s11032-004-0238-y

Bharadwaj, C., Jorben, J., Rao, A., Roorkiwal, M., Patil, B. S., Jayalakshmi, Ahammed, S. K., Saxena, D. R., Yasin, M., Jahagirdar, J. E., Sontakke, P. L., Pithia, M. S., Chudasama, M. K., Swarup, I., Singh, R. K., Nitesh, S. D., Chitikineni, A., Singh, S., Singh, I., Pratap, A., Dixit, G. P., Srivastava, A. K., & Varshney, R. K. (2022). Development of high yielding Fusarium wilt resistant cultivar by pyramiding of "genes" through marker-assisted backcrossing in Chickpea (*Cicer arietinum* L.). *Frontiers in Genetics*. https://doi.org/10.3389/fgene.2022.924287

Gautami, B., Pandey, M. K., Vadez, V., et al. (2012). Quantitative trait locus analysis and construction of consensus genetic map for drought tolerance traits based on three recombinant inbred line populations in cultivated groundnut (*Arachis hypogaea* L.). *Molecular Breeding, 30*, 757–772. https://doi.org/10.1007/s11032-011-9660-0

Blair, M. W., Medina, J. I., Astudillo, C., et al. (2010). QTL for seed iron and zinc concentration and content in a Mesoamerican common bean (*Phaseolus vulgaris* L.) population. *Theoretical and Applied Genetics, 121*, 1059–1070. https://doi.org/10.1007/s00122-010-1371-0

Bohra, A., Bharadwaj, C., Radhakrishnan, T., et al. (2019). Translational genomics and molecular breeding for enhancing precision and efficiency in crop improvement programs: Some examples in legumes. *Indian Journal Of Genetics And Plant Breeding, 79*, 227–2403.

Bohra, A., Sahrawat, K. L., Kumar, S., et al. (2015). Genetics- and genomics-based interventions for nutritional enhancement of grain legume crops: Status and outlook. *Journal of Applied Genetics, 56*, 151–161. https://doi.org/10.1007/s13353-014-0268-z

Bohra, A., Saxena, K. B., Varshney, R. K., & Saxena, R. K. (2020). Genomics-assisted breeding for pigeonpea improvement. *Theoretical and Applied Genetics, 133*, 1721–1737. https://doi.org/10.1007/s00122-020-03563-7

Chaitieng, B., Kaga, A., Han, O. K., et al. (2002). Mapping a new source of resistance to powdery mildew in mungbean. *Plant Breeding, 121*, 521–525. https://doi.org/10.1046/j.1439-0523.2002.00751.x

Chapman, M. A. (2015). Transcriptome sequencing and marker development for four underutilized legumes. *Applications in Plant Sciences, 3*, apps.1400111. https://doi.org/10.3732/apps.1400111

Cho, S., Chen, W., & Muehlbauer, F. J. (2004). Pathotype-specific genetic factors in chickpea (*Cicer arietinum* L.) for quantitative resistance to ascochyta blight. *Theoretical and Applied Genetics, 109*, 733–739. https://doi.org/10.1007/s00122-004-1693-x

Choi, H.-K. (2018). Translational genomics and multi-omics integrated approaches as a useful strategy for crop breeding. *Genes & Genomics, 41*, 133–146. https://doi.org/10.1007/s13258-018-0751-8

Chu, Y., Wu, C. L., Holbrook, C. C., et al. (2011). Marker-assisted selection to pyramid nematode resistance and the high oleic trait in peanut. *The Plant Genome, 4*, 110–117. https://doi.org/10.3835/plantgenome2011.01.0001

Cobos, M. J., Fernández, M. J., Rubio, J., et al. (2005). A linkage map of chickpea (*Cicer arietinum* L.) based on populations from Kabuli × Desi crosses: Location of genes for resistance to fusarium wilt race 0. *Theoretical and Applied Genetics, 110*, 1347–1353. https://doi.org/10.1007/s00122-005-1980-1

Collard, B. C. Y., & Mackill, D. J. (2007). Marker-assisted selection: An approach for precision plant breeding in the twenty-first century. *Philosophical Transactions of the Royal Society of London. Series B, Biological Sciences, 363*, 557–572. https://doi.org/10.1098/rstb.2007.2170

Deavours, B. E., & Dixon, R. A. (2005). Metabolic engineering of isoflavonoid biosynthesis in alfalfa. *Plant Physiology, 138*, 2245–2259. https://doi.org/10.1104/pp.105.062539

Deeplanaik, N., Kumaran, R. C., Krishna, V., et al. (2013). Expression of drought responsive genes in pigeonpea and *in silico* comparison with soybean cDNA library. *Journal of Crop Science and Biotechnology, 16*, 243–251. https://doi.org/10.1007/s12892-013-0069-7

Dhaliwal, S. K., Gill, R. K., Sharma, A., et al. (2022). A large-effect QTL introgressed from ricebean imparts resistance to *Mungbean yellow mosaic India virus* in blackgram (*Vigna mungo* (L.) Hepper). *Theoretical and Applied Genetics, 135*, 4495–4506. https://doi.org/10.1007/s00122-022-04234-5

Diapari, M., Sindhu, A., Bett, K., et al. (2014). Genetic diversity and association mapping of iron and zinc concentrations in chickpea (*Cicer arietinum* L.). *Genome, 57*, 459–468. https://doi.org/10.1139/gen-2014-0108

Díaz, L. M., Ricaurte, J., Tovar, E., et al. (2018). QTL analyses for tolerance to abiotic stresses in a common bean (*Phaseolus vulgaris* L.) population. *PLoS One, 13*, e0202342–e0202342. https://doi.org/10.1371/journal.pone.0202342

Dirlewanger, E., Isaac, P. G., Ranade, S. A., et al. (1994). Restriction fragment length polymorphism analysis of loci associated with disease resistance genes and developmental traits in *Pisum sativum* L. *Theoretical and Applied Genetics, 88*, 17–27. https://doi.org/10.1007/bf00222388

Dubey, A., Farmer, A., Schlueter, J. A., et al. (2011). Defining the transcriptome assembly and its use for genome dynamics and transcriptome profiling studies in Pigeonpea (*Cajanus cajan* L.). *DNA Research, 18*, 153–164. https://doi.org/10.1093/dnares/dsr007

Ender, M., Terpstra, K., & Kelly, J. D. (2007). Marker-assisted selection for white mold resistance in common bean. *Molecular Breeding, 21*, 149–157. https://doi.org/10.1007/s11032-007-9115-9

Ford, R., Pang, E. C. K., & Taylor, P. W. J. (1999). Genetics of resistance to ascochyta blight (*Ascochyta lentis*) of lentil and the identification of closely linked RAPD markers. *Theoretical and Applied Genetics, 98*, 93–98. https://doi.org/10.1007/s001220051044

Gao, X., Chai, H. H., Ho, W. K., et al. (2023). Deciphering the molecular basis for photosynthetic parameters in Bambara groundnut (*Vigna subterranea* L. Verdc) under drought stress. *BMC Plant Biology, 23*, 10.1186/s12870-023-04293-w.

Gnanasambandam, A., Paull, J., Torres, A., et al. (2012). Impact of molecular technologies on Faba bean (*Vicia faba* L.) breeding strategies. *Agronomy, 2*, 132–166. https://doi.org/10.3390/agronomy2030132

Gnanesh, B. N., Bohra, A., Sharma, M. C., et al. (2011). Genetic mapping and quantitative trait locus analysis of resistance to sterility mosaic disease in pigeonpea [*Cajanus cajan* (L.) Millsp.]. *Field Crops Research, 123*, 53–61. https://doi.org/10.1016/j.fcr.2011.04.011

Gresta, F., Mercati, F., Santonoceto, C., et al. (2016). Morpho-agronomic and AFLP characterization to explore guar (*Cyamopsis tetragonoloba* L.) genotypes for the Mediterranean environment. *Industrial Crops and Products, 86*, 23–30. https://doi.org/10.1016/j.indcrop.2016.03.038

Grusak, M. A. (2002). Photochemicals in plants: Genomics-assisted plant improvement for nutritional and health benefits. *Current Opinion in Biotechnology, 13*, 508–511.

Guindon, M. F., Cazzola, F., Palacios, T., et al. (2021). Biofortification of pea (*Pisum sativum* L): A review. *Journal of the Science of Food and Agriculture, 101*, 3551–3563. https://doi.org/10.1002/jsfa.11059

Gutierrez, N., Avila, C. M., Duc, G., et al. (2006). CAPs markers to assist selection for low vicine and convicine contents in faba bean (*Vicia faba* L.). *Theoretical and Applied Genetics, 114*(1), 59–66. https://doi.org/10.1007/s00122-006-0410-3. Epub 2006 Sep 30.

He, X.-Z., & Dixon, R. A. (2000). Genetic manipulation of isoflavone 7-O-methyltransferase enhances biosynthesis of 4′-O-methylated isoflavonoid phytoalexins and disease resistance in alfalfa. *Plant Cell, 12*, 1689–1702. https://doi.org/10.1105/tpc.12.9.1689

Humphry, M., Magner, T. F., Lynne McIntyre, C., et al. (2003). Identification of a major locus conferring resistance to powdery mildew (*Erysiphepolygoni* DC) in mungbean (*Vigna radiata* L. Wilczek) by QTL analysis. *Genome, 46*, 738–744. https://doi.org/10.1139/g03-057

Idrissi, O., Udupa, S. M., De Keyser, E., et al. (2016). Identification of quantitative trait loci controlling root and shoot traits associated with drought tolerance in a lentil (*Lens culinaris* Medik.) recombinant inbred line population. *Frontiers in Plant Science, 7*, 1174. https://doi.org/10.3389/fpls.2016.01174

Iruela, M., Castro, P., Rubio, J., et al. (2007). Validation of a QTL for resistance to ascochyta blight linked to resistance to fusarium wilt race 5 in chickpea (*Cicer arietinum* L.). *European Journal of Plant Pathology, 119*, 29–37. https://doi.org/10.1007/s10658-007-9121-0

Izquierdo, P., Astudillo, C., Blair, M. W., et al. (2018). Meta-QTL analysis of seed iron and zinc concentration and content in common bean (*Phaseolus vulgaris* L.). *Theoretical and Applied Genetics, 131*, 1645–1658. https://doi.org/10.1007/s00122-018-3104-8

Jaganathan, D., Thudi, M., Kale, S., et al. (2015). Genotyping-by-sequencing based intra-specific genetic map refines a "QTL-hotspot" region for drought tolerance in chickpea. *Molecular Genetics and Genomics, 290*, 559–571. https://doi.org/10.1007/s00438-014-0932-3

Janila, P., Pandey, M. K., Manohar, S. S., Variath, M. T., Nallathambi, P., Nadaf, H. L., Sudini, H., & Varshney, R. K. (2016). Foliar fungal disease-resistant introgression lines of groundnut (*Arachis hypogaea* L.) record higher pod and haulm yield in multilocation testing. *Plant Breeding, 135*, 355–366.

Jung, G., Skroch, P. W., Coyne, D. P., et al. (1997). Molecular-marker-based genetic analysis of tepary bean-derived common bacterial blight resistance in different developmental stages of common bean. *Journal of the American Society for Horticultural Science, 122*, 329–337. https://doi.org/10.21273/jashs.122.3.329

Kale, S. M., Jaganathan, D., Ruperao, P., et al. (2015). Prioritization of candidate genes in "QTL-hotspot" region for drought tolerance in chickpea (*Cicer arietinum* L.). *Scientific Reports, 5*, 10.1038/srep15296.

Katoch, M., & Chahota, R. K. (2021). Mapping of quantitative trait locus (QTLs) that contribute to drought tolerance in a recombinant inbred line population of horsegram (*Macrotyloma uniflorum*). *bioRxiv*. (Cold Spring Harbor Laboratory. https://doi.org/10.1101/2021.03.09.434553

Keerthi, C. M., Ramesh, S., Byregowda, M., & Vaijayanthi, P. V. (2018). Simple sequence repeat (SSR) marker assay-based genetic diversity among Dolichos bean (*Lablab purpureus* L. sweet) advanced breeding lines differing for productivity per se traits. *International Journal of Current Microbiology and Applied Sciences, 7*, 3736–3744. https://doi.org/10.20546/ijcmas.2018.705.433

Kelly, J. D., Gepts, P., Miklas, P. N., & Coyne, D. P. (2003). Tagging and mapping of genes and QTL and molecular marker-assisted selection for traits of economic importance in bean and cowpea. *Field Crops Research, 82*, 135–154. https://doi.org/10.1016/s0378-4290(03)00034-0

Khazaei, H., Podder, R., Caron, C. T., et al. (2017). Marker–trait association analysis of iron and zinc concentration in lentil (*Lens culinaris* Medik.) seeds. *Plant Genome, 10*, 1–8. https://doi.org/10.3835/plantgenome2017.02.0007

Khedikar, Y. P., Gowda, M. V. C., Sarvamangala, C., et al. (2010). A QTL study on late leaf spot and rust revealed one major QTL for molecular breeding for rust resistance in groundnut (*Arachis hypogaea* L.). *Theoretical and Applied Genetics, 121*, 971–984. https://doi.org/10.1007/s00122-010-1366-x

Kolkman, J. M., & Kelly, J. D. (2003). QTL conferring resistance and avoidance to white mold in common bean. *Crop Science, 43*, 539–539. https://doi.org/10.2135/cropsci2003.0539

Kumar, S., Palve, A., Patel, S., et al. (2020). Development of genomic microsatellite markers in cluster bean using next-generation DNA sequencing and their utility in diversity analysis. *Current Plant Biology, 21*, 100134–100134. https://doi.org/10.1016/j.cpb.2019.100134

Kushwah, A., Bhatia, D., Rani, U., et al. (2021). Molecular mapping of quantitative trait loci for ascochyta blight and botrytis grey mould resistance in an inter-specific cross in chickpea (*Cicer arietinum* L.) using genotyping by sequencing. *Breeding Science, 71*, 229–239. https://doi.org/10.1270/jsbbs.20085

Kushwah, A., Bhatia, D., Barmukh, R., Singh, I., Singh, G., Bindra, S., Vij, S., Chellapilla, C., Pratap, A., Roorkiwal, M., Kumar, S., Varshney, R. K., & Singh, S. (2022). Genetic mapping of QTLs for drought tolerance in chickpea (*Cicer arietinum* L.). *Frontiers in Genetics, 13*. https://doi.org/10.3389/fgene.2022.953898

Langridge, P., & Chalmers, K. (2004). The principle: Identification and application of molecular markers. In H. Lörz & G. Wenzel (Eds.), *Biotechnology in agriculture and forestry* (Molecular markers systems) (Vol. 55, pp. 3–22). Springer.

Lee, G. J., Carter, T., Villagarcia, M. R., et al. (2004). A major QTL conditioning salt tolerance in S-100 soybean and descendent cultivars. *Theoretical and Applied Genetics, 109*, 1610–1619. https://doi.org/10.1007/s00122-004-1783-9

Liu, C., Wu, J., Wang, L., et al. (2017). Quantitative trait locus mapping under irrigated and drought treatments based on a novel genetic linkage map in mungbean (*Vigna radiata* L.). *Theoretical and Applied Genetics, 130*, 2375–2393. https://doi.org/10.1007/s00122-017-2965-6

Liu, S., Yu, K., & Park, S. J. (2008). Development of STS markers and QTL validation for common bacterial blight resistance in common bean. *Plant Breeding, 127*(1), 62–68. https://doi.org/10.1111/j.1439-0523.2007.01423.x

Lucas, M. R., Ehlers, J. D., Huynh, B.-L., et al. (2013). Markers for breeding heat-tolerant cowpea. *Molecular Breeding, 31*, 529–536. https://doi.org/10.1007/s11032-012-9810-z

Luo, H., Guo, J., Yu, B., et al. (2021). Construction of ddRADseq-based high-density genetic map and identification of quantitative trait loci for trans-resveratrol content in peanut seeds. *Frontiers in Plant Science, 12*, 438. https://doi.org/10.3389/fpls.2021.644402

Ma, Y., Coyne, C. J., Grusak, M. A., et al. (2017). Genome-wide SNP identification, linkage map construction and QTL mapping for seed mineral concentrations and contents in pea (*Pisum sativum* L.). *BMC Plant Biology, 17*, 43. https://doi.org/10.1186/s12870-016-0956-4

Mahmoud, G. A.-E. (2021). Biotic stress to legumes: Fungal diseases as major biotic stress factor. In *Sustainable agriculture reviews 51: Legume agriculture and biotechnology* (Vol. 2, p. 181). https://doi.org/10.1007/978-3-030-68828-8_7

Miklas, P. N., Smith, J. R., & Singh, S. P. (2006). Registration of common bacterial blight resistant dark red kidney bean germplasm line USDK-CBB-15. *Crop Science, 46*, 1005–1007. https://doi.org/10.2135/cropsci2005.06-0110

Mir, R. R., & Varshney, R. K. (2013). Future prospects of molecular markers in plants. In R. J. Henry (Ed.), *Molecular markers in plants* (pp. 169–190). Blackwell Publishing Ltd.

Mir, R. R., Hiremath, P. J., Riera-Lizarazu, O., & Varshney, R. K. (2013). *Evolving molecular marker technologies in plants: From RFLPs to GBS* (pp. 229–247). Springer eBooks. https://doi.org/10.1007/978-94-007-5687-8_11

Muchero, W., Ehlers, J. D., Close, T. J., & Roberts, P. A. (2009). Mapping QTL for drought stress-induced premature senescence and maturity in cowpea [*Vigna unguiculata* (L.) Walp.]. *Theoretical and Applied Genetics, 118*, 849–863. https://doi.org/10.1007/s00122-008-0944-7

Mugabe, D., Coyne, C. J., Piaskowski, J., et al. (2019). Quantitative trait loci for cold tolerance in chickpea. *Crop Science, 59*, 573–582. https://doi.org/10.2135/cropsci2018.08.0504

Myers, G. O., Fatokun, C., & Young, N. D. (1996). RFLP mapping of an aphid resistance gene in cowpea (*Vigna unguiculata* L. Walp). *Euphytica, 91*, 181–187. https://doi.org/10.1007/bf00021068

Nawaz, M. A., & Chung, G. (2020). Genetic improvement of cereals and grain legumes. *Genes, 11*, 1255. https://doi.org/10.3390/genes11111255

Ning, L., Sun, P., Wang, Q., Ma, D., Hu, Z., Zhang, D., & Yu, D. (2015). Genetic architecture of biofortification traits in soybean (*Glycine max* L. Merr.) revealed through association analysis and linkage mapping. *Euphytica, 204*(2), 353–369.

Nzungize, J., Gepts, P., Buruchara, R. A., et al. (2011). Introgression of Pythium root rot resistance gene into Rwandan susceptible common bean cultivars. *African Journal of Plant Science, 5*, 193–200. https://doi.org/10.5897/ajps.9000150

Obala, J., Mukankusi, C., Rubaihayo, P., et al. (2012). Improvement of resistance to Fusarium root rot through gene pyramiding in common bean. *African Crop Science Journal, 20*, 1–13.

Ojiewo, C., Monyo, E., Desmae, H., et al. (2019). Genomics, genetics and breeding of tropical legumes for better livelihoods of smallholder farmers. *Plant Breeding, 138*(4), 487–499.

Palmer, C. M., & Guerinot, M. L. (2009). Facing the challenges of CuFe and Zn homeostasis in plants. *Nature Chemical Biology, 5*, 333–340. https://doi.org/10.1038/nchembio.166

Palmgren, M. G., Clemens, S., Williams, L. E., et al. (2008). Zinc biofortification of cereals: Problems and solutions. *Trends in Plant Science, 13*, 464–473. https://doi.org/10.1016/j.tplants.2008.06.005

Pandey, A. K., & Basandrai, A. K. (2020). Will Macrophomina phaseolina spread in legumes due to climate change? A critical review of current knowledge. *Journal of Plant Diseases and Protection, 128*, 9–18. https://doi.org/10.1007/s41348-020-00374-2

Pandey, M. K., Khan, A. J., Singh, V., et al. (2017). QTL-seq approach identified genomic regions and diagnostic markers for rust and late leaf spot resistance in groundnut (*Arachis hypogaea* L.). *Plant Biotechnology Journal, 15*, 927–941. https://doi.org/10.1111/pbi.12686

Pandey, M. K., Roorkiwal, M., Singh, V. K., et al. (2016). Emerging genomic tools for legume breeding: Current status and future prospects. *Frontiers in Plant Science, 7*, 455. https://doi.org/10.3389/fpls.2016.00455

Park, C., Dwiyanti, M. S., Nagano, A. J., Liu, B., Yamada, T., & Abe, J. (2019). Identification of quantitative trait loci for increased α-tocopherol biosynthesis in wild soybean using a high-density genetic map. *BMC Plant Biology, 19*, 1–5. https://doi.org/10.1186/s12870-019-2117-z

Paul, P., Samineni, S., Thudi, M., et al. (2018). Molecular mapping of QTLs for heat tolerance in chickpea. *International Journal of Molecular Sciences, 19*, 2166. https://doi.org/10.3390/ijms19082166

Pratap, A., Chaturvedi, S. K., Tomar, R., Rajan, N., Malviya, N., Thudi, M., Saabale, P. R., Prajapati, U., Varshney, R. K., & Singh, N. P. (2017). Marker-assisted introgression of resistance to fusarium wilt race 2 in Pusa 256, an elite cultivar of *desi* chickpea. *Molecular Genetics and Genomics, 292*, 1237. https://doi.org/10.1007/s00438-017-1343-z

Pazhamala, L., Saxena, R. K., Singh, V. K., et al. (2015). Genomics-assisted breeding for boosting crop improvement in pigeonpea (*Cajanus cajan*). *Frontiers in Plant Science, 6*, 50. https://doi.org/10.3389/fpls.2015.00050

Priyanka, B., Sekhar, K., Sunita, T., et al. (2010). Characterization of expressed sequence tags (ESTs) of pigeonpea (*Cajanus cajan* L.) and functional validation of selected genes for abiotic stress tolerance in *Arabidopsis thaliana*. *Molecular Genetics and Genomics, 283*, 273–287. https://doi.org/10.1007/s00438-010-0516-9

Ragot, M., & Lee, M. (2007). Marker-assisted selection in maize: current status, potential, limitations and perspectives from the private and public sectors. In E. P. Guimarães et al. (Eds.), *Marker-assisted selection, current status and future perspectives in crops, livestock, forestry, and fish* (pp. 117–150). FAO.

Raju, N. L., Gnanesh, B. N., Lekha, P., et al. (2010). The first set of EST resource for gene discovery and marker development in pigeonpea (*Cajanus cajan* L.). *BMC Plant Biology, 10*, 1–22. https://doi.org/10.1186/1471-2229-10-45

Ramalingam, J., Alagarasan, G., Savitha, P., et al. (2020). Improved host-plant resistance to *Phytophthora* rot and powdery mildew in soybean (*Glycine max* (L.) Merr.). *Scientific Reports, 10*, 13928. https://doi.org/10.1038/s41598-020-70702-x

Ravi, K., Vadez, V., Isobe, S., et al. (2010). Identification of several small main-effect QTLs and a large number of epistatic QTLs for drought tolerance related traits in groundnut (*Arachis hypogaea* L.). *Theoretical and Applied Genetics, 122*, 1119–1132. https://doi.org/10.1007/s00122-010-1517-0

Rehman, A. U., Malhotra, R. S., Bett, K., et al. (2011). Mapping QTL associated with traits affecting grain yield in chickpea (*Cicer arietinum* L.) under terminal drought stress. *Crop Science, 51*, 450–463. https://doi.org/10.2135/cropsci2010.03.0129

Román, B., Šatović, Z., Ávila, C. M., et al. (2003). Locating genes associated with *Ascochyta fabae* resistance in *Vicia faba*. *Australian Journal of Agricultural Research, 54*, 85–90. https://doi.org/10.1071/ar02034

Roorkiwal, M., Pandey, S., Thavarajah, D., Hemalatha, R., & Varshney, R. K. (2021). Molecular mechanisms and biochemical pathways for micronutrient acquisition and storage in legumes to support biofortification for nutritional security. *Frontiers in Plant Science, 12*, 682842. https://doi.org/10.3389/fpls.2021.682842

Sabbavarapu, M. M., Sharma, M., Chamarthi, S. K., et al. (2013). Molecular mapping of QTLs for resistance to *Fusarium* wilt (race 1) and *Ascochyta* blight in chickpea (*Cicer arietinum* L.). *Euphytica, 193*, 121–133. https://doi.org/10.1007/s10681-013-0959-2

Saha, G. C., Sarker, A., Chen, W., et al. (2010a). Inheritance and linkage map positions of genes conferring resistance to stemphylium blight in lentil. *Crop Science, 50*, 1831–1839. https://doi.org/10.2135/cropsci2009.12.0709

Saha, G. C., Sarker, A., Chen, W., et al. (2010b). Identification of markers associated with genes for rust resistance in *Lens culinaris* Medik. *Euphytica, 175*, 261–265. https://doi.org/10.1007/s10681-010-0187-y

Sallam, A., Arbaoui, M., El-Esawi, M. A., et al. (2016). Identification and verification of QTL associated with frost tolerance using linkage mapping and GWAS in winter faba bean. *Frontiers in Plant Science, 7*, 1098. https://doi.org/10.3389/fpls.2016.01098

Sampaio, A. M., Araújo, S. d. S., Rubiales, D., & Vaz Patto, M. C. (2020). *Fusarium* wilt management in legume crops. *Agronomy, 10*, 1073. https://doi.org/10.3390/agronomy10081073

Saxena, R. K., Singh, V. K., Kale, S. M., et al. (2017). Construction of genotyping-by-sequencing based high-density genetic maps and QTL mapping for fusarium wilt resistance in pigeonpea. *Scientific Reports, 7*, 1911. https://doi.org/10.1038/s41598-017-01537-2

Sayama, T., Nakazaki, T., Ishikawa, G., et al. (2009). QTL analysis of seed-flooding tolerance in soybean (*Glycine max* [L.] Merr.). *Plant Science, 176*, 514–521. https://doi.org/10.1016/j.plantsci.2009.01.007

Sen Gupta, D., McPhee, K., & Kumar, S. (2017). Development of molecular markers for iron metabolism related genes in lentil and their expression analysis under excess iron stress. *Frontiers in Plant Science, 8*. https://doi.org/10.3389/fpls.2017.00579

Sharma, K., Winter, P., Kahl, G., & Muehlbauer, F. J. (2004). Molecular mapping of *Fusarium oxysporum* f. sp. *ciceris* race 3 resistance gene in chickpea. *Theoretical and Applied Genetics, 108*, 1243–1248. https://doi.org/10.1007/s00122-003-1561-0

Shasidhar, Y., Variath, M. T., Vishwakarma, M. K., et al. (2020). Improvement of three popular Indian groundnut varieties for foliar disease resistance and high oleic acid using SSR markers and SNP array in marker-assisted backcrossing. *Crop Journal, 8*, 1–15. https://doi.org/10.1016/j.cj.2019.07.001

Singh, D., Chaudhary, P., Taunk, J., et al. (2021). *Fab* advances in *Fabaceae* for abiotic stress resilience: From "omics" to artificial intelligence. *International Journal of Molecular Sciences, 22*, 10535. https://doi.org/10.3390/ijms221910535

Singh, V., Yadav, R. K., Yadav, N. R., Yadav, R., Malik, R. S., & Singh, J. (2017). Identification of genomic Regions/genes for high iron and zinc content and cross transferability of SSR markers in mungbean (*Vigna radiata* L.). *Legume Research-An International Journal, 40*(6), 1004–1011.

Subramaniyan, R., Manivannan, N., Iyanar, K., et al. (2022). Novel and stable QTL regions conferring resistance to MYMV disease and its inheritance in blackgram (*Vigna mungo* (L.) Hepper). *Journal of Genetics, 101*, 18. https://doi.org/10.1007/s12041-022-01359-w

Sujay, V., Gowda, C., Pandey, M. K., et al. (2012). Quantitative trait locus analysis and construction of consensus genetic map for foliar disease resistance based on two recombinant inbred line populations in cultivated groundnut (*Arachis hypogaea* L.). *Molecular Breeding, 30*, 773–788. https://doi.org/10.1007/s11032-011-9661-z

Tar'an, B., Buchwaldt, L., Tullu, A., et al. (2003). Using molecular markers to pyramid genes for resistance to ascochyta blight and anthracnose in lentil (*Lens culinaris* Medik). *Euphytica, 134*, 223–230. https://doi.org/10.1023/b:euph.0000003913.39616.fd

Tar'an, B., Michaels, T. E., & Pauls, K. P. (2001). Mapping genetic factors affecting the reaction to *Xanthomonas axonopodis* pv. *phaseoli* in *Phaseolus vulgaris* L. under field conditions. *Genome, 44*, 1046–1056. https://doi.org/10.1139/g01-099

Thudi, M., Upadhyaya, H. D., Rathore, A., et al. (2014). Genetic dissection of drought and heat tolerance in chickpea through genome-wide and candidate gene-based association mapping approaches. *PLoS One, 9*(5), e96758. https://doi.org/10.1371/journal.pone.0096758

Upadhyaya, H. D., Bajaj, D., Das, S., et al. (2016). Genetic dissection of seed-iron and zinc concentrations in chickpea. *Scientific Reports, 6*(1), 1–12.

Urva, P. (2021). Detection and classification of grain crops and legumes disease: A survey. *Sparklinglight Transact Artif Intell Quantum Comp, 01*, 41–55. https://doi.org/10.55011/staiqc.2021.1105

Urwat, U., Ahmad, S. M., Masi, et al. (2021). Fe and Zn stress induced gene expression analysis unravelled mechanisms of mineral homeostasis in common bean (*Phaseolus vulgaris* L.). *Scientific Reports, 11*, 24026. https://doi.org/10.1038/s41598-021-03506-2

Varshney, R. K. (2016). Exciting journey of 10 years from genomes to fields and markets: Some success stories of genomics-assisted breeding in chickpea, pigeonpea and groundnut. *Plant Science, 242*, 98–107. https://doi.org/10.1016/j.plantsci.2015.09.009

Varshney, R. K., Mohan, S. M., Gaur, P. M., Gangarao, N. V. P. R., Pandey, M. K., Bohra, A., ... & Gowda, C. L. L. (2013). Achievements and prospects of genomics-assisted breeding in three legume crops of the semi-arid tropics. *Biotechnology Advances, 31*(8), 1120–1134.

Varshney, R. K., Pandey, M. K., Bohra, A., et al. (2019). Toward the sequence-based breeding in legumes in the post-genome sequencing era. *Theoretical and Applied Genetics, 132*, 797–816. https://doi.org/10.1007/s00122-018-3252-x

Varshney, R. K., Pandey, M. K., Janila, P., et al. (2014a). Marker-assisted introgression of a QTL region to improve rust resistance in three elite and popular varieties of peanut (*Arachis hypogaea* L.). *Theoretical and Applied Genetics, 127*, 1771–1781. https://doi.org/10.1007/s00122-014-2338-3

Varshney, R. K., Mohan, S. V., Gaur, P. M., et al. (2014b). Marker-assisted backcrossing to introgress resistance to fusarium wilt race 1 and ascochyta blight in C 214, an elite cultivar of chickpea. *Plant Genome, 7*, 11. https://doi.org/10.3835/plantgenome2013.10.0035

Wang, D.-g., Lin, Z., Kai, L., et al. (2017). Marker-assisted pyramiding of soybean resistance genes Rsc4, Rsc8, and Rsc14Q to soybean mosaic virus. *Journal of Integrative Agriculture, 16*, 2413–2420.

Wang, H., Pandey, M. K., Qiao, L., et al. (2013). Genetic mapping and quantitative trait loci analysis for disease resistance using F_2 and F_5 generation-based genetic maps derived from "Tifrunner" × "GT-C20" in peanut. *Plant Genome, 6*, 3. https://doi.org/10.3835/plantgenome2013.05.0018

Wu, X., Islam, A. S., Limpot, N., et al. (2020). Genome-wide SNP identification and association mapping for seed mineral concentration in mung bean (*Vigna radiata* L.). *Frontiers in Genetics, 11*, 656. https://doi.org/10.3389/fgene.2020.00656

Xiong, H., Kobayashi, T., Kakei, Y., et al. (2012). AhNRAMP1 iron transporter is involved in iron acquisition in peanut. *Journal of Experimental Botany, 63*, 4437–4446. https://doi.org/10.1093/jxb/ers117

Ye, H., Song, L., Schapaugh, W. T., et al. (2019). The importance of slow canopy wilting in drought tolerance in soybean. *Journal of Experimental Botany, 71*, 642–652. https://doi.org/10.1093/jxb/erz150

Yu, K., Park, S. J., Poysa, V., & Gepts, P. (2000). Integration of simple sequence repeat (SSR) markers into a molecular linkage map of common bean (*Phaseolus vulgaris* L.). *The Journal of Heredity, 91*, 429–434.

Yu, K., Park, S. J., Zhang, B., et al. (2004). An SSR marker in the nitrate reductase gene of common bean is tightly linked to a major gene conferring resistance to common bacterial blight. *Euphytica, 138*, 89–95. https://doi.org/10.1023/b:euph.0000047077.75285.60

Yundaeng, C., Somta, P., Chen, J., et al. (2021). Fine mapping of QTL conferring Cercospora leaf spot disease resistance in mungbean revealed TAF5 as candidate gene for the resistance. *Theoretical and Applied Genetics, 134*, 701–714. https://doi.org/10.1007/s00122-020-03724-8

Zate, Z. Z. (2021). Quantitative Trait Loci mapping of agronomic traits in a Cowpea (*Vigna unguiculata* L.) bi-parental cross. *Journal of Sustainable Development, 9*, 39–58. https://doi.org/10.20372/au.jssd.9.1.2021.0258

Zhang, W. B., Qiu, P. C., Jiang, H., et al. (2012). Dissection of genetic overlap of drought and low-temperature tolerance QTLs at the germination stage using backcross introgression lines in soybean. *Molecular Biology Reports, 39*, 6087–6094. https://doi.org/10.1007/s11033-011-1423-9

Zwart, R. S., Thudi, M., Channale, S., et al. (2019). Resistance to plant-parasitic nematodes in chickpea: Current status and future perspectives. *Frontiers in Plant Science, 10*, 966. https://doi.org/10.3389/fpls.2019.00966

Plant Protease Inhibitors Towards Modulating Resistance Response to Insect-Pests in Legumes

Vishal Chugh, Vigya Mishra, Shalini Purwar, and Akhilesh Kumar Singh

Abstract Plant proteases are enzymes that play a critical role in various plant physiological processes, including resistance responses to biotic stresses. In legumes, the presence of proteases has been shown to modulate the resistance response to insect-pests. These enzymes have been shown to be involved in the regulation of defence mechanisms such as phytohormone synthesis, cell wall reinforcement, and the production of defence-related compounds. Moreover, the manipulation of protease activity through genetic engineering or chemical modulation has been shown to influence the susceptibility of legumes to insect-pest damage. Harnessing the potential of plant protease inhibitors offers environmentally friendly alternatives to synthetic pesticides, contributing to the sustainability of agricultural systems. This chapter underscores the importance of understanding plant-insect interactions and the mechanisms underlying resistance to effectively leverage protease inhibitors for enhancing legume crop productivity and ensuring food security in a rapidly changing world.

Keywords Protease inhibitors, Biochemical defences, Cystatins, Gene regulation, Trypsin inhibitors, Kunitz-type inhibitors, Bowman-Birk inhibitors (BBIs), Proteinase activity

---

V. Chugh (✉)
Department of Basic and Social Sciences, College of Horticulture, Banda University of Agriculture and Technology, Banda, Uttar Pradesh, India

V. Mishra
Department of Post Harvest Technology, College of Horticulture, Banda University of Agriculture and Technology, Banda, Uttar Pradesh, India

S. Purwar
Department of Basic and Social Sciences, College of Forestry, Banda University of Agriculture and Technology, Banda, Uttar Pradesh, India

A. K. Singh
Department of Entomology, College of Agriculture, Rani Lakshmi Bai Central Agricultural University, Jhansi, India

A. Pratap et al. (eds.), *Integrated Improvement of Food Legumes*,
https://doi.org/10.1007/978-3-031-82629-0_4

1 Introduction

If the earliest land plants and insects are taken into account, they have coexisted for the last 350 million years. During this time, they have developed a number of interactions that have an impact on the organisms at all levels, from basic biochemistry to population genetics. The most frequent interaction between the two phyla involves insect predation of plants and plant defences against herbivorous insects, even though some of these associations, like pollination, are mutually beneficial. Given this historical relationship, it is not surprising that plants use a variety of tactics to try to fend off or elude their insect herbivorous predators. Some species accumulate high levels of compounds that serve as biochemical defences through their toxicity, or their physical properties, while some other plants do not devote resources to the accumulation of defensive compounds but instead try to minimize herbivore damage through rapid growth and development, dispersion, or habitat choice. Different genotypes, even within a species, employ slightly distinct coexistence approaches with insect-pests, which can impact how resources are distributed between growth and defence (Jander et al., 2001). Given this diversity, it may be even more surprising that there appears to be a defence mechanism shared by all plants, based on the plant wounding response, and that this defence mechanism appears to be active even in species like Arabidopsis (*Arabidopsis thaliana*), which have low levels of constitutive defence and might be assumed to avoid herbivores rather than defend themselves. According to Thaler et al. (2001), the induced resistance mechanism works effectively against a range of insect herbivores on a particular plant species. Researchers have been exploring the production and control of particular compounds linked to pest defence in plants for a number of decades. Several studies have established that a number of secondary plant compounds that were originally thought to be toxic have since been identified as protective substances that help plants endure insects (Polya, 2003; Janzen et al., 1977, 1986; Rhodes, 1979). The complex molecules such as antibiotics, alkaloids, and terpenes, as well as proteins such as enzymes, enzyme inhibitors, and lectins, are among the inducible chemicals, which are also produced constitutively. These compounds are produced in response to insect or pest attack, which triggers intracellular and intercellular signals which activate on the gene that produces them.

Many inducible chemicals are the products of complex biochemical pathways, such as the synthesis of phytoalexins and lignin, which require the induced synthesis of several biosynthetic enzymes (Dixon, 1986). Chitinases, beta-glucanases, other PR-proteins, and protease inhibitors (PI) are examples of inducible defensive proteins (Chen & Varner, 1985; Corbin et al., 1987; Keller et al., 1988; Kombrink et al., 1988; Legrand et al., 1987; Vera & Conejero, 1988; Ryan, 1989). Among the protective compounds in plant tissues that are both developmentally controlled and produced in response to insect and pathogen attacks are protease inhibitor proteins. The structures, roles, and mechanisms of action of plant protease inhibitors have been discussed in a number of reviews (Rodríguez-Sifuentes et al., 2020; Habib & Khalid, 2007; Hellinger & Gruber, 2019; Fernández-Fernández et al., 2023; Boisen,

1983; Liener et al., 1988; Ryan, 1973, 1981). This chapter reviews a few recent and older studies that are pertinent to the uses of protease inhibitors in plant protection, with a focus on legumes. Because of the harm caused by the insects through their direct feeding as well as the fact that they serve as a vector or infection site for plant pathogens, insects are significant pests of legumes around the world. A particular crop species and its location have a significant impact on how important insect-pests are to legumes in comparison to other biotic pressures (such as diseases and weeds). When compared to crops grown in cold-season Mediterranean regions, tropical crops—such as beans in Central America and pulses in Asia and Africa—typically experience greater direct feeding damage from a wider array of insect-pests (Qaim & Zilberman, 2003). Aphids and the viruses carried by them as vectors are more likely to damage crop and pasture legumes produced in temperate climates or during colder seasons (Heie, 1994). The life stage, tissues, and feeding strategy of insect herbivores will all have a significant impact on how economically significant they are. The legume pests that cause the greatest economic harm feed on stored or unplanted seeds (Cardona & Kornegay, 1999; Clement et al., 1999). Although some legume species are able to effectively correct for pod and seed damage, insects that attack the pods can still result in significant economic loss (Russin et al., 1987; Brier & Rogers, 1991). Insects that feed on foliage or sap have the potential to cause significant harm if the environment encourages rapid population expansion.

2 Classification of Plant Protease Inhibitors

There are many proteins in the nature that interact with proteases to block their proteolytic function. The plant kingdoms have a number of non-homologous families of protease inhibitors (Laskowski & Kato, 1980). As an illustration of their

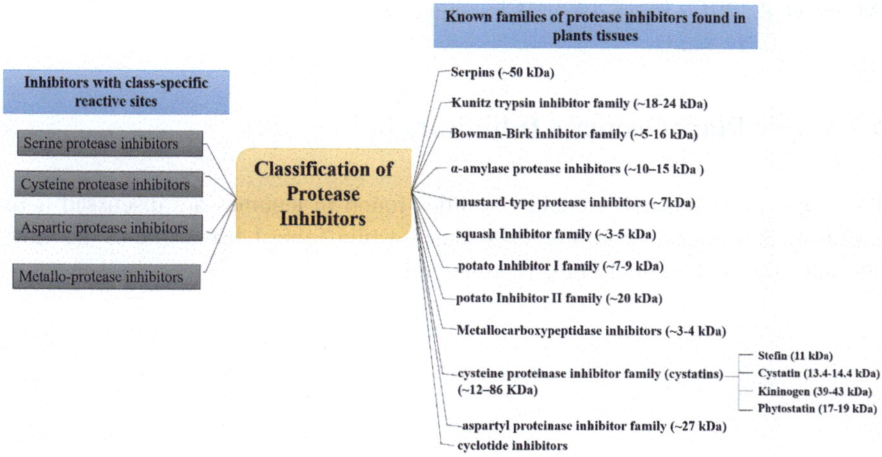

Fig. 1 Classification of protease inhibitors

significance for plant physiology and defence, protease inhibitors (PIs) make up 1–10% of the total soluble protein content in the seeds of legumes (Ryan, 1990). Plants use protease inhibitors to prevent the proteolytic breakdown of storage proteins in seeds, allowing for the controlled mobilization of nutrients in the form of amino acids or small peptides, as well as protection from herbivores by inhibiting their ability to digest food or by starving them through the gastrointestinal tract (Gomes et al., 2011). According to Hellinger and Gruber (2019), there are about 6700 plant-derived proteinaceous PIs, which may be divided into at least 12 different families based on structural similarity or sequence homology (Fig. 1). For each of the four mechanistic types of proteolytic enzymes—serine, cysteine, aspartic, and metallo-proteases—inhibitor families have been identified. Serpins, Kunitz trypsin inhibitor family, Bowman-Birk inhibitor family, mustard-type, potato Inhibitor I family, potato Inhibitor II family, squash Inhibitor family, cyclotide inhibitors, carboxypeptidase A, B inhibitor family, cysteine proteinase inhibitor family (cystatins), and aspartyl proteinase inhibitor family are the known families of protease inhibitors found in plant tissues.

The majority of these compounds have stabilizing motifs since the activity of the proteinaceous plant PIs typically necessitates well-structured domains (Bateman & James, 2011). Greater proteinogenic plant PIs have more than one domain and one or more inhibitory units. In contrast, many plant PIs have a molecular weight typical of peptides (5–100 amino acids), which makes them more bioactive. According to Arnison et al. (2013), they are also known as ribosomally produced and post-translationally modified peptides. Plant PIs frequently have a single inhibiting domain and one or more of the following: (i) secondary structural elements of protein, which are typically less susceptible to proteolytic degradation than unstructured domains (i.e. α-helices or β-sheets); (ii) post-translational modifications (i.e. pyroglutamate to protect the vulnerable N-terminus from aminopeptidases); (iii) cysteine-stabilizing motifs (e.g. cystine knot); or (iv) cyclization (e.g. cyclotides), which protects the peptide ends from aminopeptidases as well as carboxypeptidases (Mylne et al., 2012; Ryan, 1990; Volpicella et al., 2011).

3 Major Plant Protease Inhibitors in Legumes

The major plant protease inhibitor families found in legumes are discussed here with respect to their role in insect resistance, while Table 1 lists some of the most relevant studies pertaining to PIs reported in legumes:

Table 1 Protease inhibitors (PIs) in different legumes effective against targeted insect-pests

Sr. No.	Legume	Targeted enzyme	Target insect-pest	Molecular weight	Reference
1.	Soybean (*Glycine max*)	Trypsin and chymotrypsin	*Spodoptera littoralis*	17.9 kDa	El-Latif (2015)
2.	Pigeonpea (*Cajanus cajan, C. platycarpus*)	Trypsin and chymotrypsin	*Helicoverpa armigera*	Not available	Swathi et al. (2015)
3.	Soybean (*Glycine max*)	Trypsin	*B. cucurbitae*	Not available	Kaur et al. (2017)
4.	Chickpea (*Cicer arietinum, Cicer cuneatum, C. bijugum, C. chrossanicum, C. reticulatum*)	Trypsin and chymotrypsin	*Helicoverpa armigera*	Not available	Golla et al. (2018)
5.	Winged bean (*Psophocarpus tetragonolobus*) seed	Trypsin and chymotrypsin	*Helicoverpa armigera*	About 6–28 kDa	Shibata et al. (1986); Roy and Dutta (2009); Giri et al. (2003)
6.	Chickpea (*C. arietinum*) seeds	Trypsin	*Helicoverpa armigera*	30 kDa	Giri et al. (1998) and Kansal et al. (2008)
7.	Black-eyed pea (*Vigna unguiculata*)	Trypsin and chymotrypsin	*Cotton boll weevil (Anthonomus grandis)*	10 kDa	Franco et al. (2003)
8.	Mung bean (*V. radiata*)	Trypsin	*Helicoverpa armigera*	47 kDa	Kansal et al. (2008); Prasad, Dutta-Gupta, and Padmasree (2010a)
9.	Black gram (*V. mungo*)	Trypsin and chymotrypsin	*Helicoverpa armigera*	8 kDa	Kansal et al. (2008); Prasad, Dutta-Gupta, and Padmasree (2010a)
10.	Common bean (*Phaseolus vulgaris*)	Trypsin	*Spodoptera litura*	Not available	Negi et al. (2018)

(continued)

Table 1 (continued)

Sr. No.	Legume	Targeted enzyme	Target insect-pest	Molecular weight	Reference
11.	Uruvalheira (*Platypodium elegans*)	Trypsin and chymotrypsin	*Spodoptera frugiperda*	19.7 kDa	Ramalho et al. (2018)
12.	Soybean (*G. max*)	Trypsin	*Spodoptera litura*	Not available	Vasudev and Sohal (2016)
13.	Faba bean (*Vicia faba*)	Cysteine protease	*Halyomorpha halys*	Not available	Rondoni et al. (2018)
14.	Chickpea (*C. arietinum*)	Trypsin	*Helicoverpa armigera*	20 kDa	Srinivasan et al. (2005)
15.	*P. coccineus*	α-Amylase	*Hypothenemus hampei* *Tecia solanivora*	14, 18, and 32 kDa, 9–17 kDa	Valencia-Jiménez et al., 2008); de Azevedo et al. (2006)
16.	*P. vulgaris*	α-Amylase	Coffee berry borer pest, *Callosobruchus maculatus, Callosobruchus chinensis, Zabrotes subfasciatus, Sitophilus oryzae, Acanthoscelides obtectus, Cryptolestes ferrugineus, Cryptolestes pusillus, Oryzaephilus surinamensis, Sitophilus granaries, Tribolium castaneum, T. Castaneum, Drosophila melanogaster, Sarcophaga bullata, Aedes aegypti, Monomorium pharaonic, Apis mellifica, Venturia canescens, Ephestia cautella, E. Elutella, E. Kuehniella, Manduca sexta, Ostrinia nubilalis, Blattella germanica, Liposcelis decolor, Acheta domesticus, Eurydema oleracea, Graphosoma lineatum*	16–19 kDa, 15 kDa	Barbosa et al. (2010); Solleti et al. (2008); Nishizawa et al. (2007); Ignacimuthu and Prakash (2006); Kluh et al. (2005)
17.	*Vigna unguiculata*	α-Amylase	*Acanthoscelides obtectus, Callosobruchus Maculatus* and *Zabrotes subfasciatus*	25 kDa	Pelegrini et al. (2008); Wisessing (2012)
18.	*Pisum sativum*	α-Amylase	*Bruchus pisorum*	Not available	De Sousa-Majer et al. (2007)

3.1 Serpins

According to Rawling et al. (2018), the serpin family is the largest and most common superfamily of inhibitor family I4 PIs. Nearly every form of life, including viruses, bacteria, plants, and mammals, has been found to contain serpin-like genes. Large multicellular eukaryotic organisms have a number of serpin genes, but prokaryotes often only have one. Twenty-nine serpin genes have been identified in the thale cress (*Arabidopsis thaliana*) genome (Silverman et al., 2001). These are large PIs with molecular masses between 330 and 500 amino acids, with a common scaffold made up of three β-sheets and as many as nine α-helices (Gettins, 2002). Because they create a covalent compound with cysteine or serine proteases, serpins are known as "suicide inhibitors" (Bateman et al., 2011; Gettins, 2002). The primary function of serpins is the deactivation of endogenous proteases. Thus, these peptides play a key regulatory role in maintaining plant homeostasis. However, they are also said to act as a deterrent to insects and pathogens by suppressing their digestive proteases (Gomes et al., 2011). Leu-Arg-Xaa, where Xaa is a (Ala, Cys, Gly, and Ser), is the most prevalent reactive centre present in virtually all plant serpins at P2-P1-P1′. Serpins with this reactive centre are referred to as "LR serpins" (Roberts & Hejgaard, 2008). The inhibitor's reactive centre loop hooks into the protease's catalytic site. The inhibitor's reactive centre loop attaches into the protease's catalytic site. An acyl ester intermediate is formed between the inhibitor and the γ-oxygen atom of a serine residue of the protease upon breaking of the scissile link (P1-P1′). As a result, the protease structure is irreversibly altered as the serpin reactive centre loop adopts a 'superstable' β-sheet conformation (Huntington, 2011). Serpins have been reported to inhibit caspases and papain-like cysteine proteases in addition to serine proteases (Irving et al., 2002). Chymotrypsin serine proteases are used by insect orders including Lepidoptera and Diptera as digesting enzymes, making them likely targets for plant serpins. Seven serpin genes were found and described in the genomes of pigeonpea, soybean, common bean, cowpea, mungbean, and adzuki bean (Mainkar et al., 2020). Based on conserved reactive centres (P2-P1′), domains, and exon-intron boundaries in the gene structure, authors reported that phylogenetic analysis of legume serpins divided them into two major clades (Clade-I: VanSerpin, VraSerpin, VunSerpin, PvuSerpin, and Clade-II: GmaSerpin1, GmaSerpin2, and CcaSerpin). All seven serpins were shown to share a reactive core, indicating that they can target cognate proteases and are crucial in the immune system's response to fungi and insects.

3.2 Phytocystatins

Phytocystatins, a superfamily of plant PIs, are known to have molecular weights ranging from 12 to 86 KDa and consist of four families that block the activity of cysteine and thiol proteinases, including Family-1 cystatins (stefin family), Family-2 cystatins (cystatin family), Family-3 cystatins (kininogen family), and Family-cystatins (phytocystatins) (Lima et al., 2015; Habib & Khalid, 2007). These PIs, which are the second most investigated class of inhibitors, control both endogenous and exogenous proteases. These proteins act as competitive inhibitors to regulate several cysteine protease-dependent developmental processes. A core α-helix is surrounded by antiparallel β-sheets in the phytocystatin structure. Two hairpin loops with conserved motifs are part of the inhibitory domain; they are located in the centre [Gln-Xaa-Val-Xaa-Gly (Xaa can be any amino acid)] and the C-terminal region (Pro/Leu-Trp), respectively (Benchabane et al., 2010). They do not bind to cysteine proteases in a manner that is strictly substrate-like manner; rather, they inhibit them in a way that is not catalytically competent (i.e. they still prevent access to the catalytic site). The inhibitor's two hairpin motifs engage the active sites of proteases in conversation. Although it does not bind to the active site cleft, the N-terminal conserved region is crucial for the tight binding and cysteine protease specificity of these inhibitors. In addition to acting *ex planta* to stop cysteine proteases in feeding enemies, phytocystatins also serve as endogenous regulators of plant proteases. According to Haq et al. (2004) and Martinez et al. (2016), they are capable of inhibiting papain-like cysteine proteases found in a variety of insects and nematodes. According to van Wyk et al. (2016), since cystatins are active at the acidic pH values observed in Coleopteran insects, they could work in synergy with *Bt* to control Coleopteran insects and possibly prevent the development of *Bt* resistance in these insects. Western corn rootworm (WCR) growth and development are significantly inhibited by soybean cysteine proteinase inhibitor N (soyacystatin N, scN) feeding bioassay results, which attenuate digestive proteolysis (Zhao et al., 1996). Recombinant scN, however, was found to be more inhibiting than the powerful and broadly selective cysteine proteinase inhibitor E-64, according to Koiwa et al. (2000). The proteinase activities of each fraction were shown to be primarily scN-sensitive based on in-gel experiments, which the authors used to split the WCR digestive proteolytic activity into two fractions using mildly denaturing SDS-PAGE. They also extracted five peptides (28–33 kDa) from WCR gut extracts using scN affinity chromatographic separation, since binding affinity to the target proteinase determines how effective scN is as a proteinase inhibitor and an insecticide. These peptides resemble a cathepsin L-like cysteine proteinase (DvCAL1, *Diabrotica virgifera* cathepsin L), which is encoded by a WCR cDNA, according to analysis of their N-terminal sequence. Authors established that cathepsin L orthologs are pivotal digestive proteinases of WCR larvae and are targets of plant defensive cystatins (phytocystatins), like scN.

Since insects have evolved strategies to avoid any anti-digestive effects brought on by their plant host through coevolution, such as changes in gene expression in the gut, the actual application potential of cystatins against insect-pests is currently restricted (Zhu-Salzman & Zeng, 2015). A library of 20,352 cDNAs was prepared from the midgut tissue of cowpea bruchid larvae (*Callosobruchus maculatus*) given both a standard diet and diets with antinutritional substances in order to obtain insight into the nature and range of genes involved in adaptation to dietary stress. 1756 scN-responsive cDNAs were found to be clustered into 967 contigs, of which

653 were singletons, according to cDNA microarray analysis and quantitative real-time PCR (RT-PCR) confirmation with chosen genes of the transcript responses of the larvae to sustained and acute high doses of dietary soybean cystatin (scN). The proteins that were predicted to be encoded by the found differentially regulated sequences were involved in metabolism, structure, development, signalling, defence, and stress response. A group of genes whose altered expression was caused on by acute (lasting 24 h), high-level exposure to dietary scN largely overlapped with the transcript profile brought on by chronic, lower-level exposure. While structural, defence, and stress-related genes were mostly down-regulated by scN, protein and carbohydrate hydrolases were usually up-regulated. These findings demonstrate that in order to lessen the effects of a digestive protease inhibitor, insects actively deploy genetic resources in the gastrointestinal tract.

3.3 *Kunitz-Type Protease Inhibitors*

Among the serine PIs, particularly two types, Kunitz and Bowman-Birk, have are the most studied and are commonly present in legumes. According to Srinivasan et al. (2005), the Kunitz type contains one or two 18–24 kDa polypeptide chains with four cysteine residues that form two intra-chain disulfide bonds and block trypsin activity. One to multiple inhibitory domains are present in Kunitz-type inhibitors. Several 8 kDa-sized peptides have also been identified in plants and animals (Bendre et al., 2018). According to Bateman and James (2011), the universal β-trefoil structure, made up of up to 12 β-antiparallel strands arranged in a barrel-like domain with up to three disulfide connections connecting this subdomain, dominates the distinctive Kunitz motif. They were first found to be effective in soybean as serine PIs (such as those from the S01 chymotrypsin superfamily) and for the inhibition of aspartic acid proteases (such as human cathepsin D) (Gomes et al., 2011). With their traditional 'key-to-lock' binding to proteases, Kunitz inhibitors work as competitive inhibitors. Kunitz-type inhibitors are known as physiological homeostasis regulators in plants and also serve as pathogen protease inhibitors. One of the most destructive insect-pests to affect legumes is *Helicoverpa armigera.* The *Cajanus platycarpus*, a wild-type relative of pigeonpea (*Cajanus cajan*), has a serine PI that has been shown to have a significant inhibitory effect on the *H. armigera* gut trypsin-like protease (HGPs) (Swathi et al., 2015). The inhibitor contains a number of isoinhibitors with high homology to Kunitz-type PIs and demonstrated high stability at 50 °C in basic or acidic circumstances. These isoinhibitors have also demonstrated efficiency against HGPs. The scientists also added that the wild nature of the substance may be the cause of the formation of multiple isoinhibitors, whereas domesticated kinds may have lost their genetic variety. Golla et al. (2018) showed lesser pod damage by *H. armigera* in wild relatives of chickpea (*Cicer arietinum*) compared to cultivated chickpea due to the stronger insect gut trypsin and chymotrypsin inhibition induced by PIs found in the wild-type plants. In vitro inhibitory action against the gut protease of *Helicoverpa armigera* (HGP) was reported in the seed flour extracts of eight chosen varieties of mungbean by Kansal et al. (2008a). According to the study, mungbean trypsin inhibitor has insecticidal properties that could be used to grow transgenic plants. The same research group also discovered Kunitz trypsin inhibitor activity in seed flour extracts of 10 chosen kinds of chickpeas against gut protease of *Helicoverpa armigera* (Kansal et al., 2008b). The inhibitory effect of the chickpea trypsin inhibitor against *Helicoverpa armigera* was validated by the authors both in vitro and in vivo. According to an insect bioassay investigation, feeding *H. armigera* larvae with a meal supplemented with increasing doses of chickpea trypsin inhibitor resulted in a progressive fall in larval weight, growth, and survival as well as a temporal prolongation of larval growth. The adult emergence was said to have been negatively impacted by the inhibitor. Insecticidal effectiveness of soybean Kunitz trypsin inhibitor against the legume pod borer, *Helicoverpa armigera*, was also reported by Shukla et al. (2005). In a recent study, Gujjarlapudi et al. (2023) used mild trichloroacetic acid (TCA) extraction, trypsin-affinity chromatography, and further confirmation with immunodetection and MALDI-TOF studies to purify Kunitz inhibitor from the mature seeds of *V. radiata* (VrKI) and *C. platycarpus* (CpKI). Upon characterization of their structural (CD & fluorescence spectroscopy) and functional properties (temperature and DTT stability), authors established that Kunitz inhibitors are effective in the management of pod borer '*Helicoverpa armigera*'.

3.4 Bowman-Birk Inhibitors (BBIs)

BBIs comprise one or two inhibitory domains and have a molecular weight between 5 and 16 kDa. Two inhibitory loops are present in an antiparallel sheet domain on the opposite ends of the molecule. For BBI proteins, seven disulfide linkages serve as stabilizing components, while two active domains work as trypsin or chymotrypsin inhibitors. The inhibitor's two-domain structure can join forces with two different protease molecules to produce a stochiometric 1:1:1 complex (Bateman & James, 2011). According to the established mechanism for substrate-like binding to the protease's catalytic core, they work as competitive protease inhibitors (Sharma, 2015). In vitro application of a soybean trypsin-chymotrypsin inhibitor was reported to have a negative impact on *Batrocera cucurbitae* growth and development, a significant pest of cucurbit crops (Kaur et al., 2017). This study found a 66% drop in pupation and a 53% decrease in emergence. The scientists showed that the observed reduction in insect growth and development was caused by the total inhibition of insect trypsin activity by the Bowman-Birk inhibitor.

Red gram proteinase inhibitor (RgPI) was isolated and further characterized from pigeonpea (*Cajanus cajan*) seeds using ammonium sulfate precipitation, ion-exchange, affinity, and gel filtration chromatography (Prasad et al., 2010a). The authors reported two protein bands with molecular masses of ~8.5 and ~ 16.5 kDa corresponding to monomeric and dimeric forms of RgPI, along with other oligomeric forms: trimer, tetramer, and pentamer. This information was based on

SDS-PAGE under nonreducing conditions and further confirmation by matrix-assisted laser desorption ionization time-of-flight (MALDI-TOF) mass spectrometry. RgPI was further reduced with DTT, and the results of native-PAGE and two-dimensional gel electrophoresis showed that isoinhibitors with pI values of 5.95, 6.25, 6.50, 6.90, and 7.15 were present. It was inferred from the isolated RgPI's MALDI-TOF-TOF mass spectrum and N-terminal sequence ('DQHHSSKACC') that it belonged to the family of Bowman-Birk inhibitors. When supplied in the diet, *Manduca sexta* larvae showed less growth and development due to a decrease in the activity of trypsin-like proteinases in the midgut.

The insecticidal potential of isolated Bowman-Birk proteinase inhibitors from red gram (RgPI) and black gram (BgPI) towards lepidopteran insect-pests was examined in a different study by Prasad et al. (2010b). Both RgPI and BgPI exhibited impressive inhibitory efficacy against the midgut trypsin-like proteinases of *Achaea janata*, according to the authors. RgPI showed low to negligible inhibitory action, while BgPI displayed moderate to low inhibitory activity against the midgut trypsin-like proteinases of *Helicoverpa armigera, Spodoptera litura, Papilio demoleus, Amsacta albistriga, Corcyra cephalonica, Bombyx mori,* and *Daphnis nerii.* According to the study, the midgut trypsin-like proteinases of *A. janata* and *H. armigera,* as well as midgut trypsin- and chymotrypsin-like proteinases of *S. litura,* were susceptible to inhibition by RgPI and BgPI. As the concentration of the inhibitors in the diet increased, feeding studies showed that the body weight and survival rate of these larvae decreased in a dose-dependent manner. When compared to BgPI, RgPI significantly reduced the body weight and survival rate of *A. janata* larvae. In contrast, when compared to RgPI, BgPI revealed a considerable decrease in the body weight of *S. litura* larvae. Using the same methods as for KI, Gujjarlapudi et al. (2023) isolated and analyzed Bowman-Birk inhibitor from ripe seeds of *Vigna radiata* (VrBBI) and *Cajanus platycarpus* (CpBBI). Authors determined that BBI(s) are successful in managing castor semi-looper, '*Achaea Janata,*' after characterizing their structural (CD & fluorescence spectroscopy) and functional (temperature & DTT stability) capabilities.

3.5 α-Amylase Inhibitors (α-AIs)

α-AIs are abundant in microorganisms, higher plants, and animals (Kasar et al., 2022). According to Svensson et al. (2004), there are six different proteinaceous α-amylase inhibitor (AAI) subtypes found in plants: thionin-like, chloroform-methanol (CM) extractable proteins, knottin-like, Kunitz-type, thaumatin-like, and legume lectin-like inhibitors (Fig. 2). Several plant species, such as legumes (Ishimoto et al., 1996; Marshall & Lauda, 1975) and cereals (Franco et al., 2000; Iulek et al., 2000; Yamagata et al., 1998), can be used to extract -amylase inhibitors. Bifunctional α-amylase/TIs (AATI), which have been found in significant starch-containing agricultural plants such as barley, amaranth, wheat, and maize, are 10–15 kDa in size. These molecules include numerous a-helical segments and

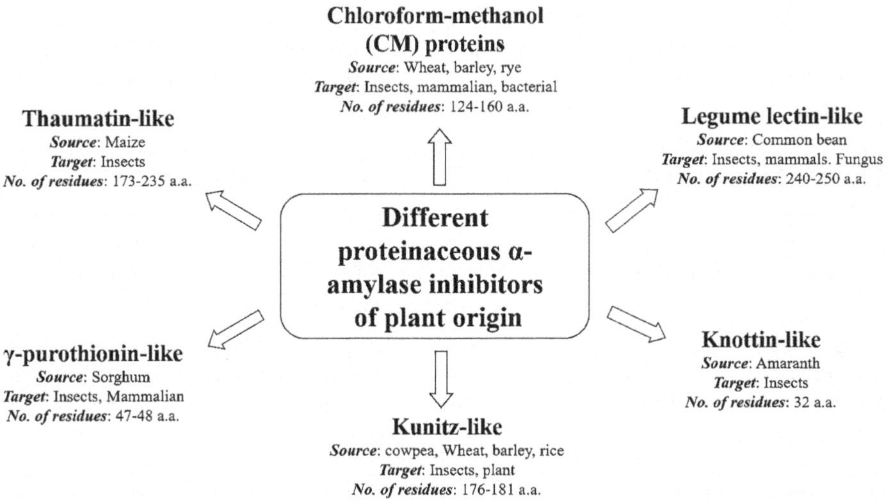

Fig. 2 Classification of plant α-amylase inhibitors based on structural similarity. (Mehrabadi et al., 2012; Richardson, 1991)

cystine-spanning loops. These peptides act as competitive PIs because their inhibitory loops acquire confirmations that are similar to those of conventional substrates.

The control of endogenous α-amylase activity is a physiological function of α-amylase inhibitors. Additionally, plants naturally utilize α-AIs to protect themselves from insect-pests (Ishimoto et al., 1989; Kluh et al., 2005). There is also a lot of interest in using α-AIs to produce transgenic plants that are resistant to insect-pests and to utilize them to manage them (Chrispeels et al., 1998; Gatehouse & Gatehouse 1998; Valencia-Jiménez et al., 2008; Yamada et al., 2001). It has already been established that inhibitors of insect α-amylase are a crucial biological mechanism in the management of insect-pests (Franco et al., 2005; Carlini & Grossi-de-Sa' 2002; Svensson et al., 2004). The expression of the cDNA encoding αAl-I into some plants, such as pea (*Pisum sativum* L.) and azuki bean (*Vigna angularis* L.) against bruchid beetle pests (*Coleoptera: Bruchidae*), has been well documented for exhibiting the ability of these inhibitors to be used as plant resistance factors against some species of insect-pests (Ishimoto et al., 1989; Yamada et al., 2001; Kluh et al., 2005). Pea and azuki bean transgenic plants expressing α-amylase inhibitors from common beans (α-AI) were completely resistant to the *Bruchus pisorum* and *Callosobruchus chinensis* weevils (Morton et al., 2000).

The common white, red, and black kidney beans were found to contain the lectin-like inhibitors, αAI-1 and αAI-2 (Ishimoto & Chrispeels, 1996). They showed different specificity against α-amylases because of the mutation in their primary structure (Grossi de Sa et al., 1997). Although αAI-1 inhibits a number of insect and mammalian α-amylases, it is ineffective against the Mexican bean weevil (*Zabrotes subfasciatus*). However, αAI-2 inhibits the α-amylase of *Z. subfasciatus* but does not inhibit the α-amylases recognized by αAI-1 (Ishimoto & Chrispeels, 1996; Kluh et al., 2005).

4 Transgenic Interventions to Address Insect Resistance to Protease Inhibitors

Insects can adjust to the presence of PIs through a variety of processes, including raising the activity of enzymes that break down PIs, producing fewer enzymes that are susceptible to inhibitors, digesting PIs, reducing the susceptibility of proteases to PIs, and producing more enzymes that are PI-resistant. The ability of endosymbiotic insect bacteria to create non-sensitive PI proteases, which aid insects in digesting protein in their guts, has also been demonstrated (Wielkopolan et al., 2018). In particular, how *H. armigera* and *C. maculatus* regulate gene expression in response to PIs, including the upregulation of presumably insensitive proteases (Bown et al., 2004; Moon et al., 2004), has shed light on how insects can overcome plant proteinase inhibitors (PIs) (Patankar et al., 2001; Zhu-Salzman et al., 2003). In addition, insects have a large number of detoxification enzymes, including cytochrome P450s and glutathione S-transferases, which detoxify plant secondary metabolites. For some time, it has been understood that certain detoxifying enzymes are produced at higher rates by insects as a result of plant defence responses (Schuler, 1996). The discovery that *Helicoverpa zea* really activates specific cytochrome P450 genes linked to the detoxification of hazardous chemicals produced by plants using the same signals plants employ to coordinate their defence response, SA and JA, is an interesting advance (Li et al., 2002b, b).

The wild-type relatives of common legumes have a collection of genes that are active against insects and other pests, making them resistant to their attack (Rodríguez-Sifuentes et al., 2020). Recombinant PIs have been created in microbes using these genes. Since multiple PIs might be applied simultaneously to the crop, using wild-type genes might end up being a preferable tactic. The efficiency of PI is ensured by this tactic by preventing insect adaptability. Two recombinant phytocystatins from cowpea (*Vigna unguiculata*) were developed in *E. coli* by Monteiro Jnior et al. (2017). The proteins exhibited great stability, effective papain and chymopapain inhibition, and a reduction of up to 80% in the overall proteolytic activity of *C. maculatus* midgut proteases. On the contrary, none of the recombinant PIs altered the percentages of larval mortality, adult emergence, the average weight of emerged individuals, or the mean development time of *C. maculatus*. According to the scientists, the beetle adapted to the diet and as a result, developed proteases that could break down the inhibitor (Zhu-Salzman et al., 2003) and cysteine protease resistant to phytocystatins (Koo et al., 2008). On the other hand, Cruz et al. (2015) observed that certain cultivars of Brazilian cowpea were resistant to infestation and harm by *C. maculatus*.

The successful emergence of insect resistance has been greatly influenced by transgenic plants that produce insecticidal crystal proteins (ICPs). The ICPs are created by the *B. thuringiensis* (*Bt*) during the sporulation process. In an insect's midgut, proteinases can solubilize and activate *Bt* toxins, ICPs, δ-endotoxins, or crystalline proteins. A pore is made in the epithelial cells of the insect gut when the toxin interacts to a particular receptor, killing the insect (Whalon & Wingerd, 2003). When *Bt* toxin-producing ICP genes were inserted into several crop species under the control of tissue-specific or constitutive promoters, resistance to insect-pests was primarily seen (Chang et al., 2017; Katta et al., 2020; Soliman et al., 2017; Zhong et al., 2019). Latin America has authorized the use of transgenic soybeans that express *Cry1Ac* + *Cry1Ab* for commercial purposes (Koch et al., 2015). By transferring the *cry2IIAa* gene with Mendelian segregation at 15% transformation efficiency, an effective transformation technique was devised in chickpeas that demonstrated resistance to pod borer (Sawardekar et al., 2017). By introducing *cry1Ac* and *cry2Aa* constructions, Ghosh et al. (2017) created a line of transgenic pigeon peas that are independent from one another. These transgenes in transgenic lines were passed down over three generations and shown resistance to *H. armigera*, which causes approximately 80–100% larval mortality. The *cry2Aa* gene was transferred using an *Agrobacterium*-mediated transformation procedure, which produced 17 putative transformants with fresh leaf protein expression levels ranging from 0.04 lg/g to 1.45 lg/g. Pigeonpea transgenics were developed using this method. After feeding on detached leaves from transgenic plants, *H. armigera* drastically reduced the weight and length of the larvae (Baburao & Sumangala, 2018). In a different work, *cry2Aa* gene in-planta transformation was used to create transgenic pigeonpea with 0.8% transformation frequency in T1 generation. According to Singh et al. (2018), the Cry2Aa protein accumulated in these transgenic plants at a rate of 25–80 lg/g FW, which resulted in 80–100% larval mortality to the pod borer, *H. armigera. Holotrichia parallela*, a Coleopteran pest, was resistant to soybean when the *Cry8-like* gene from *B. thuringenesis* was expressed, and the protein was predominantly expressed at the root stage, where insects feed on roots (Qin et al., 2019).

Mamta and Rajam (2017) present new perspectives for eco-friendly insect-pest management in plants using an RNAi-based insecticidal method. Insect-pest critical gene-specific double-stranded RNA (dsRNA) is converted by the RNase III enzyme Dicer to small interfering RNAs (siRNAs), which direct the Argonaut protein of the RNA-induced silencing complex (RISC) to degrade complementary mRNA (Scott et al., 2013). In order to promote disease resistance in plants, dsRNAs are frequently utilized to interfere with specific gene silencing through genetic changes. Even at T3 generation, the soybean pod borer larval mortality rate decreased as a result of soybean pod borer dsRNA encoding SpbP0 expression (Meng et al., 2017). Targeting TREH, ATPD, ATPE, and CHSI genes that caused a high percentage of soybean aphid (*Aphis glycines*) death in soybean was done using a dsRNA/Nano carrier formulation (Yan et al., 2020).

A bacterial defence system against invading bacteriophages is called clustered regularly interspaced short palindromic repeats (CRISPR)/CRISPR-associated 9

(CRISPR/Cas9). For CRISPR-Cas9-mediated genome editing, Cas9 and adaptable single guide RNA (sgRNA) are two essential components. In order to disrupt the function of a gene, Cas9 creates a frameshift in the targeted region by cleaving complementary and non-complementary DNA strands, respectively (Doudna & Charpentier, 2014). Cas9 is a monomeric RNA-guided DNA endonuclease with two domains called RuvC and HNH nucleases. The CRISPR/Cas9 technique has been used to disable numerous insect genes, including those of *H. armigera* and *S. exigua*. In both *H. armigera* and *S. exuigua*, spinosyn resistance was demonstrated by knockout mutations in the a-6-nicotinic acetylcholine receptor (nAchR) by CRISPR/Cas9 (Wang et al., 2020; Zuo et al., 2020). In a different study, resistance to both *cry2Aa* and *cry2Ab* was conferred on *H. armigera* by CRISPR/Cas9-mediated deletion of the ABC transporter gene, *HaABCA2* (Wang et al., 2018).

5 Conclusion

Legumes have a variety of defences against insect and pest attacks. Among these, PIs are naturally occurring proteins that are produced when certain herbivores are present and have the ability to inhibit the activity of proteolytic enzymes. There has been substantial discussion of the effectiveness of PIs in preventing or controlling insect damage. Their usage in transgenic plants is intended as a result. Introducing PIs in recombinant microbes is the best way to use them in agriculture. Using multiple PIs simultaneously in crops is an option here that could prevent the emergence of insect resistance. An excellent alternative for developing inhibitors with the potential to be used as co-adjutants in crop development is the research of wild-type legume plants as PI sources. In conclusion, plant protease inhibitors have emerged as a promising tool for modulating the resistance response to insect-pests in legumes. These natural compounds play a crucial role in plant defence mechanisms by inhibiting digestive proteases in insects, thereby interfering with their ability to obtain essential nutrients from the host plant. Through their multifaceted effects, protease inhibitors not only act as direct toxins but also disrupt insect growth and development. Moreover, they exhibit a wide range of insecticidal activity against diverse pest species, making them valuable candidates for sustainable pest management strategies. By harnessing the potential of plant protease inhibitors, researchers and agriculturalists can pave the way for environmentally friendly and effective approaches to protect legume crops from insect damage, reducing reliance on synthetic pesticides and promoting the sustainability of agricultural systems. As our understanding of plant-insect interactions and the mechanisms underlying resistance continues to expand, the use of protease inhibitors holds great promise in enhancing legume crop productivity and ensuring food security in a rapidly evolving world.

References

Arnison, P. G., Bibb, M. J., Bierbaum, G., et al. (2013). Ribosomally synthesized and post-translationally modified peptide natural products: Overview and recommendations for a universal nomenclature. *Natural Product Reports, 30*, 108–160. https://doi.org/10.1039/c2np20085f

Baburao, T. M., & Sumangala, B. (2018). Development and molecular characterization of transgenic pigeon pea carrying *cry2Aa* for pod borer resistance. *Journal of Pharmacognosy and Phytochemistry, 7*, 1581–1585.

Barbosa, A. E., Albuquerque, É. V., Silva, M. C., et al. (2010). *α-Amylase inhibitor-1* gene from *Phaseolus vulgaris* expressed in *Coffea arabica* plants inhibits α-amylases from the coffee berry borer pest. *BMC Biotechnology, 10*. https://doi.org/10.1186/1472-6750-10-44

Bateman, K. S., & James, M. N. G. (2011). Plant protein proteinase inhibitors: Structure and mechanism of inhibition. *Current Protein & Peptide, 12*, 341–347. https://doi.org/10.2174/138920311796391124

Benchabane, M., Schlüter, U., Vorster, J., et al. (2010). Plant cystatins. *Biochimie, 92*, 1657–1666. https://doi.org/10.1016/j.biochi.2010.06.006

Bendre, A. D., Ramasamy, S., & Suresh, C. G. (2018). Analysis of Kunitz inhibitors from plants for comprehensive structural and functional insights. *International Journal of Biological Macromolecules, 113*, 933–943. https://doi.org/10.1016/j.ijbiomac.2018.02.148

Boisen, S. (1983). Protease inhibitors in cereals. *Acta Agriculturae Scandinavica, 33*, 369–381. https://doi.org/10.1080/00015128309435377

Bown, D. P., Wilkinson, H. S., & Gatehouse, J. A. (2004). Regulation of expression of genes encoding digestive proteases in the gut of a polyphagous lepidopteran larva in response to dietary protease inhibitors. *Physiological Entomology, 29*, 278–290. https://doi.org/10.1111/j.0307-6962.2004.00402.x

Carlini, C. R., & Grossi-de-Sá, M. F. (2002). Plant toxic proteins with insecticidal properties. A review on their potentialities as bioinsecticides. *Toxicon, 40*, 1515–1539. https://doi.org/10.1016/S0041-0101(02)00240-4

Chang, X., Lu, Z., Shen, Z., et al. (2017). Bitrophic and tritrophic effects of transgenic *cry1Ab/cry2Aj* maize on the beneficial, nontarget *Harmonia axyridis* (Coleoptera: Coccinellidae). *Environmental Entomology, 46*, 1171–1176. https://doi.org/10.1093/ee/nvx113

Chen, J., & Varner, J. E. (1985). An extracellular matrix protein in plants: Characterization of a genomic clone for carrot extensin. *The EMBO Journal, 4*, 2145–2151. https://doi.org/10.1002/j.1460-2075.1985.tb03908.x

Chrispeels, M. J., Montes, M., & Thomas. (1998). Genetic engineering with α-amylase inhibitors makes seeds resistant to bruchids. *Seed Science Research, 8*, 257–264. https://doi.org/10.1017/s0960258500004153

Corbin, D. R., Sauer, N., & Lamb, C. R. (1987). Differential regulation of a hydroxyproline-rich glycoprotein gene family in wounded and infected plants. *Journal of Molecular Cell Biology, 7*, 4337–4344. https://doi.org/10.1128/mcb.7.12.4337-4344.1987

Cruz, L. P., de Sá, L. F. R., Santos, L. A., et al. (2015). Evaluation of resistance in different cowpea cultivars to *Callosobruchus maculatus* infestation. *Journal of Pest Science, 89*, 117–128. https://doi.org/10.1007/s10340-015-0657-z

de Azevedo, P. R., Nogueira Batista, J. A., Mattar da Silva, M. C., et al. (2006). An α-amylase inhibitor gene from *Phaseolus coccineus* encodes a protein with potential for control of coffee berry borer (*Hypothenemus hampei*). *Phytochemistry, 67*, 2009–2016. https://doi.org/10.1016/j.phytochem.2006.06.029

de Sa, M. F. G., Mirkov, T. E., Ishimoto, M., et al. (1997). Molecular characterization of a bean α-amylase inhibitor that inhibits the α-amylase of the Mexican bean weevil *Zabrotes subfasciatus*. *Planta, 203*, 295–303. https://doi.org/10.1007/s004250050195

De Sousa-majer, M. J., Hardie, D. C., Turner, N. C., & Higgins, T. J. (2007). Bean α-amylase inhibitors in transgenic peas inhibit development of pea weevil larvae. *Journal of Economic Entomology, 100*(4), 1416–1422.

Dixon, R. A. (1986). The phytoalexin response: Elicitation, signalling and control of host gene expression. *Biological Reviews, 61*, 239–291. https://doi.org/10.1111/j.1469-185x.1986.tb00719.x

Doudna, J. A., & Charpentier, E. (2014). The new frontier of genome engineering with CRISPR-Cas9. *Science, 346*, 1258096–1258096. https://doi.org/10.1126/science.1258096

El-latif, A. O. A. (2015). Protease purification and characterization of a serine protease inhibitor from Egyptian varieties of soybean seeds and its efficacy against *Spodoptera littoralis*. *Journal of Plant Protection Research, 55*, 16–25. https://doi.org/10.1515/jppr-2015-0003

Fernández-Fernández, Á. D., Stael, S., & Van Breusegem, F. (2023). Mechanisms controlling plant proteases and their substrates. *Cell Death and Differentiation, 30*, 1047–1058. https://doi.org/10.1038/s41418-023-01120-5

Franco, O. L., Rigden, D. J., R. Melo, F., et al. (2000). Activity of wheat α-amylase inhibitors towards bruchid α-amylases and structural explanation of observed specificities. *European Journal of Molecular Biology, 267*, 2166–2173. https://doi.org/10.1046/j.1432-1327.2000.01199.x

Franco, O. L., Santos, B. J., et al. (2003). Effects of black-eyed pea trypsin/chymotrypsin inhibitor on proteolytic activity and on development of *Anthonomus grandis*. *Phytochemistry, 63*, 343–349. https://doi.org/10.1016/s0031-9422(03)00108-0

Franco, O. L., Melo, F. R., Mendes, P., et al. (2005). Characterization of two *Acanthoscelides obtectus* α-amylases and their inactivation by wheat inhibitors. *Journal of Agricultural and Food Chemistry, 53*, 1585–1590. https://doi.org/10.1021/jf049343x

Gatehouse, A. M. R., & Gatehouse, J. A. (1998). Identifying proteins with insecticidal activity: Use of encoding genes to produce insect-resistant transgenic crops. *Pesticide Science, 52*, 165–175. https://doi.org/10.1002/(sici)1096-9063(199802)52:2%3C165::aid-ps679%3E3.0.co;2-7

Gettins, P. G. W. (2002). Serpin structure, mechanism, and function. *Chemical Reviews, 102*, 4751–4804. https://doi.org/10.1021/cr010170+

Ghosh, G., Ganguly, S., Purohit, A., et al. (2017). Transgenic pigeonpea events expressing *Cry1Ac* and *Cry2Aa* exhibit resistance to *Helicoverpa armigera*. *Plant Cell Reports, 36*, 1037–1051. https://doi.org/10.1007/s00299-017-2133-0

Giri, A. P., Harsulkar, A. M., Ku, M. S. B., et al. (2003). Identification of potent inhibitors of *Helicoverpa armigera* gut proteinases from winged bean seeds. *Phytochemistry, 63*, 523–532. https://doi.org/10.1016/S0031-9422(03)00181-X

Golla, S. K., Rajasekhar, P., Akbar, D., & Hl, S. (2018). Proteolytic activity in the Midgut of *Helicoverpa armigera* (Noctuidae: Lepidoptera) larvae fed on wild relatives of chickpea, *Cicer arietinum*. *Journal of Economic Entomology, 111*, 2409–2415. https://doi.org/10.1093/jee/toy160

Gomes, M. T., Oliva, M. L., Lopes, M. T., & Salas, C. E. (2011). Plant proteinases and inhibitors: An overview of biological function and pharmacological activity. *Current Protein & Peptide Science, 12*, 417–436. https://doi.org/10.2174/138920311796391089

Gujjarlapudi, M., Kotarya, B., Mohanraj, S. S., et al. (2023). Development of a rapid process for purification of Bowman-Birk and Kunitz inhibitors from legume seeds, and evaluation of their biophysical, insecticidal, and antimicrobial properties. *International Journal of Biological Macromolecules, 238*, 124050. https://doi.org/10.1016/j.ijbiomac.2023.124050

Habib, H., & Fazili, K. M. (2007). Plant protease inhibitors: A defense strategy in plants. *Biotechnology and Molecular Biology Reviews, 2*, 68–85. https://doi.org/10.5897/bmbr2007.0006

Haq, S. K., Atif, S. M., & Khan, R. H. (2004). Protein proteinase inhibitor genes in combat against insects, pests, and pathogens: Natural and engineered phytoprotection. *Archives of Biochemistry and Biophysics, 431*, 145–159. https://doi.org/10.1016/j.abb.2004.07.022

Heie, O. E. (1994). Why are there so few aphid species in the temperate areas of the southern hemisphere? *European Journal of Entomology, 91*, 127–133.

Hellinger, R., & Gruber, C. W. (2019). Peptide-based protease inhibitors from plants. *Drug Discovery Today, 24*, 1877–1889. https://doi.org/10.1016/j.drudis.2019.05.026

Huntington, J. A. (2011). Serpin structure, function and dysfunction. *Journal of Thrombosis and Haemostasis, 9*, 26–34. https://doi.org/10.1111/j.1538-7836.2011.04360.x
Ignacimuthu, S., & Prakash, S. (2006). *Agrobacterium*-mediated transformation of chickpea with α-amylase inhibitor gene for insect resistance. *Journal of Biosciences, 31*, 339–345. https://doi.org/10.1007/bf02704106
Ishimoto, M., & Kitamura, K. (1989). Growth inhibitory effects of an α-amylase inhibitor from the kidney bean, *Phaseolus vulgaris* (L.) on three species of Bruchids (Coleoptera: Bruchidae). *Applied Entomology and Zoology, 24*, 281–286. https://doi.org/10.1303/aez.24.281
Ishimoto, M., Sato, T., Chrispeels, M. J., & Kitamura, K. (1996). Bruchid resistance of transgenic azuki bean expressing seed α-amylase inhibitor of common bean. *Entomologia Experimentalis et Applicata, 79*, 309–315. https://doi.org/10.1111/j.1570-7458.1996.tb00838.x
Iulek, J., Franco, O. L., Silva, M., et al. (2000). Purification, biochemical characterisation and partial primary structure of a new α-amylase inhibitor from *Secale cereale* (rye). *The International Journal of Biochemistry & Cell Biology, 32*, 1195–1204. https://doi.org/10.1016/S1357-2725(00)00053-4
Jander, G., Cui, J. P., Nhan, B., Pierce, N. E., & Ausubel, F. M. (2001). The *TASTY* locus on chromosome 1 of Arabidopsis affects feeding of the insect herbivore *Trichoplusia ni*. *Plant Physiology, 126*, 890–898.
Janzen, D. H., Juster, H. B., & Arthur Bell, E. (1977). Toxicity of secondary compounds to the seed-eating larvae of the bruchid beetle *Callosobruchus maculatus*. *Phytochemistry, 16*, 223–227. https://doi.org/10.1016/s0031-9422(00)86790-4
Janzen, D. H., Ryan, C. A., Liener, I. E., & Pearce, G. F. (1986). Potentially defensive proteins in mature seeds of 59 species of tropical *leguminosae*. *Journal of Chemical Ecology, 12*, 1469–1480. https://doi.org/10.1007/bf01012365
Júnior, J. E., Valadares, N. F., Pereira, H. D., Dyszy, F. H., et al. (2017). Expression in Escherichia coli of cysteine protease inhibitors from cowpea (*Vigna unguiculata*): The crystal structure of a single-domain cystatin gives insights on its thermal and pH stability. *International Journal of Biological Macromolecules, 102*, 29–41.
Kansal, R., Gupta, R. B., Koundal, K. R., et al. (2008a). Purification, characterization and evaluation of insecticidal potential of trypsin inhibitor from mungbean (*Vigna radiata* L. Wilczek) seeds. *Acta Physiologiae Plantarum, 30*, 761–768. https://doi.org/10.1007/s11738-008-0178-y
Kansal, R., Kumar, M., Kuhar, K., et al. (2008b). Purification and characterization of trypsin inhibitor from *Cicer arietinum* L. and its efficacy against *Helicoverpa armigera*. *Brazilian Journal of Plant Physiology, 20*, 313–322. https://doi.org/10.1590/s1677-04202008000400007
Kasar, S. S., Maheshwari, V. L., & Pawar, P. K. (2022). Bioactive α-amylase inhibitors: Sources, mechanism of action, biochemical characterization, and applications. *Natural Products as Enzyme Inhibitors: An Industrial Perspective, 59–104*. https://doi.org/10.1007/978-981-19-0932-0_3
Katta, S., Talakayala, A., Reddy, M. K., et al. (2020). Development of transgenic cotton (Narasimha) using triple gene *Cry2Ab-Cry1F-Cry1Ac* construct conferring resistance to lepidopteran pest. *Journal of Biosciences, 45*. https://doi.org/10.1007/s12038-020-0006-0
Kaur, H., Kaur, A., Kaur, A. P., et al. (2017). Assessment of soybean inhibitor as a biopesticide against melon fruit fly, *Bactrocera cucurbitae* (Coquillett). *Journal of Plant Diseases and Protection, 124*, 445–451. https://doi.org/10.1007/s41348-017-0108-6
Keller, B., Sauer, N., & Lamb, C. R. (1988). Glycine-rich cell wall proteins in bean: Gene structure and association of the protein with the vascular system. *The EMBO Journal, 7*, 3625–3633. https://doi.org/10.1002/j.1460-2075.1988.tb03243.x
Kluh, I., Horn, M., Hyblova, J., et al. (2005). Inhibitory specificity and insecticidal selectivity of α-amylase inhibitor from. *Phytochemistry, 66*, 31–39. https://doi.org/10.1016/j.phytochem.2004.11.001
Koch, M. S., Ward, J. M., Levine, S. L., et al. (2015). The food and environmental safety of *Bt* crops. *Frontiers in Plant Science, 06*, 283. https://doi.org/10.3389/fpls.2015.00283

Koiwa, H., Shade, R. E., Zhu-Salzman, K., et al. (2000). A plant defensive cystatin (soyacystatin) targets cathepsin L-like digestive cysteine proteinases (DvCALs) in the larval midgut of western corn rootworm (*Diabrotica virgifera virgifera*). *FEBS Letters, 471*, 67–70. https://doi.org/10.1016/s0014-5793(00)01368-5

Kombrink, E., Schroder, M., & Hahlbrock, K. (1988). Several "pathogenesis-related" proteins in potato are 1,3—glucanases and chitinases. *Proceedings. National Academy of Sciences. United States of America, 85*, 782–786. https://doi.org/10.1073/pnas.85.3.782

Koo, Y. M., Ahn, J. K., Salzman, R. A., et al. (2008). Functional expression of an insect cathepsin B-like counter-defence protein. *Insect Molecular Biology, 17*, 235–245. https://doi.org/10.1111/j.1365-2583.2008.00799.x

Laskowski, M., & Kato, I. (1980). Protein inhibitors of proteinases. *Annual Review of Biochemistry, 49*, 593–626. https://doi.org/10.1146/annurev.bi.49.070180.003113

Legrand, M., Kauffmann, S., Geoffroy, P., & Fritig, B. (1987). Biological function of pathogenesis-related proteins: Four tobacco pathogenesis-related proteins are chitinases. *Proceedings. National Academy of Sciences. United States of America, 84*, 6750–6754. https://doi.org/10.1073/pnas.84.19.6750

Li, X., Berenbaum, M. R., & Schuler, M. A. (2002a). Plant allelochemicals differentially regulate *Helicoverpa zea* cytochrome *P450* genes. *Insect Molecular Biology, 11*, 343–351. https://doi.org/10.1046/j.1365-2583.2002.00341.x

Li, X., Schuler, M. A., & Berenbaum, M. R. (2002b). Jasmonate and salicylate induce expression of herbivore cytochrome *P450* genes. *Nature, 419*, 712–715. https://doi.org/10.1038/nature01003

Liener, I. E., Goodale, R. L., Deshmukh, A., et al. (1988). Effect of a trypsin inhibitor from soybeans (Bowman-Birk) on the secretory activity of the human pancreas. *Gastroenterol, 94*, 419–427. https://doi.org/10.1016/0016-5085(88)90431-3

Lima, A., dos Reis, S., & de Souza, C. (2015). Phytocystatins and their potential to control plant diseases caused by fungi. *Protein and Peptide Letters, 22*, 104–111. https://doi.org/10.2174/0929866521666140418101711

Mainkar, P. S., Manoj, M. L., Jayaswal, D., Agarwal, Y., Prajapat, R. K., & Kansal, R. (2020). Identification and in-silico characterization of Serpin genes in legumes genomes. *Indian Journal of Agricultural Sciences, 90*, 1763–1768. https://doi.org/10.56093/ijas.v90i9.106624

Mamta, B., & Rajam, M. V. (2017). RNAi technology: A new platform for crop pest control. *Physiology and Molecular Biology of Plants, 23*, 487–501. https://doi.org/10.1007/s12298-017-0443-x

Marshall, J., & Lauda, C. (1975). Purification and properties of phaseolamin, an inhibitor of alpha-amylase, from the kidney bean, *Phaseolus vulgaris*. *The Journal of Biological Chemistry, 250*, 8030–8037. https://doi.org/10.1016/s0021-9258(19)40811-9

Martinez, M., Santamaria, M., Diaz-Mendoza, M., et al. (2016). Phytocystatins: Defense proteins against phytophagous insects and Acari. *International Journal of Molecular Sciences, 17*, 1747. https://doi.org/10.3390/ijms17101747

Mehrabadi, M., Franco, O. L., & Bandani, A. R. (2012). Plant proteinaceous alpha-amylase and proteinase inhibitors and their use in insect pest control. *New Perspectives in Plant Protection, 11*, 229–246.

Meng, F., Li, Y., Zang, Z., et al. (2017). Expression of the double-stranded RNA of the soybean pod borer *Leguminivora glycinivorella* (Lepidoptera: Tortricidae) ribosomal protein P0 gene enhances the resistance of transgenic soybean plants. *Pest Management Science, 73*, 2447–2455. https://doi.org/10.1002/ps.4637

Moon, J. C., Salzman, R. A., Ahn, J. K., et al. (2004). Transcriptional regulation in cowpea bruchid guts during adaptation to a plant defence protease inhibitor. *Insect Molecular Biology, 13*, 283–291. https://doi.org/10.1111/j.0962-1075.2004.00485.x

Morton, R. L., Schroeder, H. E., Bateman, K. S., et al. (2000). Bean alpha-amylase inhibitor 1 in transgenic peas (Pisum sativum) provides complete protection from pea weevil (*Bruchus piso-*

rum) under field conditions. *Proceedings of the National Academy of Sciences, 97*, 3820–3825. https://doi.org/10.1073/pnas.070054597

Mylne, J. S., Chan, L. Y., Chanson, A. H., et al. (2012). Cyclic peptides arising by evolutionary parallelism via Asparaginyl-endopeptidase–mediated biosynthesis. *Plant Cell, 24*, 2765–2778. https://doi.org/10.1105/tpc.112.099085

Negi, P., Chand, S., Thakur, N., & Nath, A. K. (2018). Biological activity of serine protease inhibitor isolated from the seeds of *Phaseolus vulgaris*. *Agricultural Research, 7*, 265–270. https://doi.org/10.1007/s40003-018-0304-z

Patankar, A. G., Giri, A. P., Harsulkar, A. M., et al. (2001). Complexity in specificities and expression of *Helicoverpa armigera* gut proteinases explains polyphagous nature of the insect pest. *Insect Biochemistry and Molecular Biology, 31*, 453–464. https://doi.org/10.1016/s0965-1748(00)00150-8

Pelegrini, P. B., Lay, F. T., Murad, A. M., et al. (2008). Novel insights on the mechanism of action of α-amylase inhibitors from the plant defensin family. *Proteins: Structure, Function, and Bioinformatics, 73*, 719–729. https://doi.org/10.1002/prot.22086

Polya, G. M. (2003). Protein and non-protein protease inhibitors from plants. *Natural Product Chemistry, 29*, 567–641. https://doi.org/10.1016/s1572-5995(03)80015-7

Prasad, E. R., Dutta-Gupta, A., & Padmasree, K. (2010a). Insecticidal potential of Bowman–Birk proteinase inhibitors from red gram (*Cajanus cajan*) and black gram (*Vigna mungo*) against lepidopteran insect-pests. *Pesticide Biochemistry and Physiology, 98*, 80–88. https://doi.org/10.1016/j.pestbp.2010.05.003

Prasad, E. R., Merzendorfer, H., Madhurarekha, C., et al. (2010b). Bowman−Birk proteinase inhibitor from *Cajanus cajan* seeds: Purification, characterization, and insecticidal properties. *Journal of Agricultural and Food Chemistry, 58*, 2838–2847. https://doi.org/10.1021/jf903675d

Qaim, M., & Zilberman, D. (2003). Yield effects of genetically modified crops in developing countries. *Science, 299*, 900–902.

Qin, D., Liu, X.-Y., Miceli, C., et al. (2019). Soybean plants expressing the *Bacillus thuringiensis cry8*-like gene show resistance to *Holotrichia parallela*. *BMC Biotechnology, 19*, 66. https://doi.org/10.1186/s12896-019-0563-1

Ramalho, S. R., Bezerra, S., Araújo, D., et al. (2018). Novel peptidase Kunitz inhibitor from *Platypodium elegans* seeds is active against *Spodoptera frugiperda* larvae. *Journal of Agricultural and Food Chemistry, 66*, 1349–1358. https://doi.org/10.1021/acs.jafc.7b04159

Rhoades, D. F. (1979). *Evolution of plant chemical defense against herbivores. Herbivores: Their interaction with secondary plant metabolites* (pp. 3–54). Academic Press.

Richardson, M. (1991). Seed storage proteins: The enzyme inhibitor. In *Methods in plants. Biochemistry*.

Rodríguez-Sifuentes, L., Marszalek, J. E., Chuck-Hernández, C., & Serna-Saldívar, S. O. (2020). Legumes protease inhibitors as biopesticides and their defense mechanisms against biotic factors. *International Journal of Molecular Sciences, 21*, 3322. https://doi.org/10.3390/ijms21093322

Rondoni, G., Bertoldi, V., Malek, R., et al. (2018). *Vicia faba* plants respond to oviposition by invasive *Halyomorpha halys* activating direct defences against offspring. *Journal of Pest Science, 91*, 671–679. https://doi.org/10.1007/s10340-018-0955-3

Roy, S., & Dutta, S. K. (2009). Genomic and cDNA cloning, expression, purification, and characterization of chymotrypsin-trypsin inhibitor from winged bean seeds. *Bioscience, Biotechnology, and Biochemistry, 73*, 2671–2676. https://doi.org/10.1271/bbb.90519

Ryan, C. A. (1973). Proteolytic enzymes and their inhibitors in plants. *Annual Review of Plant Physiology, 24*, 173–196. https://doi.org/10.1146/annurev.pp.24.060173.001133

Ryan, C. A. (1981). Proteinase inhibitors. In A. Marcus (Ed.), *The biochemistry of plants* (Vol. 6, pp. 357–370). Academic.

Ryan, C. A. (1989). Proteinase inhibitor gene families: Strategies for transformation to improve plant defences against herbivores. *BioEssays, 10*, 20–24. https://doi.org/10.1002/bies.950100106

Ryan, C. A. (1990). Protease inhibitors in plants: Genes for improving defenses against insects and pathogens. *Annual Review of Phytopathology, 28*, 425–449. https://doi.org/10.1146/annurev.py.28.090190.002233

Sawardekar, S. V., Katageri, I. S., Salimath, P. M., Kumar, P. A., & Kelkar, V. G. (2017). Standardization of in-vitro genetic transformation technique in chickpea (Cicer arietinum L.) for pod-borer resistance. *Advances Agricultural Research & Technology Journal, 1*, 1–2.

Schuler, M. A. (1996). The role of cytochrome P450 monooxygenases in plant-insect interactions. *Plant Physiology, 112*, 1411–1419. https://doi.org/10.1104/pp.112.4.1411

Scott, J. G., Michel, K., Bartholomay, L., et al. (2013). Towards the elements of successful insect RNAi. *Journal of Insect Physiology, 59*, 1212. https://doi.org/10.1016/j.jinsphys.2013.08.014

Sharma, K. (2015). Protease inhibitors in crop protection from insects. *International Journal of Current Research and Academic Review, 3*(2), 55–70.

Shibata, H., Hara, S., Ikenaka, T., & Abe, J. (1986). Purification and characterization of proteinase inhibitors from winged bean (*Psophocarpus tetragonolobus* (L.) DC.) seeds. *Journal of Biochemistry, 99*, 1147–1155. https://doi.org/10.1093/oxfordjournals.jbchem.a135578

Shukla, S., Arora, R., & Sharma, H. C. (2005). Biological activity of soybean trypsin inhibitor and plant lectins against cotton bollworm/legume pod borer, *Helicoverpa armigera. Plant Biotechnology, 22*, 1–6. https://doi.org/10.5511/plantbiotechnology.22.1

Singh, S., Kumar, N. R., Maniraj, R., et al. (2018). Expression of *Cry2Aa*, a *Bacillus thuringiensis* insecticidal protein in transgenic pigeon pea confers resistance to gram pod borer, *Helicoverpa armigera. Scientific Reports, 8*. https://doi.org/10.1038/s41598-018-26358-9

Soliman, H. I. A., Abo-El-Hasan, F. M., El-seedy, A. S., & Mabrouk, Y. M. (2017). *Agrobacterium*-mediated transformation of tomato (*Lycopersicon esculentum* mill.) using a synthetic *cry1ab* gene for enhanced resistance against *Tuta absoluta* (meyrick). *Journal of Microbiology, Biotechnology and Food Sciences, 7*, 67–74. https://doi.org/10.15414/jmbfs.2017.7.1.67-74

Solleti, S. K., Bakshi, S., Purkayastha, J., et al. (2008). Transgenic cowpea (*Vigna unguiculata*) seeds expressing a bean *α-amylase inhibitor 1* confer resistance to storage pests, bruchid beetles. *Plant Cell Reports, 27*, 1841–1850. https://doi.org/10.1007/s00299-008-0606-x

Srinivasan, A., Giri, A. P., Harsulkar, A., et al. (2005). A Kunitz trypsin inhibitor from chickpea (*Cicer arietinum* L.) that exerts anti-metabolic effect on podborer (*Helicoverpa armigera*) larvae. *Plant Molecular Biology, 57*, 359–374. https://doi.org/10.1007/s11103-004-7925-2

Svensson, B., Fukuda, K., Nielsen, P. K., & Bønsager, B. C. (2004). Proteinaceous α-amylase inhibitors. *Biochimica et Biophysica Acta, Proteins and Proteomics, 1696*, 145–156. https://doi.org/10.1016/j.bbapap.2003.07.004

Swathi, M., Mohanraj, S. S., Swaroop, V., et al. (2015). Proteinase inhibitors from *Cajanus platycarpus* accessions active against pod borer *Helicoverpa armigera. Acta Physiologiae Plantarum, 37*. https://doi.org/10.1007/s11738-015-1991-8

Thaler, J. S., Stout, M. J., Karban, R., & Duffey, S. S. (2001). Jasmonate-mediated induced plant resistance affects a community of herbivores. *Ecological Entomology, 26*, 312–324.

Valencia-Jiménez, A., Alberto, J., & Grossi-de-Sa, M. F. (2008). Activity of α-amylase inhibitors from *Phaseolus coccineus* on digestive α-amylases of the coffee berry borer. *Journal of Agricultural and Food Chemistry, 56*, 2315–2320. https://doi.org/10.1021/jf0733564

van Wyk, S. G., Kunert, K. J., Cullis, C. A., et al. (2016). Review: The future of cystatin engineering. *Plant Science, 246*, 119–127. https://doi.org/10.1016/j.plantsci.2016.02.016

Vasudev, A., & Sohal, S. K. (2016). Partially purified Glycine max proteinase inhibitors: Potential bioactive compounds against tobacco cutworm, *Spodoptera litura* (Fabricius, 1775) (Lepidoptera: Noctuidae). *Turkish Journal of Zoology, 40*, 379–387. https://doi.org/10.3906/zoo-1508-20

Vera, P., & Conejero, V. V. (1988). Pathogenesis-related (PR) proteins of tomato. P69 as an alkaline endoproteinase. *Plant Physiology, 87*, 58–63.

Volpicella, M., Leoni, C., Costanza, A., et al. (2011). Cystatins, serpins and other families of protease inhibitors in plants. *Current Protein & Peptide Science, 12*, 386–398. https://doi.org/10.2174/138920311796391098

Wang, H., Shi, Y., Wang, L., et al. (2018). *CYP6AE* gene cluster knockout in *Helicoverpa armigera* reveals role in detoxification of phytochemicals and insecticides. *Nature Communications, 9*, 4820. https://doi.org/10.1038/s41467-018-07226-6

Wang, J., Ma, H., Zuo, Y., et al. (2020). CRISPR-mediated gene knockout reveals nicotinic acetylcholine receptor (nAChR) subunit α6 as a target of spinosyns in *Helicoverpa armigera*. *Pest Management Science, 76*, 2925–2931. https://doi.org/10.1002/ps.5889

Whalon, M. E., & Wingerd, B. A. (2003). Bt: Mode of action and use. *Archives of Insect Biochemistry and Physiology, 54*, 200–211. https://doi.org/10.1002/arch.10117

Wielkopolan, B., Krawczyk, K., & Obrępalska-Stęplowska, A. (2018). Gene expression of serine and cysteine proteinase inhibitors during cereal leaf beetle larvae feeding on wheat: The role of insect-associated microorganisms. *Arthropod-Plant Interactions, 12*, 601–612. https://doi.org/10.1007/s11829-018-9608-y

Yamada, T., Hattori, K., & Ishimoto, M. (2001). Purification and characterization of two α-amylase inhibitors from seeds of tepary bean (*Phaseolus acutifolius* A. Gray). *Phytochemistry, 58*, 59–66. https://doi.org/10.1016/s0031-9422(01)00178-9

Yamagata, H., Kunimatsu, K., Kamasaka, H., et al. (1998). Rice Bifunctional α-amylase/subtilisin inhibitor: Characterization, localization, and changes in developing and germinating seeds. *Bioscience, Biotechnology, and Biochemistry, 62*, 978–985. https://doi.org/10.1271/bbb.62.978

Yan, S., Qian, J., Cai, C., et al. (2020). Spray method application of transdermal dsRNA delivery system for efficient gene silencing and pest control on soybean aphid Aphis glycines. *Journal of Pest Science, 93*, 449–459. https://doi.org/10.1007/s10340-019-01157-x

Zhao, Y., Botella, M. A., Subramanian, L., et al. (1996). Two wound-inducible soybean cysteine proteinase inhibitors have greater insect digestive proteinase inhibitory activities than a constitutive homolog. *Plant Physiology, 111*, 1299–1306. https://doi.org/10.1104/pp.111.4.1299

Zhong, Y., Ahmed, S., Deng, G., et al. (2019). Improved insect resistance against *Spodoptera litura* in transgenic sweetpotato by overexpressing *Cry1Aa* toxin. *Plant Cell Reports, 38*, 1439–1448. https://doi.org/10.1007/s00299-019-02460-8

Zhu-Salzman, K., & Zeng, R. (2015). Insect response to plant defensive protease inhibitors. *Annual Review of Entomology, 60*, 233–252. https://doi.org/10.1146/annurev-ento-010814-020816

Zhu-Salzman, K., Koiwa, H., Salzman, R. A., et al. (2003). Cowpea bruchid *Callosobruchus maculatus* uses a three-component strategy to overcome a plant defensive cysteine protease inhibitor. *Insect Molecular Biology, 12*, 135–145. https://doi.org/10.1046/j.1365-2583.2003.00395.x

Zuo, Y., Xue, Y., Lu, W., et al. (2020). Functional validation of nicotinic acetylcholine receptor(nAChR) α6as a target of spinosyns in *Spodoptera exigua* utilizing the CRISPR/Cas9system. *Pest Management Science, 76*, 2415–2422. https://doi.org/10.1002/ps.5782

Brier, H. B., & Rogers, D. J. (1991). Susceptibility of soybeans to damage by Nezara viridula (L.)(Hemiptera: Pentatomidae) and Riptortus serripes (F.)(Hemiptera: Alydidae) during three stages of pod development. *Australian Journal of Entomology, 30*(2), 123–128.

Cardona, C., & Kornegay, J. (2019). Bean germplasm resources for insect resistance. In Global plant genetic resources for insect-resistant crops (pp. 85–99). CRC Press.

Clement, S. L., Cristofaro, M., Cowgill, S. E., et al. (2019). Germplasm resources, insect resistance, and grain legume improvement. In Global plant genetic resources for insect-resistant crops (pp. 131–148). CRC Press.

Giri, A. P., Harsulkar, A. M., Deshpande, V. V., et al. (1998). Chickpea defensive proteinase inhibitors can be inactivated by podborer gut proteinases. *Plant Physiology, 116*(1), 393–401.

Roberts, T. H., & Hejgaard, J. (2008). Serpins in plants and green algae. *Functional & integrative genomics, 8*, 1–27.

Irving, J. A., Steenbakkers, P. J., Lesk, A. M., et al. (2002). Serpins in prokaryotes. *Molecular biology and evolution, 19*(11), 1881–1890.

Kansal, R., Kumar, M., Kuhar, K., et al. (2008). Purification and characterization of trypsin inhibitor from Cicer arietinum L. and its efficacy against Helicoverpa armigera. *Brazilian Journal of Plant Physiology, 20*, 313–322.

Nishizawa, K., Teraishi, M., Utsumi, S., et al. (2007). Assessment of the importance of α-amylase inhibitor-2 in bruchid resistance of wild common bean. *Theoretical and Applied Genetics, 114*, 755–764.

Rawlings, N. D., Barrett, A. J., Thomas, P. D., et al. (2018). The MEROPS database of proteolytic enzymes, their substrates and inhibitors in 2017 and a comparison with peptidases in the PANTHER database. *Nucleic acids research, 46*(D1), D624-D632.

Russin, J. S., Layton, M. B., Orr, D. B., et al. (1987). Within-plant distribution of, and partial compensation for, stink bug (Heteroptera: Pentatomidae) damage to soybean seeds. *Journal of economic entomology, 80*(1), 215-220.

Silverman, G. A., Bird, P. I., Carrell, R. W., et al. (2001). The serpins are an expanding superfamily of structurally similar but functionally diverse proteins. *Journal of Biological Chemistry, 276*(36), 33293–33296.

Wisessing, A., & Choowongkomon, K. (2012). Amylase inhibitors in plants: structures, functions and applications. *Functional Plant Science and Biotechnology, 6*, 31–41.

Accelerated Breeding for Nutritional and Grain Quality Traits in Pulses: Conceptualization and Path Ahead

Biswajit Mondal, Yogesh Kumar, Vaibhav Kumar, Prasoon Verma, Aneeta Yadav, Ankita Sagar, Mehul Misra, Nandan Singh, P. S. Basu, Farindra Singh, and N. P. Singh

Abstract Pulse crops have been an essential component of South Asian food habits and a major source of protein for the vegan population. The global production of pulses has increased significantly with the pulses revolution in India, which shares 24–25% of it. From 82.56 million tons in 2014, global pulses production recorded an all-time high in 2017 with 98.72 million tons (a 19.56% increase) and restored to 96.83 million tons in 2018 (17.28% increase). Likewise, in India, the pulses production increased from 17.15 million tons in 2014–15 to 26.05 million tons in 2022–23. Genetic enhancement and breeding for nutritional and grain quality is going to be a major avenue of research after the considerable achievement in production increase of pulse crops. Breeding for nutritional and grain quality in pulse crops needs wide-scale screening of genotypes with high-throughput phenotyping protocols to identify the trait-specific donors for their utilization in hybridization program. Early generation selection of seed-based traits being an issue, selection strategies need to be worked out. Rapid generation advancement and attainment of homozygosity may help to address this issue of phenotyping the segregating generation and selection. Policy support in the form of setting relevant criteria for nutritionally superior or biofortified genotypes in the variety release system and ensuring premium price to the growers to maximize their returns are essential. Considerable research progress has been made for protein content enhancement in chickpea and pigeonpea relating to identification of the high-protein donors, working out the inheritance pattern, and genetic mapping. Nutrition profiles of the major pulse crops have been studied by the conventional biochemists; this information will be the initial benchmark for the genetic enhancement programs. Traits like the concentration of beta-carotene, Fe, Zn, folate, etc., in seeds, along with grain quality traits, have been

B. Mondal (✉) · Y. Kumar · V. Kumar · P. Verma · A. Sagar · M. Misra
N. Singh · P. S. Basu · F. Singh · N. P. Singh
ICAR-Indian Institute of Pulses Research, Kanpur, UP, India

A. Yadav
Faculty of Agricultural Sciences and Allied Industries, RAMA University, Kanpur, UP, India

A. Pratap et al. (eds.), *Integrated Improvement of Food Legumes*,
https://doi.org/10.1007/978-3-031-82629-0_5

studied in different pulse crops. High-throughput nondestructive phenotyping protocol and selection strategies for handling segregating generation are major bottlenecks for breeding for nutritional or grain quality in pulse crops. Nutritional and grain quality breeding for pulses is an interdisciplinary approach involving the plant breeders, biochemists, processing engineers, and food technologists; a holistic approach is needed to address the nutritional security issue through this approach.

Keywords Genotype screening · Genetic mapping · Pre-breeding · Protein Digestibility Corrected Amino Acid Score (PDCAAS) · Protease inhibitors · Spectral assays (Near infrared, X-ray fluorescence)

1 Introduction

In the 68th session of the United Nations General Assembly, the Food and Agriculture Organization (FAO) adopted a historical resolution on 20 December, 2013; it was the declaration of the year 2016 as the International Year of Pulses. Two major aspects were focused in the declaration, namely, the role of pulses in providing nutritional security to humans and sustainability to the cropping system (FAO, 2014). Traditionally, pulse crops have been an essential component of South Asian food habits, and in India, they are the major source of protein for the vegan population. A wide variety of food legumes are consumed across the globe. In India, food legumes are majorly consumed in the form of pulses (husked/unhusked split cotyledons of mature grains) and whole grains in different forms (soups, flour, and processed products). Several leguminous vegetables are also used in South Asian cuisines. In Europe, North and South America, common beans have been an important component of the diet and are consumed with many non-vegetarian delicacies. In Africa, cowpea (grain and vegetable), beans, and chickpea are components of daily diets, along with cassava, millets, and other dryland cereals. Out of the different food habits related to legumes (grain and vegetable) across the globe, pulses provide considerable variation in taste and flavor specific to different crops (chickpea, pigeonpea, lentil, green gram, black gram, etc.). Besides adding different options to the food, pulse crops belonging to the *Leguminosae* family fix atmospheric nitrogen to the soil through bacterial symbiosis in their roots and thrive in marginal tracts with limited resources. Thus, pulses add a sustainability component to the cropping cycle and stability to the overall farming system.

The then Director General, FAO, Jose Graziano da Silva, officially launched the International Year of Pulses on 11th November, 2015, with a note, "Nutritious seeds for a sustainable future," and emphasized on the nutritional quality of pulse crops and their role in a sustainable farming system for small and marginal farmers of South Asia and Africa. The International Year of Pulses 2016 was marked with several discourses and discussions worldwide, in different forums, bringing together farmers, marketing channel personnel, millers, food processers, and scientists, upholding the importance of pulse crops. The legacy of the International Year of

Pulses 2016 had a deep impact in the pulses research, funding, and policymaking. The year 2016 witnessed a significant success story as well, the pulse revolution in India. In the cropping season 2016–17, pulse production in India reached an all-time high of 23.13 million tons with a productivity of 786 kg/ha, from 14.76 million tons of production in 2007–08 with a productivity of 625 kg/ha (MoAFW, 2019a). During the cropping season 2017–18, the growth was sustained, and pulse production in India further increased to 25.23 million tons with a productivity of 841 kg/ha (MoAFW, 2019a). Consecutively for two years, the record pulses production in India was sustained, and thereby the achievement of this milestone was declared as the pulses revolution (MoAFW, 2019b). Chickpea had a lion's share in upscaling the pulses production in India (44.5% in 2017–18). With the considerable achievement in increasing the yield, nutritional quality is going to be a major thrust area in breeding pulse crops. Being the largest consumer of pulses and housing a major portion of the vegan population, India depends on the pulse crops for food and nutritional security. Pulse crops serve as major suppliers of quality protein in diets and have a high amino acid score based on lysine content (WHO, 2007). Pulses provide several beneficial nutrition components, like protein, fibers, Fe, and Zn, along with several other dietary elements. Theose crops have been a part of traditional diets across South Asia and Africa, thereby making them ideal candidate crops to be taken over in research programs for genetic enhancement for nutritional quality. India has been ranked 102 in the Global Hunger Index out of 117 countries which qualified for the rating (GHI, 2019), which is indeed alarming and indicated the hidden hunger looming large on one of the largest democracies in the world. Forbes India has also stressed on the widespread protein deficiency in India, which needs action (Khandelwal & Kavishwar, 2018)..

Enhancing the nutrition parameters and other quality parameters in pulses is the need of the hour to address nutrition security. Research in this field is in the primary stage to deliver nutritionally enhanced products to the consumers as well. Low-throughput classical biochemical protocols (for assays of protein, fat, carbohydrate, fiber, folate, beta-carotene, and elements) that lack a breeding strategy to handle the early generation segregation (in F_2 to F_4) of traits expressed in seeds are major challenges in the way of a quality breeding program in pulses. Efficiency is related to the time required or speed of a plant breeding program to develop candidate advanced breeding lines for varietal release. Rapid generation advancement/ speed breeding adopts generation advancement by single seed decent method, on the way, therer is a probability of losing favourable recobinants or transgressive segregants which are important while breeding for rare alleles. Therefore, efficient breeding for nutrition and quality parameters in pulses is a relevant topic of discourse with respect to the thrust areas of contemporary plant breeding. Existing reviews (Simon et al., 2009; Burstin et al., 2011; Jukanti et al., 2012; Robinson et al., 2019 etc.) have summarized the nutrition profiles of the pulse crops and biochemical characterization of the protein fractions; however, data on the plant breeding perspective is limited in the literature. In this chapter, the breeding principles and perspectives for quality and nutrition parameter breeding in pulses (especially chickpeas, being the major pulse commodity) will be focused on and discussed..

2 Production and Global Market

Plant breeding is a commercially driven subject. The contemporary global market scenario has witnessed changes in consumer behavior, and the share of pulse crops in trade is increasing steadily. The global production of pulses has increased significantly with the pulses revolution in India. From 82.56 million tons in 2014, global pulses production recorded an all-time high in 2017 with 98.72 million tons (FAOSTAT, 2020), which was a significant increase of 19.56%. In 2018, 96.83 million tons (FAOSTAT, 2020) of global pulses production was recorded, which was a 17.28% increase with respect to the production of 2014. India is the leading pulses producer of the world and percent share of India in global pulses production has ranged from 20 to 25% between 2014 and 2018 (FAOSTAT, 2020). The last decade saw an unprecedented increase in pulses production in India which was recorded at 26.06 million tons in 20,22–23 and is expected to decrease slightly to 24.49 million tonnes during 2023–24 (DES, 2024). Among different pulse crops, chickpea (*desi* and *Kabuli* types) is the most important commodity in the global market. The pulses revolution in India largely attributes to the hike in chickpea production, which enabled the country towards self-sufficiency in the crop (Dixit et al., 2019). This event had a significant impact on the global pulses market. From a trade volume of 15.61 million tons of pulses worth US$ 11697.97 million in 2016, a significant reduction was observed in 2018, when 13.7 million tons of pulses worth US$ 7929.88 million were imported (APEDA, 2021). A prominent fall in the import of pulses in India was observed, from 6.12 million tons (worth US$ 3988.66 million) in 2016, the import reduced to 2.33 million tons (worth US$ 1040.49 million) during 2018 (APEDA, 2021). A fall in the import of peas from Canada and Australia in response to hike in chickpea production in India was the major cause behind it. However, the global chickpea market is predicted to grow steadily. Market research agency IMARC, which is based in the USA, UK, and India, projects the chickpea market to reach 21 million tons by 2024 with a compound annual growth rate (CAGR) of 4.1% between 2019 and 2024 (IMARC, 2019). Another marker research agency, Technavio, has predicted the global chickpea market to grow by US$ 3.5 billion between 2018 and 2022 with a CAGR of 4%. Irish market research organization Fact.MR has studied the consumer patterns in Europe and the USA and suggests an increase in demand for *desi* chickpeas along with the conventional preference of *kabuli* types. The organization predicts above-average growth in the global chickpea market, generating US$ 10679.6 million revenue to the USA by the end of 2022, with the dried chickpea market surpassing US$ 5000 million (Fact.MR, 2017). The market value of direct consumption of chickpeas alone has been projected to exceed US$ 3500 million, making the commodity immensely important for the days to come (Fact.MR, 2017), along with all the primary and secondary processed food products.

3 Role of Pulses in Health and Nutrition

In the contemporary era, with the change in lifestyle and food habits, nutritional quality has been a matter of consideration. Thereby, along with yield, nutritional aspects have become a significant thrust area for plant breeding. Pulses share only 5% of total protein consumption globally (Joshi & Rao, 2016), but it is significantly higher at the country level. In India, 6–24% of proteins in people's diets come from pulses (NSSO, 2014; NNMB, 2006; Minocha et al., 2017; Kupad & Minocha, 2017). Another study from ICRISAT, Hyderabad, reported the role of pulses in protein intake of up to 72% based on a survey conducted in rural areas of Telangana and Maharashtra (Padmaja et al., 2016). Swaminathan et al. (2012) also highlighted the dietary imbalance in the Indian population with respect to protein intake, which is at large cereal-based, and discussed criteria like amino acid score based on lysine content, protein digestibility corrected amino acid score (PDCAAS; with lysine as the first limiting amino acid), and protein: energy (PE) ratio, which provides the realistic estimate of the population nutrition parameters. Legumes have been scored >100 along with animal-based proteins for amino acid score based on lysine content (WHO, 2007), which were conventionally grouped as second-class proteins by dieticians with respect to the animal-based protein, classified as first-class protein. Animal protein foods contain cholesterol, which is not advantageous for humans after a certain age, but no such disadvantages are associated with plant-based protein foods. Food legumes contain several anti-nutrition factors (ANFs) like protease inhibitors, amylase inhibitors, lectins, saponins, phytic acid, and long-chain oligosaccharides (raffinose family)/polysaccharides (Singh & Pratap, 2016, 2018). Plant biochemistry research of the 1980s and 1990s concentrated majorly on quantification and characterization of those ANFs in food legume crops. Roy et al. (2010) highlighted the duality regarding the ANFs, which are interesting and intriguing. ANF like the raffinose family of oligosaccharides (RFO), which consist of oligosaccharides like raffinose, stachyose, and verbascose, are commonly found in pulse crops and are often associated with flatulence. However, recent research suggests RFOs reduce the risk of intestinal cancer and promote bowel clearance. During germination, RFOs serve as a carbon source. The long-chain oligosaccharides or polysaccharides help in reducing the glycemic index by reducing the rate of digestion. It reduces hankering for carbohydrates and is utilized by dieticians for weight management diets. Another ANF lectin has been reported as hemagglutinogen, whereas its role in arresting metastasis, killing cancerous cells, activating body defense mechanisms, and reducing obesity has also been reported (Roy et al., 2010). Saponins are another group of ANF present in legume crops, which has been reported to trigger RBC hemolysis in higher concentrations. Recent researches reports saponins to decrease the serum lipid and glucose levels and also lower cancer risk. Diets rich in saponins have been reported to prevent dental caries as well. Amylase inhibitors have the potential to be used in treatment strategy to control diabetes for its property to reduce the digestion of starch and alter response of sugar to insulin. Hemolytic disorder-causing ANFs from faba bean—vicine and convicine—have also been reported to reduce cholesterol. Roy et al. (2010) reviewed the beneficial attributes of protease inhibitor ANFs (trypsin and chymotrypsin inhibitors) and reported their anti-cancerous and anti-inflammatory properties. The authors also reported that ANF angiotensin I-converting enzyme (ACE) inhibitors to be associated with the control of hypertension. Findings of authors like Ha et al. (2014) and Halton et al. (2006) advocate replacing animal protein foods with legume proteins to limit the intake of saturated fats and increase the intake of dietary fibers, which helps lower the risk of cardiovascular disorders. A report from the WHO (WHO, 2008) stated the fact that adoption of pulse-based diets prevents up to 80% of heart disease, stroke, type 2 diabetes, and some cancers.

Pulses are being considered as the food of the future based on the contemporary lifestyle of sedentary behavior and less physical activity. They have a low glycemic index and have been reported to indirectly control the blood glucose levels, thereby preventing the risk of diabetes (Lunde et al., 2011; Marinangeli & Jones, 2012). The indirect control of blood glucose levels is achieved through the feeling of satiety (feeling of a full stomach), thereby reducing hunger and increasing the food intake gap. The ANFs can be reduced through processing, as they are either thermo-labile or water-soluble. Soaking the seeds before cooking and draining the water (for water-soluble ANFs) or stirring the seeds in dry heat (for thermo-labile ANFs) before cooking makes pulses/grain legumes ready for consumption. Thus, ANFs are a minor point of consideration in the way of quality breeding for pulses, and traits like the accumulation of protein, Fe, Zn, folate, beta-carotene, etc., can be addressed initially.

4 Considerations for Quality Breeding in Pulses

Quality breeding in pulses or food legumes is an area where research is in the primary stage. Considerable research has been done on soybeans and groundnuts which are majorly used for oil purposes. Identification of the trait-specific donor lines, working out environmental effects on the concerned trait, tracing the inheritance pattern and genetics of the trait, working out the selection parameters (with correlated traits), and handling of the segregating populations are major challenges for a quality breeding program. Cumbersome phenotyping assays and seed-based traits make breeding for quality parameters different from a conventional breeding program for yield enhancement. Considerations of breeding programs for quality parameters are discussed hereby.

4.1 Selection of Parents

A quality breeding program is very precise, and the selection of parents is very crucial since the breeding materials generated out of the crossing program must not segregate for major biotic (disease) or abiotic (drought/heat/salinity/cold/frost) stress susceptibility. Those parental lines (released varieties, advanced breeding lines, trait-specific donors) should have yield at par with the check variety and have desirable plant architecture. Trade-offs between quality parameters and basic yield parameters should be worked out during the crossing program or early generation selection. However, without desirable yield parameters, varieties with quality parameters cannot be popular with the farmers. This is a primary consideration of a breeding program for quality parameters.

4.2 *Phenotyping Strategy*

Diversity for the trait existing in the germplasm accessions is a prior prerequisite for any plant breeding program. Studies regarding diversity for quality and nutrition traits in germplasm accessions of pulse crops are limited. Conventional biochemical assays of protein, fat, carbohydrate, elements, etc., parameters involve separate steps for sample preparation and prevent handling of larger sample numbers. Reproducibility of the result is limited by the system of detection and the concerned machine. All the basic protocols available for biochemical analysis of protein, fat, carbohydrates, and elements are destructive and low throughput in nature. This bottleneck has prevented identification of stable donor lines for quality parameters. With the advent of nondestructive spectral assays (near-infrared and X-ray fluorescence quantification), this hurdle can be overcome. The germplasm accessions and landraces are homogeneous heterozygous in genetic makeup, and in pulse crops, there are considerable variations related to seed traits in such populations. Such populations need acclimatization and cycles of pre-breeding for proper characterization. Seed scanners available will also have a significant role in constituting subpopulations from a base population on the basis of seed traits prior to their field evaluation.

4.3 *Standardization of the Phenotyping Protocol*

Non-destructive spectral assays, like near-infrared assay (NIR) and X-ray fluorescence (XRF), are promising high-throughput phenotyping techniques for nutrition and quality parameters scored in seeds. NIR can generate data for proximate principles of food (protein, fats/oils, carbohydrate), and XRF can generate data for protective principles of food (macro- and micronutrient elements). For those techniques, also elaborate standard curves need to be developed based on the biochemical nutrition component and seed samples (of contrasting genotypes) on the basis of which the spectral data of unknown samples will be generated. For example, for protein phenotyping in seeds, separate standard curves with bovine serum albumin (or any pure plant protein extract), ammonium nitrate, and contrasting genotypes (high-low protein) are to be generated and utilized for standardization of the machine. The quantity of seeds needed for non-destructive assay is an issue depending on the type of platform used in the machine (cup or static/rotating plate). It is based on the threshold distance needed for the spectra to travel through the media (seed tissue) for detection. For grains with small shapes and uniform coordinate geometry (pigeonpea, lentil, green gram, black gram, etc.), there are lesser interstitial spaces between the grains. The spectra are interrupted less by the air, so the data generated is reproducible. However, in crops like chickpea, where the test weight range is high (12 g/100 seeds to 62 g/100 seeds), air interrupts the incident spectral path, and often the data is not reproducible. Grinding the seeds and using flour for detection is a way out, but from a breeding point of view, loss of seeds cannot be afforded (in pre-breeding or segregating populations). In a major pulse crop like chickpea (and other pulse crops with large seeds, for example, beans), standardization and selection of phenotyping protocol is a major consideration. Moisture content is a key determining factor for phenotyping quality traits in seeds because, along with moisture content, data varies. Therefore, mentioning the seed moisture content while reporting the seed quality parameters is crucial for the reproducibility of the data.

4.4 Selection Strategy and Generation Advancement

High-throughput spectral phenotyping systems (near-infrared and X-ray fluorescence quantification) have evolved over time, but their utilization in early generation segregating lines is limited. Selection strategies for early generation segregating lines in pulse crops need to be worked out. Significant segregation for contrasting seed traits between parental lines is observed in the early segregating populations of pulse crops. It prevents the direct use of NIR or XRF on the segregating lines. Populations segregating for seed traits might be handled by constituting families based on morphology and seed traits for phenotyping with NIR or XRF. Research is limited relating to the early generation selection strategies. So, fixing the segregating lines is of prime importance to apply the high-throughput phenotyping systems, and here comes the importance of speed breeding strategies for fast generation advancement of the segregating populations to address the efficiency of the breeding program. High-throughput phenotyping protocols and rapid generation advancement strategies are key considerations in breeding programs for quality traits in pulses. Nutrition and quality parameters, being seed-based traits, destructive phenotyping (grinding of the seeds for sample in early generations) may lead to the loss of recombinants, which is a major challenge of a breeding program on nutrtional quality.

The researchable areas in quality breeding include identification of donor accessions with high nutrition and quality parameters by screening of germplasm accessions, studying the effect of environment on such traits, working out the heritability and inheritance patterns, genetic mapping and development of breeding populations. The above-discussed considerations are important while planning for any of those research areas.

5 Accelerated Plant Breeding

Development of a cultivar through conventional plant breeding and its release through an organized system of coordinated multi-location trials for commercial cultivation is a time-taking process. In India, a country with one of the largest national agricultural research systems, the approximate varietal development time is 12–13 years without any modification in the breeding methods. Through conventional methods of handling F_2 populations (pedigree, bulk, or backcross breeding), a minimum of five to six generation (thereby five or six years to reach F_5 or F_6) of advancement cycles with selection are needed to develop an advanced breeding line with desirable traits. The advanced breeding lines are evaluated in the preliminary yield trials and station trials for 1–2 years in the breeding station. The best candidate entries selected from those trials are subjected to multi-location coordinated trials for 3 years, following which the best-performing line (with respect to yield) is identified, released, and notified through a protocol. After notification, the cultivar comes in seed chain and finally reaches to the farmers. Thirteen years is a long time for consideration regarding varietal development when the field conditions (regarding biotic and abiotic stresses) change significantly, and there is a change in market preferences and consumer behavior as well. Under the present scenario of climate change, the frequency of bad seasons has increased, and improper field evaluation conditions are significant hurdles for a crop improvement program. Atlin et al. (2017) highlighted the importance of rapid breeding in developing countries under the climate change regime. The authors presented a case study on the yields of the maize belt of Iowa, USA, from 1964 to 2013 and indicated how climate change is shadowing the expression of genetically enhanced cultivars. The authors showed that the reduced yield in the severe drought year of 2012 was equivalent to the high yield of maize during the 1980s, where apparently negative genetic advance was observed. Atlin et al. (2017) stressed on regulated access to elite germplasm at a global level from the territories which are already facing the immediate effects of climate change in the form of vagaries of weather and the provision of large-scale testing of the advanced breeding lines from different breeding centers in those strategic locations. This setup will facilitate the uninterrupted inflow of superior genotypes into the breeding pipeline. This activity will shift the focus of the breeding program from generation advancement and handling of segregating generations to multi-location evaluation trials under a wide range of climatic conditions. Reinforcement and revisiting of the AICRP (All India Coordinated Research Project) system of the Indian Council of Agricultural Research (ICAR) are important in this aspect. Identification of strategic locations and wide-scale testing may save 6–7 years of generation advancement and handling of segregating generations at many plant breeding stations while developing cultivars for target regions. The concept of rapid breeding focuses on the rapid generation advancement, i.e., taking several generations in a year, thereby reducing the duration of 6–7 years of generation advancement.

Valentine (1984) discussed the conventional breeding methods and showed in schematic diagrams that all of those methods need six years to develop a candidate advanced breeding line and also discussed the interventions to reduce the duration. Understanding the rapid generation advancement holistically, this chapter can be considered as the starting point, which represented an era when conventional breeding gradually started adopting molecular marker technology. Contrastingly, Lenaerts et al. (2019) has vividly discussed all the approaches and technologies available presently to hasten the breeding cycle, including the tissue culture approaches (haploid breeding-doubled haploid and embryo culture approach), molecular marker-assisted interventions (specially marker-assisted backcrossing), genomic selection, and the most recently evolved rapid generation advancement method. Nutritional quality or grain quality breeding has a bottleneck for early generation selection due to the destructive phenotyping protocols. So, rapidly advancing the large F_2 and phenotyping at advanced generations might be desirable for quality breeding.

Plant breeders had been resorting to shuttle breeding and the use of off-season nurseries in varietal development, which utilize strategic locations with favorable environments to grow 2–3 crops in a year. Tissue culture-based approaches like doubled haploids drastically reduce the generation advancement time but are a tedious process with limited number of cultivars bred. Different research groups are now focusing on the alteration of photoperiod and thermoperiod regimes to hasten the life cycle of crops while addressing the speed breeding aspect. Considerable studies had been carried out by plant physiologists and breeders for standardizing growth conditions of different crops under phytotron conditions (chamber or glasshouse). Information generated in such studies needs to be consolidated, reviewed, and documented for the growth chamber- or phytotron-based speed breeding. Interventions focusing on germinating and developing green seeds have been adopted by certain groups as well. Thus, speed breeding or accelerated plant breeding has ushered in a novel avenue of research, which will facilitate plant breeding for faster product delivery.

Watson et al. (2018) have proposed three approaches to hasten the crop life cycle by intervention of the photoperiod under controlled environment conditions in enclosed growth chambers. The authors proposed the advancement of six generations in four crops, namely wheat (bread wheat and durum wheat), barley, chickpea, and pea, through speed breeding approaches in enclosed growth chambers, with respect to two to three generation advancements under glasshouse conditions. In the proposed scheme 1, the seeds were soaked with an aqueous solution of 0.5 ppm gibberellic acid, then stratified at 4 °C for 3 days and left to germinate on germination papers at room temperature for 2 days before sowing. The authors provided a 22-hour photoperiod, with a temperature of 22 °C and a 2-hour dark period with 17 °C to the crops. A program of ramping up and down of temperature and light was set in the growth chambers for 1.5 hours to mimic the dusk and dawn conditions. The chamber humidity was set to 70%. A mixture of spectra from white LED bars (6 units/7.6 m^2), far red LED lamps (12 units/7.6 m^2), and ceramic mercury quartz iodide lamps (32 units/7.6 m^2) was provided to the plants inside the growth chamber. The intensity of light provided was 360–380 μmol/m^2/s. Two weeks after anthesis, the irrigation was withheld to hasten the maturity. Up to three weeks after anthesis, the green developing seeds were harvested. Harvested seeds (spikelets/pods) were dried at 28–30 °C for 3–5 days to reduce the moisture content and make them ready for the next sowing. The authors could harvest up to six generations in bread wheat, durum wheat, barley, chickpea, and canola following this rapid generation advancement scheme.

In another rapid generation advancement scheme, Watson et al. (2018) carried out the generation advancement study in the same above-mentioned crops under temperature-controlled glasshouse conditions. The temperature-controlled glass houses (a standard phytotron facility also has temperature controlled glasshouse unit) were fitted with high-pressure sodium vapor lamps programmed to a temperature regime of 17 °C (night)/22 °C (day) with a turnover interval of 12 hours and a 22-hour photoperiod. The duration of two hours of darkness was a part of the 17 °C night cycle. The light intensity was adjusted to 440–650 μmol/m^2/s at approximately 45 cm above the bench. In this scheme, no ramping program of light and temperature changes was included in the circadian rhythm of this scheme, in contrast to the previous scheme. A weekly foliar spray supplementation of calcium nitrate (1 g/L) was provided to the crops. One spike of wheat and barley was harvested at 2 and 4 weeks after anthesis, respectively. Three pods per plant were harvested in canola and chickpea 6 weeks after anthesis. Three generations of wheat, barley, and chickpea could be advanced in one year in this protocol. However, two generations of canola could be grown in one year following this protocol.

In the third scheme, Watson et al. (2018) designed a low-cost customized enclosed room for rapid generation advancement. The dimensions of the room were 3 m × 3 m × 3 m, with insulated sandwich paneling fitted with seven LB-8 LED light boxes. Twelve hours photoperiod (darkness for 12 hours) was provided for 4 weeks to the growing crops and then increased to an 18-hours photoperiod (with 6 hours of darkness). No ramping of temperature and light was applied to the circadian rhythm. The air conditioner was set at 18 °C in darkness and 21 °C when the LED lights were on, with fluctuations <± 1 °C. Wheat, barley, oat, and triticale crops could be subjected to generation advancement under this protocol.

In the report of Ghosh et al. (2018), detailed specifications of the equipment and parts used in the experimental setup for rapid generation advancement discussed in Watson et al. (2018) are provided. Dr. Lee Hickey and his group at the John Innes Center, UK, are pioneers working in this research area, and the reports mentioned above (Watson et al., 2018; Ghosh et al., 2018) are from their laboratory. Rapid generation advancement schemes can be designed for different crops based on those reports.

A recent report of Samineni et al. (2020) regarding rapid generation advancement is quite intriguing. The authors worked under controlled glasshouse conditions and concentrated their efforts in identifying the stage of the green developing seed when it germinates. Six genotypes were taken for study, two early maturing genotypes (JG 11 and JG 14), two medium-maturity-duration genotypes (ICCV 10 and JG 16), and two late-duration genotypes (CDC-Frontier and C 235). The basic strategy was the application of an extended photoperiod to induce flowering and facilitating germination of green seeds. The authors reported that the extended photoperiod reduced the days to flowering by 8–19 days (in early maturing lines), 7–16 days (in medium-duration maturing lines), and 11–27 days (in late-maturing lines). A crucial information came out of this study that the earliest stage when the immature developing green seed is reached between 20 and 23 days after anthesis, irrespective of the maturity groups in chickpea. The seed-to-seed cycle could be reduced to 43–60, 44–64, and 52–79 days in early, medium, and late maturing genotypes. The authors faced problems to advance the late maturing genotype CDC frontier. In a year, 7, 6.2, and 6 generations were grown in the early, medium, and late-maturing-genotypes of chickpea, respectively. The results were based on six representative chickpea genotypes from three maturity-duration classes (early, medium, and late). This technology is a breakthrough for the rapid attainment of homozygosity and the development of mapping populations.

The varietal development program is a function of selecting suitable recombinants from the segregating populations based on the multi-trait selection models. A higher number of recombinations (pods) in the background of the recipient parent (segregating lines, homozygous or heterozygous) increases the frequency of suitable allelic combinations in the segregating populations, which the breeders select and advance to the next generation. Rapid generation advancement technologies discussed above which hasten the life cycle through extended photoperiod produce 2–3 pods per plant, i.e., the frequency of favorable recombinants in each generation is reduced. Maintenance and advancement of segregating lines through single seed descent or single pod descent method further reduce the frequency of favorable recombinants. However, rapid generation advancement schemes appear to be fit for breeding the seed parameter-based traits, i.e., nutritional quality, grain quality, and milling quality, where attainment of homozygosity is important prior to phenotyping. A lack of desirable recombinants may become a concern in cases of good × good crosses aimed at yield enhancement. Rapid generation advancement is an important parameter for maximizing the genetic gain, as per the breeder's equation (Lush, 1937). The breeder's equation can be written as:

$$\Delta G = \left[(\sigma_a)(i)(r)\right] / L.$$

The genetic gain (ΔG) or response to selection is estimated by multiplying three parameters, namely additive genetic variation within the population (σ_a), selection intensity (i), selection accuracy (r), and the product divided by the number of years required per breeding cycle (L, number of years required from crossing to the development of candidate advanced breeding lines). The denominator L was added by Eberhart (1970) and is the only parameter that allows scope for human intervention. The rapid generation advance technology reduces the value of the denominator L in the equation, thereby maximizing the genetic gain. Rapid generation advancement ushers in an interdisciplinary avenue of plant physiology and breeding and has an immense scope in the days to come where resources like land and labor are going to be limited.

6 Quality Traits in Pulses

Depending on the use of the end product of pulses or food legumes, there are specific traits that govern the nutritional, processing, and organoleptic quality. Thereby, quality traits in pulses can be discussed under three heads, namely, nutritional quality, processing/milling quality, and organoleptic quality.

Nutritional quality parameters include the content of proximate (protein, fat, and carbohydrate) and protective principles of food (vitamins and mineral nutrients). Pulses, being a major source of vegan protein, have been the focus of most quality enhancement research, addressing the nutritional security aspect. Both quantitative and qualitative aspects are considered for the proximate principles (protein, fat, and carbohydrate). Quantitative aspect comprises total protein/carbohydrate/lipid content data. Large-scale screening of the germplasm accessions is carried out to identify the potential accessions with high protein/carbohydrate/lipid content in seeds which might be used as donors in plant breeding programs. For example, conventional Kjeldahl's method or Duma's method or NIR assay generate the total protein content data of the pulses. Qualitative aspect comprises the component profiling of the potentially important genotypes (with high protein/fat/carbohydrate) to study the quality of the seed storage materials, i.e., amino acid profiling, polysaccharide profiling, or lipid profiling. Conventional biochemical assays or modern-day spectra-based high-throughput assays generate the quantitative data, whereas chromatography (HPLC, GC etc.) generates the qualitative profile of the proximate principles like protein, fat, and carbohydrate. Regarding amino acid profiles of high-protein genotypes, richness in sulphur-containing amino acids, which lacks in cereals and supplied from pulses in a balanced diet, is a qualitative aspect of consideration. Dietary fibre content trait is a qualitative aspect of the total carbohydrate content which is in focus relating to the potential for body weight management and blood sugar control. Parameters related to protective principles like vitamins and associated precursors (especially beta-carotene, folate, etc.), micro- and macronutrients (content of Fe, Zn, P, etc.) are quantitative only, i.e., content of beta-carotene, folate, Fe, Zn, etc. Those do not have any component traits per se.

In pulses and beans, anti-nutritional factors (ANFs) like phenolic compounds, tannins, trypsin inhibitors, saponins, and phytic acid are also important criteria for breeding for nutritional quality. Lower concentrations of those ANFs are desirable, thereby negetive selection for the content of those ANFs (i.e. selection for lesser quantity) istargeted in breeding programs. However researches have reported anticancerous properties and other health related advantages of ANFs (Agrawal, 2016). Majority of the ANFs are water soluble and can be avoided by overnight soaking. They are thermolabile as well, dry roasting of the legume seeds over hot pan also reduces the effect of ANFs on human consumption.

Grain quality parameters considered for salad and sprouts quality are desirable shape (angular), size, rate of germination, low ANFs, etc. For the consumption of immature green seeds in chickpea, traits like earliness, high podding, seeds per pod, and retention of greenness in grains till grain filling/seed setting are important. In vegetable-purpose pigeonpea, the above-mentioned traits are also important.

Grain quality parameters governing the cooking quality are hydration capacity, hydration index, swelling capacity, and swelling index. Different formulae for these parameters have been standardized by research groups for different crops which can be adopted depending on specific conditions. Cooking time (minutes) and 100-seed weight per se are also important traits for consideration.

Milling of pulses involves decortication of the grain, splitting the cotyledons to produce pulses ("*dahl*"), and grinding the cotyledons to produce flour ("*besan*"). Milling of pulses is a big industry in south Asia and till date milling properties has not been addressed in plant breeding while developing high-yielding varieties. Milling properties of pulses have been reviewed by Vishwakarma et al. (2017). Physical properties associated with the grains, namely, spatial dimensions, sphericity value, bulk density, true density, porosity, and geometric mean diameter (GMD) govern the milling recovery of pulses. Data of those parameters are needed for design of storage structures and processing equipment. Along with milling properties per se, data related to the mechanical properties of seed due to their unique coordinate geometry are also important and a prior prerequisite for designing efficient dehulling and hull separation machines.

Mechanical property parameters, namely, friction due to the testa texture of seed, force deformation of the seed, and aerodynamic parameters of the seed related to the impact of rotation mechanics of the mill, are intriguing traits which have never been addressed by the plant breeders while developing varieties. The shape of each pulse is different as may be evidenced by variation in the sphericity values as reviewed by Vishwakarma et al. (2017). The authors further reviewed that no correlations existed between size, density, angle of repose, and static coefficient of friction values of the pulses. Along with physical and mechanical properties of genotypes, information regarding seed fractions of pulse varieties is also important for the enhancement of milling qualities and breeding for better milling attributes. Those traits include the percentage share of testa, hull, cotyledons, and embryo in the whole seed. Traits like total gum content and gum profiling are also important where decortication is a problem (especially pigeonpea). Gums determine the strength of hull and cotyledon attachment and, therefore, are a deciding factor in dehulling. All those parameters are essential to developing efficient processing machines. Other important parameters include the coefficient of wholeness of kernel, performance of milling system, product recovery, dehulling index, effectiveness of dehulling, dehulling efficiency, and dehulling loss. The majority of those are secondary traits, and researchers have worked out different formulae for those parameters specific to crops and milling machines. A close discourse between the breeders and millers is needed in this interdisciplinary research field.

Flour is an important commodity of chickpeas and few other pulse crops like mung beans (green gram) and urdbeans (black gram). Rheological properties and parameters of chickpea flour, specific to the usage of end product in South Asia, are yet to be defined. Baking properties need to be standardized in chickpea flour in the light of parameters of wheat flour.

7 Principles of Efficient Breeding for Quality Traits in Pulses

Efficiency of a system is related to the time factor. Efficient breeding for quality traits targets to reduce the duration of breeding cycle and the varietal development program. The efficiency of a quality breeding program depends on four principles, presented schematically in Fig. 1. The principles of efficient quality breeding are discussed herein.

7.1 High-Throughput Phenotyping/Screening for Quality Traits in Seeds

Large-scale screening of the germplasm accessions, released varieties, and advanced breeding lines will identify quality trait-specific genotypes, which may be utilized directly or used as donor in breeding programs for cultivar development. Spectral detection techniques like NIR and XRF and other methods like ICPMS, ICPOES, and Duma's method for protein quantification can be referred to as high-throughput phenotyping methods, which generate large amounts of data in limited time..

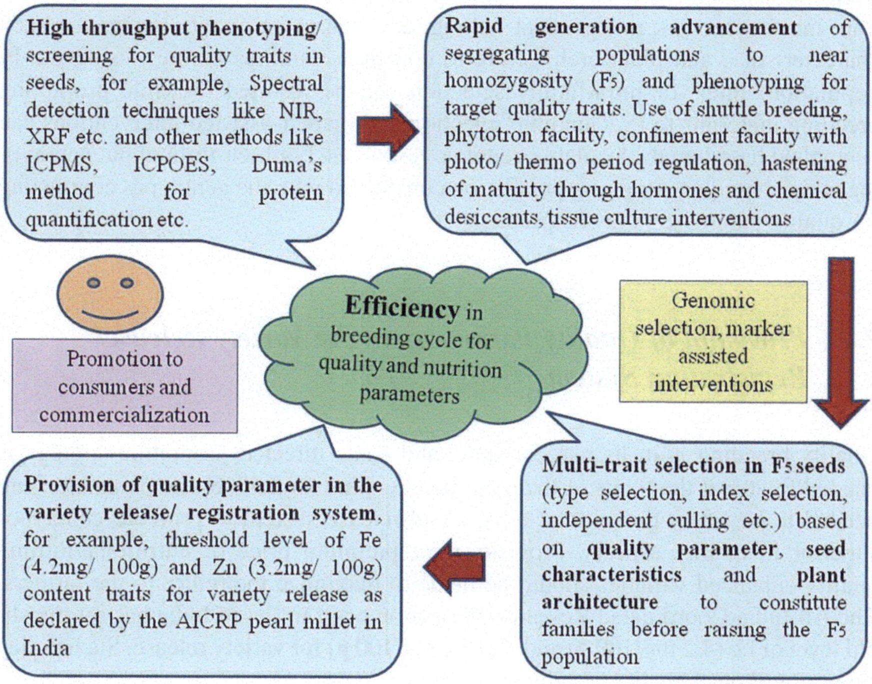

Fig. 1 Principles of efficient breeding for quality traits in pulses—phenotyping strategy, rapid generation advancement, selection strategy, and policy support

7.2 Rapid Generation Advancement of Segregating Populations

Being a seed-based trait, early generation phenotyping and selection is a problem in quality parameters, as already discussed in the considerations for nutritional quality breeding. Here, rapid advancement of the segregating populations to near homozygosity (F_5) is important to select advance-generation breeding lines for the target quality trait. Shuttle breeding (use of strategic locations for growing more than one crop per year or selection) and off-season nurseries have been used by breeders conventionally to shorten the breeding cycle. Use of phytotron facility, confinement facility with photo/thermo period regulation, hastening of maturity through hormones and chemical desiccants, and tissue culture interventions have been adopted to a limited extent in varietal breeding programs. Recently developed speed breeding and rapid generation advancement technology has immense potential to increase the efficiency of a quality breeding program.

7.3 Selection Strategy

Conventional multi-trait selection strategies may be applied in F_5 seeds (type selection, index selection, independent culling, etc.) based on quality parameter, seed characteristics, and plant architecture to constitute families before raising the F_5 population. Inter- or intra-family selection can be resorted to while promoting breeding materials to F_6. Genomic selection and marker-assisted interventions can be applied based on the available genomic resources. For high-throughput genotyping, chips based on a training population constituted with the genotypes contrasting for quality parameters may be produced.

7.4 Provision of Quality Parameter in the Variety Release/Registration System

Quality breeding is in its nascent state, and some directed promotion strategy is needed to attract the focus of the plant breeders in this field. Some parameters are needed to be set in the varietal testing and release system to promote genotypes superior in quality aspects. Provisions for premium price of nutritional/milling quality-enhanced varieties should be made to maximize the return to the farmers. The All India Coordinated Research Project on pearl millet in India has set threshold level of Fe (4.2 mg/100 g) and Zn (3.2 mg/100 g) for variety release. Such policy support will push quality breeding.

8 Breeding Program for Quality Traits in Pulses

An outline of a nutritional or grain quality breeding program in pulses has been provided in a flow chart in Fig. 2. Identification of quality trait-specific donors is the prerequisite of the breeding program. Wide-scale screening of the released varieties

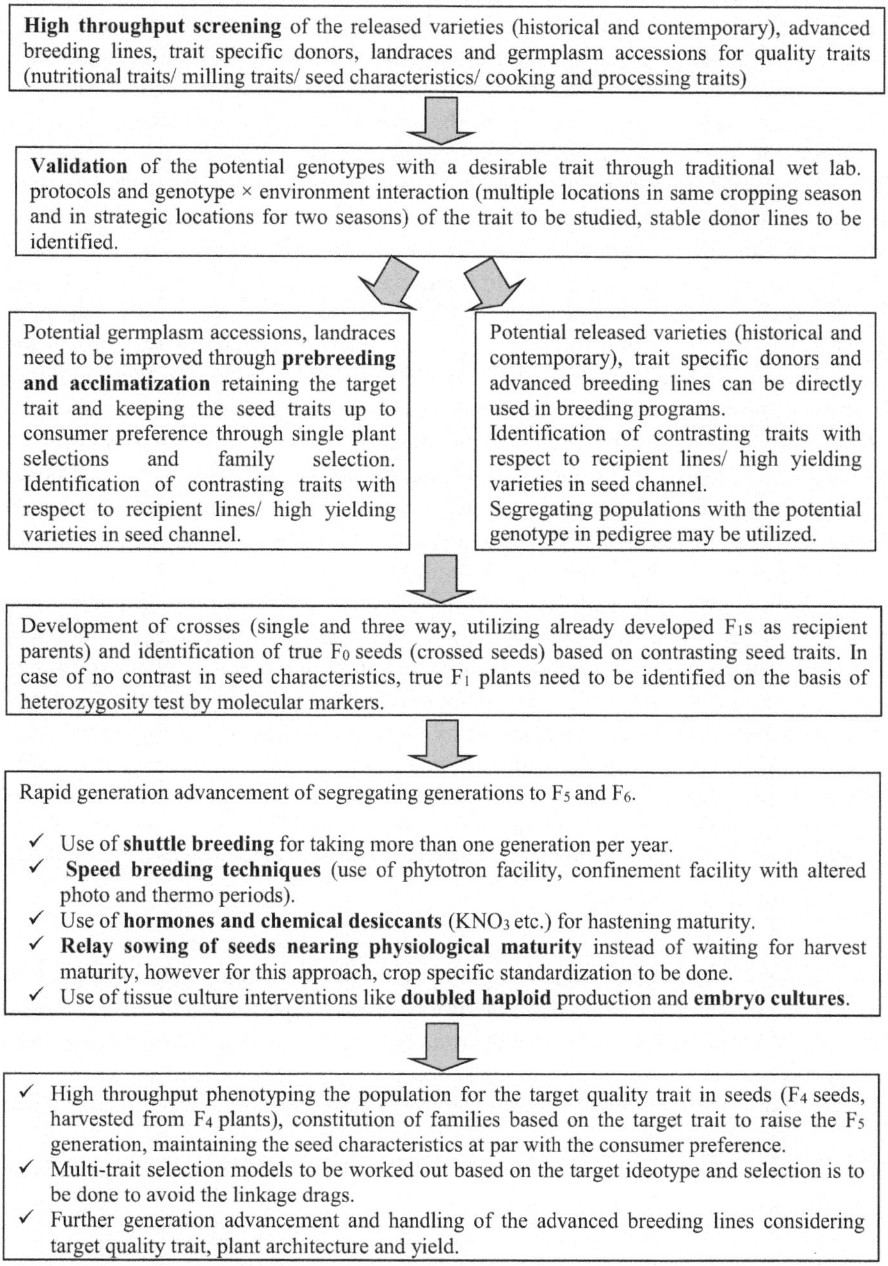

Fig. 2 Outline of a nutritional or grain quality breeding program in pulses

(timeline varieties, obsolete varieties, check varieties, and recent releases), advanced breeding lines, trait-specific donors, landraces, and germplasm accessions has to be done for quality traits for this purpose. Nutritional traits, milling traits, seed characteristics, cooking, and processing traits can be targeted for the genetic enhancement program. Nondestructive high-throughput phenotyping systems (NIR, XRF, etc.) can be utilized for wide-scale screening to identify the potential genotypes with the quality traits. In the next step, these potential genotypes with desirable quality attributes are to be confirmed through conventional biochemistry protocols. The genotype × environment interactions of those quality traits are to be studied, and stable donor lines are to be identified. Contrasting genotypes along with checks may be evaluated in multiple locations during the same cropping season and in strategic locations for different seasons. If germplasm accessions and landraces are identified as potential stable donors in the multi-location or over the season trials, they need to be improved through prebreeding and acclimatization to avoid linkage drags. Seed characteristics are a matter of concern for the wild accessions and landraces. Prebreeding in such genotypes should retain the target trait and keep the seed traits up to consumer preference through single-plant selections and family selection. Contrasting traits between the target recipient lines/high-yielding varieties in the seed channel and potential donor lines need to be identified prior to their adoption in the breeding program. If the released varieties (historical and contemporary varieties), trait-specific donors, and advanced breeding lines are identified as potential genotypes for quality traits, they can be directly used in breeding programs or re-notified through the variety release system. Available segregating populations with the potential genotype in pedigree may also be utilized to reduce the time required for the development of advanced breeding lines.

After the identification of the stable donors, the hybridization program is to be taken up (single and three-way crosses, utilizing already developed F_1s as recipient parents). True F_0 seeds (crossed seeds) are to be identified based on contrasting seed traits. In case of no contrast in seed characteristics, true F_1 plants need to be identified on the basis of a heterozygosity test by molecular markers. The F_2 population is to be raised and progenies to be maintained through single pod descent or single seed descent. Rapid generation advancement interventions may be adopted to advance the segregating populations to F_5 and F_6.

Suitable rapid generation advancement strategies are enlisted below, which may be standardized and utilized.

(a) Use of shuttle breeding for taking more than one generation per year.
(b) Confined chamber or controlled glasshouse speed breeding techniques (use of phytotron facility, confinement facility with altered photo and thermo periods).
(c) Use of hormones and chemical desiccants (KNO_3, etc.) for hastening maturity.
(d) Relay sowing of seeds nearing physiological maturity instead of waiting for harvest maturity; however, for this approach, crop-specific standardization is to be done.
(e) Use of tissue culture interventions like doubled haploid production and embryo cultures.

As the F_4 seeds are harvested from the F_4 plants, they can be phenotyped for the target trait and constituted into families for field evaluation and selection in F_5. Families may also be constituted based on seed characteristics (if segregation is observed) and phenotyped for grain quality in F_4 seeds expressing the grain characteristics of F_5. The F_5 families may be grown from the F_4 seeds, evaluated, and selected under field conditions, maintaining the seed characteristics at par with the consumer preference. Multi-trait selection models should be worked out based on the target ideotype, and probable linkage drags from the wild donor background are to be avoided. Depending on those models, further generation advancement and handling of the advanced breeding lines are carried out, considering the target quality trait, plant architecture, and yield. Rest of the steps in the nutritional or grain quality breeding program is common with a conventional breeding program for yield enhancement. The promising advanced breeding lines with the desirable expression of the target quality trait will be evaluated for yield superiority against the check at the research station and multi-location trials. The superior genotypes in yield and quality parameters, with respect to the check, will be released and notified for commercial use in the farmer's field.

9 Conclusions and Future Prospects

Genetic enhancement and breeding for nutrition and grain quality in pulses is a promising approach to deal with malnutrition and ensure nutritional security. Vegan sources of protein are becoming popular across the globe for the diversification of diets to combat with the lifestyle disorders. The South Asian and African food habits of consuming pulses and food legumes are gradually being popularized across the Europe and America, which will add a new dimension to the cuisines. Considerable research has been carried out to initiate the focused genetic enhancement and breeding programs for nutrition and grain quality in pulses, which have been highlighted and discussed above. Identification of the potential donor genotypes and study of the environmental effects for the particular quality traits reported in the literature will be the stepping stones for this interdisciplinary approach. Nutrition and grain quality parameters being seed-specific, the genetic enhancement approach will be different from the conventional approach for yield enhancement in terms of early generation selection and other factors. Wang et al. (2003) provided an outline of approaches for enhancing the nutritional quality of legume seeds with respect to protein content, amino acid profile improvement, carbohydrate enhancement, Fe and Zn content, etc. The authors stressed on both conventional breeding and pathway-targeted transgenic development, but the latter approach is more complicated for a food product development.

Conventional breeding targets the desirable ideotype and yield to develop new varieties in order to increase and sustain the production. Till 2017, yield was a major thrust area in the pulse breeding. Enhanced pulse production in India for three consecutive seasons since 2017 has provided some yield stability, and breeding for nutritional quality is gradually becoming the focus. Conventional breeding in pulse crops has led to the development of the biofortified lentil variety IPL 220 from the Indian Institute of Pulses Research, Kanpur. Lentil variety IPL 220 contains an enhanced level of Fe (73–114 mg/Kg) and Zn (51–63 mg/Kg) in seeds. Chickpea variety IPC 2005–62 (*desi* type) has been developed at the Indian Institute of Pulses Research, Kanpur, which contains an enhanced level of total protein in seeds. Protein content of IPC 2005–62 seeds ranges from 25–26%, per 100 gm dry weight, with respect to the protein content of the other chickpea varieties ranging from 18–22%, per 100 gm dry weight (based on total nitrogen content, *Kjeldahl* method). The development of the above-mentioned varieties is a result of correlated selection for enhanced yield and desirable plant architecture, which is an eye-opener and usher in a new horizon of research. The path ahead involves targeted breeding program for nutritional enhancement, enabled by research funding and policy support facilitating cultivar development for commercial use. Genetic enhancement and breeding for nutrition and grain quality is a sustainable intervention and a promising approach toward nutritional security..

References

Agarwal, A. (2016). Duality of anti-nutritional factors in pulses. *Journal of Nutritional Disorders & Therapy, 6*, 1. https://doi.org/10.4172/2161-0509.1000e12

APEDA (2021). Annual report of APEDA 2020-2021. https://apeda.gov.in/sites/default/files/files_universal_page/2024-01/Apeda_Annual_Report_English_2020-21.pdf.

Atlin, G. N., Cairns, J. E., & Das, B. (2017). Rapid breeding and varietal replacement are critical to adaptation of cropping systems in the developing world to climate change. *Global Food Security, 12*, 31–37.

Burstin, J., Gallardo, K., Mir, R. R., Varshney, R. K., & Duc, G. (2011). Improving protein content and nutrition quality. In A. Pratap & J. Kumar (Eds.), *Biology and breeding of food legumes* (pp. 314–328). CAB International.

DES (2024). Agricultural Statistics at a glance 2023. Directorate of Economics and Statistics (DES), Department of Agriculture, Cooperation and Farmers Welfare (DAC&FW), Govt. of India. https://desagri.gov.in/document-report-category/agriculture-statisticsat-a-glance/.

Dixit, G. P., Srivastava, A. K., & Singh, N. P. (2019). Marching towards self-sufficiency in chickpea. *Current Science, 116*(2), 239–242.

Eberhart, S. A. (1970). Factors affecting efficiencies of breeding methods. *Afr Soils, 15*, 655–680.

Fact.MR. (2017). *Chickpeas market forecast, trend analysis & competition tracking*. Global Market insights 2017 to 2022, Report Code: FACT296MR.

FAO. (2014). *Resolution adopted by the General Assembly on 20 December 2013*. United Nations resolution no. A/RES/68/231. https://www.un.org/en/ga/search/view_doc.asp?symbol=A/RES/68/231. Accessed on 10 Jan 2020.

FAOSTAT. (2020). *Food and agriculture data*. Food and Agriculture Organization of the United Nations.

GHI. (2019). *Global Hunger Index 2019: India*. https://www.globalhungerindex.org/pdf/en/2019/India.pdf, (https://www.globalhungerindex.org). Accessed on December, 2019.

Ghosh, S., Watson, A., Gonzalez-Navarro, O. E., Ramirez-Gonzalez, R. H., Yanes, L., Mendoza-Suárez, M., Simmonds, J., Wells, R., Rayner, T., Green, P., Hafeez, A., Hayta, S., Melton, R. E., Steed, A., Sarkar, A., Carter, J., Perkins, L., Lord, J., Tester, M., Osbourn, A., Moscou, M. J.,

Nicholson, P., Harwood, W., Martin, C., Domoney, C., Uauy, C., Hazard, B., Wulff, B. B. H., & Hickey, L. T. (2018). Speed breeding in growth chambers and glasshouses for crop breeding and model plant research. *Nature Protocols, 13*, 2944–2963.

Ha, V., Sievenpiper, J. L., de Souza, R. J., Jayalath, V. H., Mirrahimi, A., Agarwal, A., et al. (2014). Effect of dietary pulse intake on established therapeutic lipid targets for cardiovascular risk reduction: A systematic review and meta-analysis of randomized controlled trials. *Canadian Medical Association Journal, 186*(8), E252–E262.

Halton, T. L., Willett, W. C., Liu, S., Manson, J. E., Albert, C. M., Rexrode, K., & Hu, F. B. (2006). *Low-carbohydrate-diet score and the risk of coronary heart disease in women. New England Journal of Medicine 35*(19), 1991–2001.

IMARC (2019). Chickpeas Market: Global Industry Trends, Share, Size, Growth, Opportunity and Forecast 2019–2024. https://www.imarcgroup.com/prefeasibility-report-chickpea-processing-plant.

Joshi, P. K., & Rao, P. P. (2016). Global and regional pulse economies current trends and outlook. In *IFPRI Discussion Paper 01544. The International Food Policy Research Institute (IFPRI)*. South Asia Office.

Jukanti, A. K., Gaur, P. M., Gowda, C. L. L., & Chibbar, R. N. (2012). Nutritional quality and health benefits of chickpea (Cicer arietinum L.): A review. *British Journal of Nutrition, 108*, S11–S26.

Khandelwal, S., & Kavishwar, A. (2018). *Indians are protein deficient, and it needs immediate attention*. Forbes India.

Kurpad, A. V., & Minocha, S. (2017). The health-nutrition-agriculture connect for protein in India. *Bulletin of the Nutrition Foundation of India, 38*(1).

Lenaerts, B., Collard, B. C. Y., & Demont, M. (2019). Review: Improving global food security through accelerated plant breeding. *Plant Science, 287*, 110207. https://doi.org/10.1016/j.plantsci.2019.110207

Lunde, M. S. H., Hjellset, V. T., Holmboe-Ottesen, G., & Høstmark, A. T. (2011). Variations in postprandial blood glucose responses and satiety after intake of three types of bread. *Journal of Nutrition and Metabolism., 2011*, 1. https://doi.org/10.1155/2011/437587

Lush, J. (1937). *Animal breeding*. Plans. Iowa State College Press, Ames.

Marinangeli, C. P. F., & Jones, P. J. H. (2012). Pulse grain consumption and obesity: Effects on energy expenditure, substrate oxidation, body composition, fat deposition and satiety. *British Journal if Nutrition, 108*(S1), S46–S52.

Minocha, S., Thomas, T., & Kurpad, A. V. (2017). Dietary protein and the health–nutrition–agriculture connection in India. *The Journal of Nutrition, 147*, 1243–1250. https://doi.org/10.3945/jn.116.243980

MoAFW. (2019a). *Agricultural statistics at a glance 2018*. Ministry of Agriculture and Farmer's welfare (MoAFW), Department of Agriculture, Cooperation and Farmer's Welfare, Government of India.

MoAFW. (2019b). *Pulses Revolution from food to nutritional security*. Ministry of Agriculture and Farmer's welfare (MoAFW), Department of Agriculture, Cooperation and Farmer's Welfare, Crops Division, Government of India.

NNMB. (2006). *Diet and nutritional status of population and prevalence of hypertension among adults in rural areas*. National Nutrition Monitoring Bureau Technical Report No: 24. 2006; Hyderabad: National Institute of Nutrition. https://www.nin.res.in/downloads/NNMBReport06Nov20.pdf

NSSO. (2014). *Nutritional Intake in India 2011–12*. NSS 68th Round, 2011–12. Kolkata: National Sample Survey Organization office. Ministry of Statistics and Programme Implementation, Government of India. http://mospi.nic.in/sites/default/files/publication_reports/nss_report_560_19dec14.pdf. Accessed on 12 Jan 2020.

Padmaja, R., Soumitra, P., & Bantilan, M. C. S. (2016). Role of pulses in enhancing nutrition status of rural poor: Micro level evidence from semi-arid tropics of India. *Agricultural Economics Research Review, 29*, 65–74.

Robinson, G. H. J., Balk, J., & Domoney, C. (2019). Improving pulse crops as a source of protein, starch and micronutrients. *Nutrition Bulletin, 44*, 202–215.

Roy, F., Boye, J. I., & Simpson, B. K. (2010). Bioactive proteins and peptides in pulse crops: Pea, chickpea and lentil. *Food Research International, 43*, 432–442.

Samineni, S., Sen, M., Sajja, S. B., & Gaur, P. M. (2020). Rapid generation advance (RGA) in chickpea to produce up to seven generations per year and enable speed breeding. *The Crop Journal, 8*(1), 164–169.

Simon, P. W., Pollak, L. M., Clevidence, B. A., Holden, J. M., & Haytowitz, D. B. (2009). Plant breeding for human nutrition quality. *Plant Breeding Reviews, 31*, 325.

Singh, N. P., & Pratap, A. (2016). Food legumes for nutritional quality and health benefits. In U. Singh, S. S. Singh, C. S. Praharaj, & N. P. Singh (Eds.), *Biofortification of food crops*. Springer India. https://doi.org/10.1007/978-81-322-2716-8_4

Singh, N. P., & Pratap, A. (2018). Pulses for zero hunger and zero malnutrition. *Indian Farming, 68*, 33–36.

Swaminathan, S., Vaz, M., & Kurpad, A. V. (2012). Protein intakes in India. *British Journal of Nutrition, 108*, S50–S58. https://doi.org/10.1017/S0007114512002413

Valentine, J. (1984). Accelerated pedigree selection: An alternative to individual plant selection in the normal pedigree breeding method in the self-pollinated cereals. *Euphytica, 33*, 943–951.

Vishwakarma, R. K., Shivhare, U. S., Gupta, R. K., Yadav, D. N., Jaiswal, A., & Prasad, P. (2017). Status of pulse milling processes and technologies: A review. *Critical Reviews in Food Science and Nutrition., 58*, 1615. https://doi.org/10.1080/10408398.2016.1274956

Wang, T. L., Domoney, C., Hedley, C. L., Casey, R., & Grusak, M. A. (2003). Can we improve the nutritional quality of legume seeds? *Plant Physiology, 131*, 886–891.

Watson, A., Ghosh, S., Williams, M. J., et al. (2018). Speed breeding is a powerful tool to accelerate crop research and breeding. *Nature Plants, 4*, 23–29. https://doi.org/10.1038/s41477-017-0083-8

WHO. (2007). *Proteins and amino acids in human nutrition*. WHO/FAO/UNU Expert Consultation. Technical Report Serial No. 935. World Health Organization, Geneva: WHO.

WHO (World Health Organization). (2008). *Action plan for the global strategy for the prevention and control of non-communicable diseases*.

Next-Generation Breeding Approaches for Chickpea Improvement

Ravi Ranjan Kumar, Anand Kumar, Tushar Ranjan, Vikas Kumar, Juli Kumari, Kumari Rajani, Anil Kumar, Sanjay Kumar, Vinod Kumar, Bal Krishna, Jai Prakash Prasad, and Chandan Kishore

Abstract Chickpea improvement programs need to be dynamic and efficient towards rapidly developing improved varieties needed for adaptation in the existing and evolving cropping systems. Information on genetic progress achieved over time from a breeding program is essential to develop effective breeding strategies by assessing the efficiency of past improvement works in genetic yield potential and suggesting on future selection directions to facilitate further improvement. Integrating breeding approaches utilizing conventional and genomic resources would lead to speedy, precise, and efficient chickpea breeding. The availability of a

---

R. R. Kumar (✉)
Department of Molecular Biology and Genetic Engineering, Bihar Agricultural University, Bhagalpur, Bihar, India

Bhola Paswan Shastri Agricultural College, Purnea, Bihar, India

Bihar Agricultural University, Bhagalpur, Bihar, India

A. Kumar · A. Kumar · S. Kumar · C. Kishore
Department of Plant Breeding and Genetics, Bihar Agricultural University, Bhagalpur, Bihar, India

T. Ranjan · J. Kumari · V. Kumar
Department of Molecular Biology and Genetic Engineering, Bihar Agricultural University, Bhagalpur, Bihar, India

V. Kumar
Department of Seed Science and Technology, Bihar Agricultural University, Bhagalpur, Bihar, India

K. Rajani
Bhola Paswan Shastri Agricultural College, Purnea, Bihar, India

Bihar Agricultural University, Bhagalpur, Bihar, India

Department of Seed Science and Technology, Bihar Agricultural University, Bhagalpur, Bihar, India

B. Krishna · J. P. Prasad
Bhola Paswan Shastri Agricultural College, Purnea, Bihar, India

Bihar Agricultural University, Bhagalpur, Bihar, India

A. Pratap et al. (eds.), *Integrated Improvement of Food Legumes*,
https://doi.org/10.1007/978-3-031-82629-0_6

large number of molecular markers, dense genetic maps, and molecular markers associated with traits and transcriptomic resources has made it possible to integrate genomics technologies into chickpea improvement. Marker-associated selection for chickpea improvement is being followed for improving resistance to several biotic and abiotic stresses like drought tolerance and introgressing disease resistance against wilt and ascochyta blight. Pod borer continues to remain the most challenging insect-pest of chickpea due to non-availability of sources with high levels of resistance in germplasm; however with the advent of genome-wide selection, breeders will also have an opportunity to develop new varieties for complex traits. The integration of genomic technologies in chickpea breeding will greatly improve the efficiency of breeding programs in the development of better cultivars and reduce the time required for cultivar development. Genomic resources developed in the recent past have enabled trait dissection as well as trait improvement through modern molecular breeding approaches. This chapter describes the accomplishments and way forward in chickpea improvement deploying next generation breeding approaches.

Keywords Marker-assisted selection (MAS) · Genome-wide selection · Transcriptomics · Breeding efficiency · Marker-assisted backcrossing · RNA-sequence

Abbreviations

GS	Genomic selection
QTL	Quantitative trait locus
MABC	Marker-Assisted Backcrossing
MARS	Marker-Assisted Recurrent Selection
GWS	Genome Wide Selection
SNPs	Single Nucleotide Polymorphisms
NGS	Next–Generation Genotyping and Sequencing Technologies
TILLING	Targeting induced local lesions in genomes
FW	*Fusarium* wilt
AB	*Ascochyta* blight
ICGSC	International Chickpea Genome Sequencing Consortium

1 Introduction

Rapid changes in the climate, coupled with population explosion and limited arable land, are the greatest challenges before the humanity in attaining food and nutritional security across the globe. In this context, food legumes play an important role in reducing hunger and malnutrition, especially in developing countries, in combination with cereals like maize, rice, and wheat (Kaushal et al., 2024). Among more than a dozen food legumes, chickpea (*Cicer arietinum* L.), a cool-season crop, is one of the most important food legumes. The genus *Cicer* belongs to the *Leguminosae* family and to the subfamily *Papilionoideae*. Most of the *Cicer* species are diploid with $2n = 2x = 16$ chromosomes and a 738 Mb genome size (Varshney et al., 2013). There are two distinct types of chickpea, "*kabuli*" (also known as *macrosperma*) and "*desi*" (also known as *microsperma*), differing in their geographic distribution and different plant type (Mohanty et al., 2022). The *desi* types are found in central Asia and in the Indian subcontinent, while the *kabuli* types are mostly found in the Mediterranean region. *Kabuli* types are usually taller and have large beige or cream seed color and "ram's head" seeds with white flowers. *Desi* types are generally shorter, possessing small leaflets, pods, and seeds, and predominantly pink-colored flowers. *Kabuli* chickpea seems to have evolved from the *desi* types (Moreno et al., 1978; Redden et al., 2007; Thudi et al., 2017).

2 Importance of Chickpea

Chickpea is a good and cheap source of protein for people in developing countries, especially South Asia, where the majority of the population is largely vegetarian, either by choice or because of economic reasons. Additionally, chickpea is rich in minerals (phosphorus, calcium, magnesium, iron, and zinc), fiber, unsaturated fatty acids, and β-carotene. Chickpea also improves soil fertility by fixing atmospheric nitrogen, meeting up to 80% of its nitrogen (N) requirement from symbiotic nitrogen fixation (Saraf et al., 1998; Coyne et al., 2020; Bharathi et al., 2022). It plays a significant role in improving soil fertility by fixing the atmospheric nitrogen and meeting 80% of its nitrogen (N) requirement from symbiotic nitrogen fixation and can fix up to 140 kg N ha^{-1} from the air. Its leaves provide substantial amount of residual nitrogen for subsequent crops and add plenty of organic matter to maintain and improve soil health and fertility. Because of its deep tap root system, chickpea can withstand drought conditions by extracting water from deeper layers in the soil profile.

Growing interest in chickpea consumption, coupled with increased preference for vegetable-based protein, has led to an increase in its global demand. Additionally, increasing awareness of both economic (premium prices for large-seeded *kabuli*) and environmental benefits of chickpea, including human health, crop diversification, and sustainable agriculture, has increased farmers' interest in cultivating chickpea.

In India, conventional breeding approaches have yielded over 350 improved cultivars, which have contributed to improved productivity, reduced fluctuations in yield, and enhanced adaption of chickpea to new niches.

2.1 Nutritional Composition

Chickpea is an important source of protein for millions of people in the developing countries, particularly in South Asia, who are largely vegetarian either by choice or because of economic reasons. In addition to having high protein content (20–22%), chickpea is rich in fiber, minerals (phosphorus, calcium, magnesium, iron, and zinc), and β-carotene. Its lipid fraction is high in unsaturated fatty acids. It is an important source of energy, protein, and soluble and insoluble fiber (Wallace et al., 2016). Further, the seed protein contains essential amino acids like lysine, methionine, threonine, valine, isoleucine, and leucine. On an average, chickpea grains contain 60–65% carbohydrates, 6% fat, and between 12% and 31% protein—higher than any other pulse crop. It is also a good source of vitamins (rich in B vitamins) and minerals like potassium and phosphorus. Chickpea, like other legume crops, also replenishes soil fertility through biological nitrogen fixation (Wood et al., 2007).

3 Chickpea Production

3.1 International Status

Chickpea is grown mostly in South Asia and Sub-Saharan Africa, which account for more than 75% of the world's chickpea area. Global chickpea production has increased from 7.68 million tonnes (1961) to 13.73 million tonnes (2014) (FAOSTAT, 2024). It is the third-largest produced food legume globally, after common bean (*Phaseolus vulgaris* L.) and field pea (*Pisum sativum* L.). Among the annual seed crops, it ranks 14th in terms of area and 16th in production. According to Food and Agriculture Organization (FAO) statistics for 1996–2005, the chickpea area averaged nearly 11 million hectares worldwide (1.2% of total crop area), and production was slightly more than eight million tonnes (0.34%). However, among pulse crops, chickpea has consistently maintained a much more significant status, ranking second in area (15.3% of total) and third in production (14.6%). Chickpea is grown in more than 50 countries (89.7% area in Asia, 4.3% in Africa, 2.6% in Oceania, 2.9% in the Americas, and 0.4% in Europe). India is the largest chickpea-producing country, recording a production of 14.20 million tonnes during 2022–23 and accounting for 64% of the global chickpea production. The other major chickpea-producing countries include Pakistan, Turkey, Iran, Myanmar, Australia, Ethiopia, Canada, Mexico, and Iraq.

3.2 National Status

East Asia dominates the world's chickpea production, accounting for 75% during 1996–2005; India ranks first in terms of cultivated area and production. However, there is a slight increase in the chickpea cultivation area in India from 9.27 million hectares (1961) to 11.70 million hectares (2024), but production increased significantly from 6.25 million tonnes (1961) to 14.20 million tonnes (2023–24) due to a significant increase in the productivity from 0.67 t/ha (1961) to 1.24 t/ha (2024). India alone produced 65%. The majority (85%) were desi types. During 1961–2005, production remained almost static. A significant decline in crop area (−0.7% per year) was offset by a mean yield increase of 0.7% per year. The overall picture of a comparatively stable production (mean 0.1% per year increase in production) obscures a number of significant trends at the national level. The most profound change has been a substantial migration of the Indian chickpea industry away from the far north to the center and south of the country. Up to the late 1960s, chickpea was concentrated in the Indo-Gangetic Plain, principally in the states of Uttar Pradesh, Punjab, Haryana, Bihar, and West Bengal. Collectively, they contributed about 65% of the total Indian production. Then the Green Revolution began to impact the chickpea industry: the expansion of the rice industry and the development of wheat varieties suitable for late sowing made rice/wheat rotations highly profitable, and gradually, chickpeas were displaced to marginal rainfed areas and afforded minimal or no inputs. In the last 50 years, the chickpea area in these states has declined by almost five million hectares, although this has been more pronounced in some states than others: during 1960–2002, the chickpea area fell by 68% in Uttar Pradesh, the leading producer, 77% in West Bengal, and 88% in Bihar and Punjab (including Haryana). Approximately half of this "lost" area was taken up by states in central and southern India. By 2001/02, these states had accounted for 80% of the national chickpea area and production. Madhya Pradesh became the largest producer (37% of the national chickpea area), followed by Rajasthan (15%), Maharashtra (13%), Karnataka (8%), and Andhra Pradesh (4%). The increase in chickpea area in the central and southern zones was underpinned by the availability of short-duration varieties that were also resistant to diseases, especially fusarium wilt. This led to a rapid increase in annual yield in some states, e.g., 3% in Maharashtra and Andhra Pradesh. Overall, a sustained increase in yield, despite the general displacement of chickpea to short-season, low-rainfall environments, has enabled the Indian chickpea industry to offset a 30% reduction in crop area over the period 1966–2005. The APY of chickpea in India is provided in Table 1.

3.3 History of Chickpea Research

Chickpea has a long history of research in India. It started as early as 1905, when the formerly Imperial Agricultural Research Institute, Pusa (now known as the Indian Agricultural Research Institute (IARI)), made a modest beginning by taking up its breeding work. However, systematic research on chickpea started with the establishment of the All India Coordinated Pulses Improvement Project (AICPIP) in 1967. In 1972, CGIAR, an international body, established ICRISAT with a global mandate for crop improvement of select dryland crops including chickpea. ICRISAT started working globally as well as with ICAR and other partners in India. Realizing the importance of the crop and providing focused attention on every aspect of

Table 1 Area, production, and yield of chickpea in India

Year	Area (in hectares)	Production (in tonnes)	Yield (Kg/hectare)
2005	6714600	5469400	8146
2006	6896200	5575400	8085
2007	7493900	6333700	8452
2008	7543700	5748600	7620
2009	7890000	7060000	8948
2010	8170000	7480000	9155
2011	9190000	8220000	8945
2012	8320000	7700000	9255
2013	8522000	8832500	10364
2014	9927000	9530000	9600
2015	8251000	7332000	8886
2016	8399000	7058000	8403
2017	9539000	9075000	9514
2018	11899185	11380000	9564
2019	10,210,000	11,425,000	11,190
2020	10,560,000	12,780,000	12,102
2021	10,950,000	13,540,000	12,365
2022	11,230,000	13,750,000	12,244
2023	11,500,000[a]	14,200,000[a]	12,347[a]
2024	11,700,000[a]	14,500,000[a]	12,393[a]

Source: FAOSTAT (2024)
[a]Estimated

chickpea, ICAR established a separate All India Coordinated Research Project (AICRP) on Chickpea in 1993. Since then, more than 190 chickpea varieties for important traits have been released in India.

In order to attain faster genetic gains, the availability of genomic resources and their deployment in breeding programs is a prerequisite. Although chickpea was considered an "orphan legume," recent research efforts globally transformed it into a genomics-resource rich crop. Several thousands of molecular markers (Nayak et al., 2010), high-density genetic maps (Thudi et al., 2011; Gujaria et al., 2011; Kale et al., 2015), transcriptomic resources (Varshney et al., 2009; Hiremath et al., 2012; Kudapa et al., 2024), and a physical map (Varshney, et al., 2014) are now available for trait dissection and crop improvement.

In recent years, the genetics of complex abiotic stresses like drought (Varshney et al., 2014; Kale et al., 2015), heat (Paul et al., 2018), salinity (Vadez et al., 2012; Pushpavalli et al., 2016), and biotic stresses like *Fusarium* wilt and *Ascochyta* blight (Sabbavarapu et al., 2013; Yadav et al., 2023) have been understood, and the genomic regions/QTLs have been identified. Furthermore, several functional genomics approaches, such as RNA-seq and Massive Analysis of cDNA Ends (MACE), with parental genotypes of mapping populations as well as near isogenic lines (NILs), have provided some candidate genes for drought tolerance that are being validated through genetical, genomics, TILLING, and/or genome editing approaches (Kudapa et al., 2024; Singh et al., 2023). In 2013, the International Chickpea Genome Sequencing Consortium, co-led by ICRISAT, the University of California-Davis (USA), and BGI-Shenzhen (China), assembled the draft genome of the *kabuli* chickpea genotype CDC Frontier, while the Next Generation Challenge Program on Chickpea Genomics (NGCPCG), India, assembled the genome sequence of the *desi* genotype ICC 4958. After assembling the draft chickpea genomes, efforts to exploit the potential of next-generation sequencing (NGS) technology to understand the genome architecture of chickpea were initiated. As part of these initiatives, NGS-based whole genome re-sequencing (WGRS) of chickpea parental lines was undertaken, which led to the identification of two million Single Nucleotide Polymorphisms (SNPs) and more than 290 K Indels. These SNPs are a valuable resource, providing enough markers to undertake the genetics research (Thudi et al., 2017). In addition, NGS-based WGRS was used to understand the impact of breeding on genetic diversity and temporal diversity trends in chickpea. For this, WGRS was used to re-sequence more than 100 chickpea varieties released in the last five decades and identified 1.2 million SNPs. These SNPs were used to identify the genomic changes during the history of chickpea breeding, suggesting an increase in diversity in the primary gene pool as a result of recent chickpea breeding programs (Thudi et al., 2017). In addition to parental lines and varieties, the chickpea reference set (comprising 300 accessions) was also re-sequenced using the WGRS approach and led to the identification of 4.9 million SNPs. These SNPs are being used to undertake the genome-wide association study (GWAS) for identifying the markers associated with traits of interest and understanding the domestication and post-domestication divergence in chickpea. Global gene banks store a huge germplasm wealth that has the potential to contribute significantly towards the goal of enhancing the rate of genetic gain. The "Chickpea Genome Sequencing Initiative" aims to re-sequence 3000 lines from the Global Chickpea Composite Collection (Varshney, 2016). This is the first time in the chickpea research history that 3000 lines have been evaluated at six different locations in India over two seasons for several traits of agronomic importance. The genomic resources generated have been successfully deployed for developing superior lines for different traits of interest.

Trait mapping and molecular breeding, such as marker-assisted backcrossing (MABC), marker-assisted recurrent selection (MARS), and advanced backcross quantitative trait loci (AB-QTL) analysis, which are routine in breeding programs for major crops, are also being practiced in chickpea. For example, superior lines with enhanced drought tolerance (Varshney et al. 2013), *fusarium* wilt, and *ascochyta* blight (Varshney et al., 2014) have been developed. The introgression of the "*QTL-hotspot*" into several elite varieties in India, as well as in Kenya and Ethiopia, had led to the development of superior lines with enhanced tolerance to drought and increased yield under rainfed and irrigated conditions, irrespective of genetic background. Furthermore, the available genomic resources have also enabled the successful deployment of modern breeding approaches like genomic selection for faster genetic gains (Roorkiwal et al., 2016). For effective utilization of the available genomic resources in crop improvement, cost-effective genotyping platforms also play a major role. Towards this direction, for use in foreground and background selection, cost-effective SNP genotyping assays like VeraCode assays (Roorkiwal et al., 2014), KASPar assays (Hirematheta al., 2012) were developed. Recently, a precise and cost-effective SNP genotyping platform, with 50,590 high-quality non-redundant SNPs on the Affymetrix Axiom®*CicerSNP* array, has been developed and is being used for high-resolution genetic mapping (Roorkiwal et al., 2018). This array will also be useful for fingerprinting the released varieties as well as assessing their adoption in addition to genetics and breeding applications. During the last 12 years, significant progress has been made in terms of developing genomic resources, and these resources have been effectively used for attaining faster genetic gains in chickpea.

4 Crop Improvement

4.1 Conventional Approaches

4.1.1 Introduction of Cultivars

Plant introduction as a method involves obtaining and evaluating genotypes from within and outside the country and identifying desirable genotypes adapted to the local environment with high productivity or any other specific desired trait. The success of plant introduction, therefore, depends on the nature of the introduced material. Generally, the material introduced from locations with similar soil characteristics and climatic conditions is expected to be successful. Plant introduction is generally facilitated by the following ways:

1. Exchange of material with fellow plant breeders
2. Exploration of areas showing rich variation of the species
3. Obtaining generic resources from international institutes/organizations

The introduced material may be homozygous and stable, like pure lines, or it may be heterogeneous, mixed with landraces or heterozygous segregating populations. If the introduced material is a pure line, it will be immediately accepted as it shows good adaptability. But with mixtures (landraces) or segregating material, it will be necessary to identify either a productive line or a line with a specific desirable trait that will be useful in a recombination breeding program. Although plant introduction is employed at the beginning of a breeding program, it is nevertheless a continuous process. It is the cheapest and fastest way of developing cultivars. Therefore, it becomes an important breeding procedure in a country where there is financial constraint or lack of trained staff and also in those countries where the area under the crop is relatively small. There are several examples of successful introduction in chickpea. A bruchid-resistant line (G109-1) is a selection from an exotic variety introduced to India from Turkey (Saxena & Raina, 1970).

Table 2 Chickpea varieties developed by pure line selection

State	Cultivars
Punjab	Pb7, G24, S26 Uttar Pradesh T1, T2, T3, T87, K4, K468, KWR108, KGD1168, Pragati, DPC92-3
Madhya Pradesh	JGG 1, JG 5, JG 62, JG 221, JG 315, Ujjain 21, Ujjain 24, Gwalior 2, JG1
Maharashtra	Chaffa, D8, BDN9-3
Rajasthan	RS10, GNG146
Tamil Nadu	Co1, Co2
Andhra Pradesh	Warangal, Jyoti
Bihar	BR17, BR77, BR78, RAU15, RAU 52
Haryana	CSG8962
Karnataka	Annegeri
Gujarat	Dahod yellow

Source: Singh (1987), Dua et al. (2001)

4.1.2 Pure Line Selection

As the number of local landraces and exotic germplasm increased, purification and evaluation of work was initiated more systematically. These landraces were variable, and purification was made through selection. These selections were evaluated in different states of India and cultivars were developed (Table 2).

4.1.3 Hybridization

The objective of hybridization is to combine desirable traits from two or more parents into a single cultivar. This follows after the existing natural variation is exhausted. The success or failure of a hybridization program is largely determined by proper choice of the parents. When the aim is to replace the existing variety with a superior one, the existing variety with adaptation to the local environment is a logical choice as one parent. The second parent must be so chosen that it complements the first parent. If creation of variation for the desired traits or broadening of the genetic base is the objective, then diverse parents are selected. Published evaluation reports of the germplasm from national gene banks, e.g., National Bureau of Plant Genetic Resources (NBPGR) in India, international institutes, and other national research reports as well as periodicals serve as valuable sources to select parents for a crossing program based on specific objectives. To ensure a rational and scientific basis in the selection of parents, biometrical approaches are employed to analyze diversity and the combining ability of genotypes involved in hybridization. Crossing in chickpea is a tedious job, and the success of the artificial hybridization ranges from 10% to 50% depending on the weather, particularly temperature and humidity, besides the genotypes involved. Since chickpea flowers are small and delicate, emasculation and pollination generally cause damage to the floral parts, thus reducing the success rate. Artificial hybridization techniques have been reviewed by Argikar (1970), Van der Maesen (1972), Smart (1976), and Auckland and Van der Maesen (1980). Pollen grains germinate in ~30 min after pollination. The germinated pollen tube takes 4 hours to reach the base of the ovary. However, it takes 24 h for fertilization after pollination. Functional specialization of stigmatic cells has been reported by Turano et al. (1983). For emasculation, the flower bud that is due to open the following day is selected. The selected flower bud is held at the base between the thumb and forefinger. The front sepal is stripped off with the help of forceps, and the keel is pushed downward. The exposed anthers are then gently removed. After that a colored cotton thread or a tag is tied around the internodes below the emasculated bud. Pollination is generally done in the morning on the following day. Half-opened flowers are the best source of matured pollen. Such flower buds are collected from the intended male parent in a petri dish or a glass tube. At the time of pollination, the standard and wings of the selected buds are removed, and the upper portion of the keel is lightly pressed between the thumb and the forefinger to shed enough pollen on the stigma of the emasculated flower.

In order to increase the success rate of artificial hybridization, the following points are taken care of:

1. Selection of large flower buds
2. Selection of lateral buds rather than the terminal ones since the success is reported to be better on the lateral buds (Sindhu et al., 1981)
3. Avoiding mechanical injury to the floral parts at the time of emasculation and pollination
4. Attempting hybridization after the formation of the first pod (Bahl & Gowda, 1975)

Timing of pollination and fertilization is also important in deciding the success rate. Under the conditions of low temperature, emasculation in the afternoon followed by pollination in the following morning gives better results. However, when the temperatures are high, morning emasculation followed by immediate pollination or pollination in the same afternoon has been reported to be successful (Khosh-Khui & Niknejad, 1972; Singh & Auckland, 1975; Bejiga & Tessema, 1981; Pundir & Reddy, 1984). Recombination breeding in chickpea was taken up in the 1930s with the objective of transferring the desirable attributes of donors into the local Indian cultivars. The first effort was made to transfer the attribute of disease resistance, and the first *Ascochyta* blight-resistant variety, C12/34, was developed from the cross Pb7 × F8 in 1946. F8 was used as a resistant donor for this. In recombination breeding, selections are made as per pedigree/bulk methods or their modifications.

Hybridization involves single crosses, three-way crosses, or multiple crosses depending upon the objective of the program.

Single Crosses

These are extensively used in chickpea to develop new cultivars. Through single crosses, efforts were made to select the superior recombinants than parents used in the cross. The major thrust remained to transfer the resistance against biotic and abiotic stresses. However, cultivars were also developed from single crosses for wide adaptation, suitable for rainfed/irrigated/late-sown conditions. The new varieties developed in India from single crosses between 1960 and 2004 included: 26 for wilt resistance, 9 for large seed, 8 for blight resistance, 8 for wide adaptation over rainfed and sometimes irrigated conditions, 6 for high yield mainly of *kabuli* types, and 5 for late sowing, usually in combination with other key traits (Dua et al., 2001; Singh, 1987).

Three-Way Crosses

These involve three different parents, and each parent possesses a specific desirable trait, with the aim of recombining the traits together in a new cultivar. In the first step, an F_1 is developed between the parents of choice, and then the third parent is crossed with the F_1 to develop a three-way cross. These crosses provide additional opportunities for gene interaction. Three-way crosses can be managed by pedigree or bulk-breeding methods. The progenies of three-way crosses are more variable with wide genetic base than single crosses. Thus, it takes more time to isolate uniform progenies. In chickpea, a large number of three-way crosses were attempted, and some of them led to the development of new varieties (Table 3).

Multiple Crosses

In these, more than three parents are involved, which are crossed in various ways. These crosses are used to broaden the genetic base and to enhance the opportunities for recombination. The segregating generations can be managed by following pedigree or bulk-breeding methods. The cultivars developed from multiple crosses are expected to have wider adaptation for a range of environments. Details of cultivars bred by following multiple crosses are given in Table 4.

Handling of Segregating Populations

It is always desirable to attempt well-planned crosses based on the diversity and combining ability of parents. Since the heterosis of F_1 hybrids has been shown to correlate well with the performance of the segregating populations later, poor- performing F1 crosses can be rejected, which enables handling of few but promising hybrids so that sufficiently large populations of F_2 from selected F_1 populations can be handled for achieving better results. Further, the performance of F_2 populations

Table 3 Varieties developed from three-way crosses in chickpea

Name of variety	Pedigree	Year of release	Important traits/ Remarks
BG244	([850-3/27 × P922] × P9847 kabuli type)	1985	Blight and stunt-resistant, cold-tolerant
BGD72	([BG256 × E100Y] × BG256)	1999	High yielding
C214	([G24 × IP58] × G24)	1971	Drought-resistant and cold-tolerant
H208	([26 × G24] × C235)	1977	Drought-resistant
HK94-134	([H82-2 × E100Y] × Bhim)	2001	Large-seeded kabuli type
ICCC37	([P481 × JG62] × P1630)	2001	Wilt and dry root rot-resistant
JG11	([PhuleG 5 × Narsinghpur bold] × ICCC37)	1989	Wilt-resistant and moderately root rot-resistant
JG130	([PhuleG 5 × Narsinghpur bold] × JG74)	2000	Wilt-resistant, dry root rot and pod borer-tolerant
JKG-1	([ICCV2 × Surutato] × ICC7344)	2002	Large-seeded kabuli
Raj Vijay Gram 202(RVG 202)	(JAKI 9226 X DCP 20) X JG 412	2015	Late sown
Raj Vijay Gram 202(RVG 202)	(ICCV 91902 X ICCV 10) X ICCV89230	2012	For irrigated and late sown cultivation
JAKI 9218	(ICCV 37 X GW5/7) X ICCV 107	2008	Rainfed cultivation
Haryana Kabuli Chana 2(HK 2)	(H 82-5 X E100 Ym) X Bheema	2005	Irrigated cultivation
Pusa Manav (ICCV 17308)	(ICCV 92944 × ICCV 94954) × JG 11	2023	High-yielding (3.0–3.5 t/ha), wilt-resistant
KGD 1168	(KWR 108 × JG 315) × ICCV 10	2024	Heat-tolerant, bold seeds
ICCV 16105	(ICCV 2 × ICCV 92944) × ICCV 00350	2022	Biofortified (high iron & zinc)

Source: Singh (1987), Dua et al. (2001), Shrivastava et al. (2017), IARI Annual Report (2023), Annual report of AICRP in Rabi Pulses (2024)

can also be taken as a guide for their further advancement. Usually, F_1 hybrids showing high heterosis and less inbreeding depression are the ideal ones to pursue, since additive gene effects seem to govern the expression of traits in such populations, which can be fixed by simple selection. Usefulness of early generation selection is often debated. It was reported to be advantageous by Dahiya et al. (1984), and Salimath and Bahl (1985), while Rahman and Bahl (1985) reported it to be useful for seeds per pod and seed weight, but not for pod number and seed yield. Yield, which is a complex trait, is generally less heritable. Therefore, selection for yield per se, particularly in the early segregating generations, may not give better

Table 4 Varieties developed from multiple crosses in chickpea in India

Name of the variety	Pedigree	Year of release	Important traits/remarks
BG256	([JG62 × 850-3/27] ×[L550 × H208])	1984	Wilt and blight-tolerant and large-seeded
ICCV2	([K850 × GW5/6 × P48] ×[L550 × Guamchil 2])	1993	Wilt and botrytis gray mould (BGM)-resistant, early maturing
Phule G 95311	([ICCC32 × ICCL 80004] × [ICCC49 × FLIP 82K] × ICCV3)	2002	Large-seeded
SG2	([E100Y × P436] × [L550 × F378])	1987	Wilt-tolerant, double-podded
PKV Kabuli 2 (KAK 2)	(ICCC 2 × Surutato 77) × ICC 7344-ICC× 870026-PBPB-14P-BP-62AK-7AKBAK	2001	Large seeded kabuli variety for irrigated cultivations
JG 11	[(Phule G 5 × Narsinghpur bold) × ICCC 37] ICCX860263-BF-BP-91 BP	1999	For irrigated and rainfed cultivations
Phule Vikram	([ICCV10 × ICCV94954] × JG130) × ICCV17308	2020	Double-podded, BGM-resistant
Super Annigeri 3	(ICCV10 × ICCV94954) × (JG11 × ICCV2) × ICCV16105	2022	Drought-tolerant, biofortified
ICCV 16105	(ICCV2 × ICCV92944) × (ICCV00350 × ICCV10)	2022	High iron/zinc, drought-tolerant
RVG 203	(JAKI9226 × DCP20) × (JG412 × ICCV17308)	2023	Late-sown, high protein (24%)
Haryana Chana 5 (HC5)	(H208 × ICCV2) × (JG11 × ICCV10) × ICCV17308	2024	Heat-tolerant, 3.2 t/ha yield

From Singh (1987), Dua et al. (2001), Shrivastava et al. (2017), AICRP Chickpea Reports

results. However, indirect selection for component traits related to productivity has been found to be useful. Bisen et al. (1985) and Salimath and Bahl (1985) observed that selection for seed size and seeds per pod may be better. Kumar and Bahl (1992) compared the yield performance of F_6 lines derived by exercising selection pressure for four successive generations (F_2–F_5) and concluded that indirect selection for yield by pod number and seed weight was more efficient than direct selection for yield. The pedigree method of selection is the most widely used method of handling segregating populations in self-pollinated crops. However, there are various reports for and against the pedigree method of selection. Lal et al. (1973) support the pedigree method of breeding for high yield in chickpea, whereas other researchers found this method less suitable and advocated the bulk method (Byth et al., 1980) or the single-seed descent method (Singh & Auckland, 1975). While comparing different methods of handling segregating populations in chickpea, Rahman and Bahl (1985) concluded that mass selection is superior to random bulk and single-seed descent methods. Singh and Waldia (1994) found that the bulk method of breeding is advantageous for the development of high-yielding and short-duration varieties. Singh (1987) has summarized the usefulness of various breeding methods in respect of specific traits, e.g., pedigree method for resistance to biotic stresses; bulk-pedigree method for traits such as drought tolerance and winter hardiness, modified bulk method for abiotic stresses, seed size, earliness, and plant type, etc.; backcross method for interspecific hybridization and limited backcross for desi × Kabuli introgression and resistance breeding.

Most recent varieties have been developed through hybridization. Single, three-way, or multiple crosses are used, depending on the objective of the breeding program. Backcross breeding is often used to incorporate one or a few traits from a germplasm line or a wild species into a well-adapted variety. Pedigree, bulk, or different modifications of these methods are used in handling segregating generations (Table 5). There is a need to enhance the precision and efficiency of selections

Table 5 Improved varieties of chickpea with specific traits

Specific trait(s)	Released varieties
Wide adaptation	C 235, L 550, Pusa 203, Pusa 209, Pusa 256, Pusa 372, Radhey
Bold seeds	Pusa 256, Pusa 391, BGD 72, Pusa 362, Phule G5, K 850, Radhey, GNG 469, Gaurav, Avrodhi, Co 3, Co 4, Gujarat Gram 2, KAK 2, BG 1003, BG 1053, JGK 1, Virat
Lodging resistance	DCP 92-3, GNG 16, Pusa 240
Tolerance to *Fusarium* wilt	KWR 108, ICCV 10, H 82-2, CSJ 8963, DCP 92-3, GCP 101, GCP 105, JG 315, GPF 2, Vijay, KGD 1168, JG 74, GNG 663, K 850, Radhey, Pusa 256, Pusa 212, KPG 59, BG 1003, BG 1053, Annegiri 1, Mahamaya 1, Vikas, BGD 72, Gaurav, PBG 1, GNG 469, Super Annigeri 1, Improved JG74, Improved Pusa 372, Improved Pusa 391, Improved Pusa 256,
Tolerance to root rot	Alok, Co-3, Co 4, KWR 108, Pusa 209, Pusa 240, Pusa 244, Pusa 362, Pusa 372, Pusa 391, Pusa 413, Pusa 417, ICCV 5, ICCV 10, Vijay, Vardan, GNG 469, BGD 72, JG 11, L 551
Tolerance to *Asochyta* blight	GNG 469, Gaurav, PBG 1, GNG 146, C 235, Pusa 261, PBG 5
Tolerance to *Botrytis* gray mold	Pusa 256, ICCV 2
Tolerance to stunt	Kiran, Pusa 244
Tolerance to root knot nematode	Kiran, Pusa 362, BGD 72
Tolerance to *Helicoverpa* pod borer	Ujjain 24, KAK 2, ICCV 6, Vijay, Vardan, Vishal, BGD 72
Drought tolerance	Co 1, RS 10, Vikas, GNG 16, RSG 44, ICCV 10, Vijay, RSG 888, Geletu, Pusa chickpea 10,216, Improved JG 11, Improved RSG 888, Improved Pusa 362, Improved JAKI 9218, Improved DCP 92-3, Improved ICCV 10
Tolerance to salinity	CSG 8962. ICCV 6

Source: Chaturvedi et al. (2003)

in the segregating generations for higher and rapid genetic gains. Precision in selection for resistance/tolerance to stresses can be improved by screening under controlled environmental conditions or at hotspot locations. Singh (1987) summarized the usefulness of different breeding methods of breeding for specific traits in chickpea as follows:

(i) Pedigree method for resistance breeding (disease, insect, nematode)
(ii) Modified bulk method for stress characters (drought, heat, cold, iron deficiency)
(iii) Backcross method for interspecific hybridization
(iv) Limited backcross for *Desi* x *Kabuli* introgression and resistance breeding
(v) Bulk pedigree method for traits such as winter hardiness and drought tolerance

4.1.4 Mutational Breeding

The use of traditional breeding methods for a longer period might have narrowed genetic variability. Further improvement in a crop species requires increasing novel genetic variability. From this point of view, mutations may be fruitful in plant-breeding programs (Micke, 1988a, b). History: The first idea of mutation, traced back to 1900, was given by De Vries (1901) in *Die Mutations Theorie*. The oldest description of spontaneous mutants appeared around 300 BC in China and was described, especially in cereal crops, in an ancient book, *Lulan* (van Harten, 1998), while induced mutations were discovered in the fruit fly (*Drosophila melanogaster* L.) by Muller (1927). After the discovery of X-ray-induced mutations, Stadler (1928) confirmed the power of X-rays to induce mutations and showed its application to barley (*Hordeum vulgare* L.). After pioneering research, induced mutations have been successfully exploited by many scientists all over the world. But mutation works were gradually reduced towards the 1960s due to many of its negative effects (Allard, 1960). In 1964, when the Food and Agricultural Organization/International Atomic Energy Agency (FAO/IAEA) Division of Nuclear Techniques in Food and Agriculture was established, plant breeders were encouraged to use mutation techniques for beneficial purposes. Toker et al. (2005) introduced a simple and reliable method to detect gamma-irradiated Cicer seeds, including *C. bijugum* P.H. Rech. and *C. reticulatum* Ladiz. Recently, mutagenesis has received great attention for use as a promising new technique known as "targeting induced local lesions in genomes" (TILLING).

Following considerations should be taken into account before starting mutation-breeding programs:

1. Mutations are mostly recessive, and they can be selected in the second generation, M_2. Unlike recessive, dominant mutations occur in very small frequencies, and they can be selected in M_1. Selection of polygenic traits should be started in individual plant progenies of M_3.
2. Mutations are beneficial at very low frequencies (~0.1%), while treatments reduce germination, growth rate, vigor, and pollen and ovule fertility in the living organisms.

3. Mutations are randomly induced, and they might occur in any gene(s). However, some gene(s) can be more readily induced than others.
4. Mutations can be recurrent. The same gene(s) in a crop plant species may be induced again.
5. Mutations generally have pleiotropic effects due to closely linked gene(s). Selection of variety, mutagen, and dose: Mutation-breeding programs should be clearly planned and well defined, and large enough to select desirable mutations as well. The variety selected for mutagenesis should be one of the best varieties released recently and used for at least two varieties. It will be useful in improving specific characteristics of well-adapted and high-yielding varieties, which have one or two deficient traits.

Cagirgan and Toker (2004) selected cold-tolerant and *Ascochyta* blight-resistant chickpea mutants. Kharkwal (1998) used three doses of gamma rays, fast neutrons, and two different chemical mutagens on four chickpea genotypes with treatments included. It was concluded that chemical mutagens were more efficient than physical ones in inducing viable and more number of mutations. Among the chemicals, N-nitroso-N-methyl urea (NMU) was the most effective, while among the physical mutagens, gamma rays were the most effective (Kharkwal, 1999). Gamma rays are the most often used mutagen to change gene(s) in released chickpea mutants for commercial production. Micke and Donini (1993) recommended 10–16 k rad X-rays for chickpea. Similarly, the 50% growth reduction (LD50) of primary shoots and the useful dose range for chickpea were given by 18–26 and 12–18 gamma rays, respectively (Conger et al., 1977). LD50 and a dose close to LD50 are considered optimum by many researchers. This will be producing the maximum frequency of mutations with minimal hazard. An optimum dose can be determined with a preliminary treatment. Overdoses of mutagens are lethal to many plants, while underdoses will produce less mutation frequency. Some factors that influence the effects of mutagens are biological (nuclear volume, chromosome volume, and DNA content of the variety), environmental (oxygen, water status, and temperature), and chemical (Sigurbjönsson, 1983). Parts of chickpea to be treated: Although air-dried seeds are the most frequently used part of chickpea, pollen grains may be used directly, and vegetative organs for in vitro mutagenesis may be used as well. However, pollen grains are used infrequently because emasculation and pollination are very difficult, time-consuming, and pollen viability is short. Irradiation of pollen grains can be beneficial to overcome pre-fertilization and post-fertilization problems, especially in intraspecific hybridizations. Treatments of interspecific and intraspecific hybrids can produce translocations. Mutations do not induce chimeras when pollens are irradiated.

4.1.5 Advantages of Mutation Breeding

Mutation breeding not only creates variability in a crop species but also shortens the time taken for the development of cultivars through induced mutation compared with that by hybridizations. The average time elapsed from the beginning of mutation treatment to the release of the mutant cultivars was approximately 9 years, while this time was 18 years for cultivars arising from crossing programs (Brock, 1977). Moreover, mutations are induced in both qualitative and quantitative characters in a short time, altering new alleles of known and previously unknown genes and modifying linkage (Konzak, 1977). Further, desirable variability could be brought about as variability in the family *Leguminosae* (*Fabaceae*) through induced mutations (Toker & Cagirgan, 2004), whereas it is limited through hybridization. It is expected to be induced in accordance with Vavilov's law of parallel variation. It can create all kind of variation in a family. Gaur and Gour (2002) reported a mutant with nine flowers per node, while chickpea in general possesses one or two flowers per node. Furthermore, mutations, both induced and spontaneous, are one of the most important sources of evolution and will be helpful in studying the genetics of created novel characters. Special problems of chickpea have been solved through mutation breeding. For example, M16119 is both cold-tolerant and cold-resistant to *Ascochyta* blight (Haq & Singh, 1994). Some chickpea mutants (CM 88 and CM 98) could result in gains of $9.6 million in Pakistan (Ahloowalia et al., 2004).

4.1.6 Disadvantages of Mutation Breeding

The frequency of desirable mutations is very low, i.e., ~0.1%. The success of mutation breeding depends on the methodology, effective screening techniques, and the population grown in M1 and successive generations. The larger the population in M1, the greater the success in selecting desirable mutants. Breeders have to screen large populations for desirable mutations. The screening procedures in large populations will take considerable time, labor, and other resources. Some mutations involve pleiotropic effects due to linked gene(s), other mutations, chromosomal aberrations, and deletions. These mutants often have to be backcrossed to parents or adapted varieties. The number of mutant chickpea varieties officially released and recorded in the FAO/IAEA Mutant Varieties Database was 2252 before the end of 2000 (Ahloowalia et al., 2004; Kharkwal et al., 2005). From these mutant varieties, 265 mutant grain legume cultivars have been released in 32 countries (Bhatia et al., 2001). Gamma rays were the most frequently used technique to alter genes, accounting for 64% of the radiation-induced mutant varieties (Ahloowalia et al., 2004; Maluszynski et al., 2000). Sanilac, a variety of common bean (*Phaseolus vulgaris L.*), was the first released legume mutant in Michigan in 1956 (van Harten, 1998). In the case of chickpea, mutation induction works were probably initiated in the Indian subcontinent (see in Sharma & Varshney, 1999). These works have been encouraged by FAO/IAEA projects in order to create useful variation (Shaikh, 1977; Sharma & Kharkwal, 1982, 1983; Shaikh et al., 1983a, b). These works created many unique mutations for use in plant-breeding programs. Approximately 15 mutants in chickpea were released for commercial production (Singh, 1987; Bhatia et al., 2001a, b, c; Singh, 2005). A mutation frequency of some spontaneous non-nodulating mutants at the rate of 120 and 490/million plants was observed (Rupela, 1992). More than 150 chickpea mutants have been mentioned in the literature till date. In conclusion, genetic variation in available germplasm collections has been widely used to combat biotic and abiotic stresses. Apart from a few stress-resistant characteristics (Singh, 1997), desirable sources of resistance have not been found in the cultivated collections, with the exception of wilds (Croser et al., 2003; Singh et al., 1998; Toker, 2005). Therefore, mutagenesis may be considered as the effective tool to create a new, desirable genetic variability in chickpea. The varieties released for the commercial production through mutational breeding are given in Table 6.

Table 6 Mutant chickpea varieties released for commercial production in various countries

Mutant variety	Parent variety	Mutagen(s)	Main characters induced	References	Released	
					Country	Year
Hyprosola (M-699)	Faridpur-1	Gamma rays, 200 Gy	Earliness	Oram et al. (1987)	Bangladesh	1981
Binasola-3 (L-84)	G-97 (Binasolo-2)	Gamma rays, 200 Gy	Earliness	Shamsuzzaman et al. (2004)	Bangladesh	2005
Kiran (RSG-2)	RSG10	Fast neutrons, 4.5 × 1012	Erectoid type	Dua et al. (2001)	India	1984
Ajay (Pusa 408)	G-130	Gamma rays, 600 Gy	Yield	Kharkwal et al. (1988)	India	1985
Atul (Pusa 413)	G-130	Gamma rays, 600 Gy	Yield	Kharkwal et al. (1988)	India	1985
Girnar (Pusa 417)	BG-203	Gamma rays, 600 Gy	Yield	Kharkwal et al. (1988)	India	1985
C-72	6153	Gamma rays, 150 Gy	*Ascochyta* blight resistance	Haq et al. (1999)	Pakistan	1983
CM-88	C-727	Gamma rays, 100 Gy	*Ascochyta* blight and *fusarium* wilt	Haq et al. (2001)	Pakistan	1994
CM-98 (CM 31-1/85)	6153	Gamma rays, 200 Gy	Yield	Haq et al. (1999)	Pakistan	1998
TAEK-SAGEL	–	Gamma rays, 100/150 Gy	Yield	Sagel et al. (2009)	Turkey	2005
Pusa 547	BG-256		Bold seeds	Kharkwal et al. (2008)	India	2006
BGM 547	BG-256		Tolerant to wilt, root rot and shunt diseases	Kharkwal et al. (2005)	India	2006

(continued)

Table 6 (continued)

Mutant variety	Parent variety	Mutagen(s)	Main characters induced	References	Released	
					Country	Year
BARI Chola-12	Binasola-3	Gamma rays (250 Gy)	Drought tolerance, bold seeds	BARI (2015)	Bangladesh	2015
Kabuli Mutant-1	ICCV 2	Fast neutrons (5 × 10125 × 1012)	Large-seeded kabuli (>30 g/100 seeds)	ICRISAT (2016)	India	2016
Plovdiv-18	Plovdiv-8a	Gamma rays (150 Gy)	Enhanced yield (20% increase), cold tolerance	Dimitrova et al. (2018)	Bulgaria	2018
GCP 105 (M-307)	JG 11	Gamma rays (350 Gy)	Dual resistance (*Fusarium* wilt + root rot)	ICAR-IIPR (2019)	India	2019
Türkmen-7	Türkmen-3	Gamma rays (200 Gy)	Early flowering (85–90 days), heat tolerance	TARI (2020)	Turkmenistan	2020
Pusa 578 (M-2021)	Pusa 256	Gamma rays (300 Gy)	Early maturity, high yield (2.8–3.2 t/ha)	ICAR-IARI (2021)	India	2021
NIFA 2022 (M-55)	NIFA 2005	EMS (0.4%) + Gamma rays (200 Gy)	*Fusarium* wilt resistance	NARC (2022)	Pakistan	2022
AUBL 17	Annigeri 1	EMS (0.3%)	Salinity tolerance (ECe 8–10 dS/m)	AUB (2023)	Australia	2023

5 Molecular Approaches for Improvement of Chickpea

The development of molecular markers and molecular mapping of QTL/genes of interest has made it possible to initiate genomics-assisted breeding in chickpea. The introduction of genomics tools and the development of markers have made it possible to accelerate the chickpea breeding program. The following molecular method may be utilized for chickpea improvement:

1. Marker-Assisted Backcrossing (MABC)
2. Marker-Assisted Recurrent Selection (MARS)
3. Identification of Gene(s) and QTL for Introgression Program
4. Genome-Wide Selection (GWS)
5. Next—Generation Genotyping and Sequencing Technologies

5.1 Marker-Assisted Backcrossing (MABC)

MABC is one of the most promising approaches for the use of molecular markers to identify and select genes controlling resistance to those factors. It is used to incorporate a gene of interest/QTL from the inferior source/donor parent into a selected cultivar or breading line (recipient parent) without any linkage drag from the donor parent (Collard & Mackill, 2008; Moose & Mumm, 2008; Ribaut et al., 2010; Varshney et al., 2009), which is already adapted by the farmers.

There are two types of selection: one is Foreground selection (FS) and the other is Background selection. In the case of Foreground selection (FS), two flanking markers are used to select a particular gene or a QTL (Hospital & Charcosset, 1997). Whereas in Background selection (BS), many markers are used that are not associated with the desirable QTL/gene and select backcross progeny with the highest proportion of the genome of the recipient parent (Frisch et al., 1999; Hospital & Charcosset, 1997).

The major advantages of MABC in comparison to conventional breeding include:

(i) The Foreground Selection can be used for selecting heterozygous plants at the seedling stage itself
(ii) Recessive alleles can be identified in Marker-Assisted Backcrossing selection
(iii) MABC can be used to develop near-isogenic lines (NILs), which are often used for genomic analysis
(iv) Linkage drag (undesirable gene binding) can be reduced through MABC (Frisch et al., 1999; Ribaut & Hoisington, 1998; Salina, 2003, Hospital, 2005)

5.2 Marker-Assisted Recurrent Selection (MARS)

Plant breeders have traditionally and routinely used various recurrent selection methods to accumulate favorable alleles for yield and other polygenic traits in cross-pollinated crops. Recurrent selection involves alternating generations of selection and interbreeding to identify favorable genetic recombination. MARS is a novel approach in which individuals for intercrossing are selected using a selection index constructed based on QTL-associated molecular markers (Chamarthi et al., 2011; Ribaut et al., 2010; Stam, 1994; Gupta et al., 2010; Tester & Langridge, 2010). The gain from selection using such an index is expected to be higher than phenotypic selection used in conventional recurrent selection methods (Bernardo & Charcosset, 2006). QTL information is generated on the populations of interest (good-by-good crosses, in opposition to good-by-bad crosses used for QTL discovery), and QTL effects for polygenic traits (usually minor) are identified that are specific to that population. If one of the two parents presents a large QTL (such as for a quality trait or biotic stress resistance), such a QTL can also be included in the selection, and the favorable allele can be fixed at early stages of recombination. Superior lines are developed by accumulating favorable alleles through successive intercrossing using only genotypic selection.

Several factors are responsible for the effectiveness of the MARS approach, including number and size of the families, population type and reliability of marker-trait associations (Mayor & Bernardo, 2009). Knowledge of the number of minor QTL associated with the trait is directly correlated with response to MARS (Chamarthi et al., 2011; Charcosset & Moreau, 2004). ICRISAT has initiated the use of MARS in breeding chickpea for drought tolerance. Although the utility of MARS is yet to be demonstrated in chickpea breeding, it could be useful in pyramiding genes involved in complex traits, like drought (Asati et al., 2022; Bhardwaj et al., 2021; Thudi et al., 2024).

5.3 *Identification of Gene(s) and QTL for Introgression Program*

Molecular markers have been used to establish linkage maps for many crop species, and they have been utilized to determine gene numbers for particular traits and for gene tagging. Many important biotic and abiotic resistance genes have been mapped and tagged in chickpea using marker approaches. RAPD, SSR, and SNP markers have been widely employed for mapping genomes and for the identification of genes/QTLs for various traits (Table 7).

Wild species/relatives of several important crop species are the reservoirs for resistance genes to different kinds of abiotic and biotic stresses. However, there has been limited use of these sources of resistance, mainly due to the transfer of associated undesirable alleles (linkage drag). But, with rapid advancements in genomics, such as genome-wide molecular markers and genetic maps and integrative QTL analysis, it is now possible to transfer the favorable alleles into an elite germplasm more efficiently. Therefore, wild species can play a major role in varietal development by being good sources of resistance/tolerance to biotic and abiotic stresses that can be used in breeding programs.

Several approaches have been used for the introgression of genes from wild species, including introgression libraries and advanced-backcross QTL. Successive backcrossing to the recurrent parent, usually for three to four generations, generates introgression lines (ILs) that make up the introgression libraries. Molecular markers are used to monitor the introgressed fragments (Chamarthi et al., 2011). The whole donor genome is represented by a set of small, contiguous overlapping fragments. Such introgression lines (ILs) have been utilized in rice (Zhang et al., 2006, Rangel et al., 2008; Gutiérrez et al., 2010), tomato (Eshed & Zamir, 1995; Kamenetzky et al., 2010), soybean (Concibido et al., 2003), and groundnut (Foncéka et al., 2009) The advanced backcross QTL (AB-QTL) approach was proposed by (Tanksley & Nelson, 1996). AB-QTL detects additive, dominant, partially dominant and overdominant QTL. The approach uses repeated backcrossing with the elite parent. Both phenotyping and genotyping (with polymorphic markers) are done on the

Table 7 List of QTLs/genes identified for different traits in chickpea

Trait(s)	Number and name of gene/QTL	Marker	References
Ascochyta blight	AR2, ar1, ar1a, ar1b, ar2a, ar2b, Ar19, CaLG02, CaLG03, CaLG04, CaLG05 and CaLG06	SSRs, RAPDs, DAF, SNPs	Cho et al. (2004) Collard et al. (2003) Tekeoglu et al. (2002) Rakshit et al. (2003) Udupa et al. (2003) Deokar et al. (2019)
	QTLAR1, *QTLAR2*	SCARs, SSRs, RAPDs	Iruela et al. (2006, 2007)
	13 QTL	SSRs, RAPDs	Kottapalli et al. (2009) Winter et al. (2000) Flandez et al. (2003) Tekeoglu et al. (2002) Millán et al. (2003) Cobos et al. (2003)
	5 QTL (1–5)	SSRs	Anbessa et al. (2009)
	3 QTL	SSR	Aryamanesh et al. (2009)
	319 genes	1118 SNPs significantly associated with AB resistance 11.2–39.3% PVE	Alo et al. (2022)
	CaAP2 (qABR4.2) and CaCNGC1 (qABR4.3)		Singh et al. (2023)
	153 candidate genes	SNPs/SSRs	Dariva et al. (2024)
	6 QTLs	SNPs	Lin et al. (2025)
Fusarium wilt	foc-0, foc-1, foc-2, foc-3, foc-4, foc-5, foc-6, foc-7	SSRs, STSs, RAPDs, KASPs	Winter et al. (2000) Cobos et al. (2005) Tekeoglu et al. (2002) Iruela et al. (2007) Sharma et al. (2004) Bhutia et al. (2024) Faramarzpour et al. (2025)
	Fusarium wilt resistance (race 2)	SNPs	Lakmes et al. (2024)
Botrytis gray mold	3 QTL (1, 2, 3)	SSRs	Anuradha et al. (2011)
Rust	1 QTL	SSR	Madrid et al. (2008)
Seed size traits	2 QTL	SSRs	Hossain et al.(2010) Cobos et al. (2007, 2009)
Seed yield	1 QTL, pentricopeptide repeat (PPR) Gene	SSRs,SNPs	Madrid et al. (2008) Basu et al. (2018)
Seed yield components	1 QTL	SSRs	Vadez et al. (2012)
Single or double pod	*S*	SSRs	Cho et al. (2002) Cobos et al. (2005) Rajesh et al. (2002)

(continued)

Table 7 (continued)

Trait(s)	Number and name of gene/QTL	Marker	References
Flowering time	2 QTL	SSRs	Cobos et al. (2007) Lichtenzveig et al. (2006)
	2 QTL	SSRs	Aryamanesh et al. (2010)
Beta carotene, leutin, seed weight	4 QTL for carotene; 3 QTL for seed weight, genomic regions on CaLG01 (1.08 Mb) and CaLG04 (2.7 Mb) linked with 100-seed weight. Two genes (Ca_04364 and Ca_04607) for 100SDW and one gene (Ca_04586) forRTR, Genomic region on CaLG01 harboring six candidate genes for 100-seed weight	SSRs	Abbo et al. (2005) Singh et al. (2016) Das et al. (2015)
Flower color	*B/b*	SSRs	Cobos et al. (2005)
Drought tolerance	MTAs significantly associated with yield and yield-related traits under drought, "QTL-hotspot" region from ca. 29 cM to 14 cM and added 49 SNPs in the region	SNPs	Li et al. (2018) Jaganathan et al. (2015)
	QTL Hotspot	SSRs	Bhardwaj et al. (2021)
	54 QTLs for different morphophysiological traits	SNPs	Kushwah et al. (2022)
	9 QTLs	SNPs	Yadava et al. (2023)
Seed iron and zinc	Genomic loci/ genes (with 29% combined PVE) associated with seed-Fe and Zn concentrations 35 SNPs significantly associated with seed Zn, Cu, Fe and Mn concentrations	SNPs	Upadhyaya et al. (2016) Fayaz et al. (2022)
Plant vigour and canopy conductance	QTLs for plant vigor and canopy conductance traits on CaLG04 and CaLG03, respectively; transcription factor Ca4_TIFY4B	SNPs	Sivasakthi et al. (2018) Nguyen et al. (2022a, b)

segregating BC2F2 or BC2F3 population generated during backcrossing, followed by QTL analysis.

5.4 Genome-Wide Selection (GWS)

Genomic selection (GS) is a novel approach in which genetic markers covering the whole genome are used so that all (QTL) are in linkage disequilibrium with at least one marker. In this approach, both phenotyping and genotyping data are used to predict genomic estimated breeding values (GEBVs) of progenies. GEBVs are the sum of the effects of all QTL across the genome, exploiting all the genetic variation for a particular trait. Integration of genome-wide sequence information with precise phenotypic variation allows capturing accessions with low-frequency variants that may be responsible for essential phenotypes such as yield components, abiotic stress tolerance, or disease resistance. The GBS approach has been used to detect genome-wide SNPs in chickpea to understand allelic diversity and population structure and to develop high-density linkage maps, QTL analysis, GWAS, and GS (Roorkiwal et al., 2020). Superior progeny lines are selected based on the GEBVs for advancement in the breeding cycle (Bernardo & Yu, 2007; Heffner et al., 2009; Jannink et al., 2010). Best Linear Unbiased Prediction (BLUP) and geostatistical mixed models have been developed to calculate GEBVs (Robinson, 1991; Streeck & Piepho, 2010). The interesting fact about GWS is that there is no need to work out marker-trait associations as required in MABC or MARS (Tester & Langridge, 2010; Bernardo, 2010a, b). Mayor and Bernardo (2009) have shown that double haploid (DH) populations are very useful in GWS compared to F2 populations, especially for complex traits that are controlled by many QTL. Though there is little information available on the use of GWS in crop plants, it is now possible to generate genome-wide marker data (using SNPs) to start GWS in breeding programs of crops like chickpea. In addition, NGS-based WGRS was used to understand the impact of breeding on genetic diversity and temporal diversity trends in chickpea.

5.5 Next-Generation Genotyping and Sequencing Technologies

An NGS technology is the sequence-based genotyping (SbG) method, which has become a popular method of choice for understanding the genetics of complex traits in both plants and animals. So far, the SbG technologies have been deployed extensively in genetic mapping, purity testing, establishing marker-trait associations, marker-assisted selection (MAS), and genomic selection (GS) for crop improvement (Varshney et al., 2014). The next-generation sequencing approach is also used for sequencing the chickpea genome of CDC Frontier (a *kabuli* chickpea variety) by the International Chickpea Genome Sequencing Consortium (ICGSC). For sequencing the chickpea genome, the ICC4958 *desi* chickpea genotype was targeted for generating a draft genome assembly using NGS platforms along with BAC end sequences and a genetic map. An improved version of the *desi* chickpea cultivar ICC4958 with a 2.7-fold increase in the length of pseudomolecules was reported (Parween et al., 2015). This improved assembly could reduce the gaps in the existing genome assembly and predicted the presence of more than 30 K protein-coding genes (Table 8).

Table 8 Genomic resources of chickpea

Platform	Unigenes/transcripts	SSRs	SNPs	References
Illumina	53,409	4816	–	Garg et al. (2011a)
454FLX	34,760	4111	495	Garg et al. (2011b)
454FLX and Illumina	1,03,215	26,252	26,082	Hiremath et al. (2011)
454FLX	1,60,883	–	1022	Gaur et al. (2012a)
454FLX and Illumina	–	–	14,454	Hiremath et al. (2012)
454FLX	37,265	4072	36,446	Jhanwar et al. (2012)
454FLX and Illumina	43,389	5409	39,940	Agarwal et al. (2012)
Illumina NovaSeq 6000	187,452	32,618	4.2 million	Varshney et al. (2019)
PacBio SMRT	81,269 (Isoforms)	–	1.8 million	Jain et al. (2022)
Oxford Nanopore	52.02 billion reads	–	932,172	Thudi et al. (2023)
Hi-C scaffolding	TELOS v2.0 genome	40,112	6.3 million	Garg et al. (2022)

5.6 Chickpea Varieties/Lines Developed Through Molecular Methods

Although conventional breeding based exclusively on phenotypic selection remains the mainstay for most breeding programs in the legumes, but through molecular method, the ultimate aim is to increase the rate of genetic gain across target environments in less time and at lower cost compared to conventional methods.

In this context, there are series of challenges that include (i) the availability of limited predictive molecular markers for target traits, (ii) limited access to marker technologies, and (iii) limited availability of high-throughput, cost-effective, and precise phenotyping platforms.

Most importantly, the deployment of molecular breeding approaches has resulted in the development of improved lines in chickpea. The cost-effective sequencing technologies have introduced a new era in genomics and breeding by pinpointing the genes responsible for distinct phenotypes, leading to the selection of plants based on genotyping information. The transfer of targeted traits has been completed in 2–3 years through marker-assisted backcrossing (MABC), as opposed to the 6–8 years needed with conventional methods. MABC has been deployed for several traits in these legume crops. In the case of chickpea, very recently, improved lines with enhanced resistance to AB have been developed. *Fusarium* wilt (FW) and *Ascochyta* blight (AB) are two major constraints to chickpea production; therefore, a stepwise MABC approach was adopted to develop dual resistance in C 214, an elite chickpea cultivar. To develop resistant lines, the *foc1* locus for FW and two quantitative trait loci (QTL) regions were targeted (Rajput et al., 2023; Sudheesh et al., 2021). In summary, during the last 12 years, significant progress has been made in terms of developing genomic resources, and these resources have been effectively used for attaining faster genetic gains in chickpea (Tables 9 and 10).

Table 9 Elite varieties targeted for molecular breeding to enhance tolerance/resistance to drought, *Fusarium* wilt and *Ascochyta* blight in India

Trait	Variety/lines	Developer
Drought tolerance	JG 11 & ICCV 10	ICRISAT, Hyderabad
	DCP92-3 & KWR 108	IIPR, Kanpur
	Pusa 362	IARI, New Delhi
	RSG 888,JAKI 9218	ICAR-AICRP for release in India
	Pusa 372	IARI, New Delhi
	Super Annigeri 3	UAS, Dharwad
	Pusa Manav (ICCV 17308)	ICAR-IARI, New Delhi
Fusarium wilt resistance	C 214	ICRISAT, Hyderabad
	Pusa 256	ICAR-IIPR, Kanpur
	JG 74, Improved JG74 Phule G12, Annigeri-1, Super Annigeri-1	JNKVV, Jabalpur, MPKV, Rahuri, ARS- Kalaburagi
	Pusa 372,Pusa 391,Pusa 256	IARI, New Delhi
	Phule Vikram	MPKV, Rahuri
	KGD 1168	UAS, Dharwad
	RVG 203	RAU, Udaipur
Ascochyta blight resistance	C 214	ICRISAT, Hyderabad
	GNG 469	RAU, Udaipur
	Phule G 95311	MPKV, Rahuri

Source—Annual Report upto 2023–24-All India Coordinated Research Project on Chickpea and Mannur et al. (2019)

Table 10 Superior molecular breeding lines in the genetic background of JG 11 and ICCV 10 evaluated in Advanced Varietal Trials of AICRP on chickpea upto 2024

SL No	Name of entry	Pedigree	Source	Key traits	Evaluation year
1	DIBG 203	(JG 11 × ICC 4958) × 3*JG 11 – 1	UAS, Dharwad	Drought tolerance, high yield	2016–2017 (Released)
2	DIBG 204	(JG 11 × ICC 4958) × 3*JG 11 – 11	UAS, Dharwad	Enhanced root traits	2016–2017
3	DIBG 205	(ICCV 10 × ICC 4958) × 3*ICCV 10 – 21	UAS, Dharwad	Early maturity	2016–2017
4	NBeG 506	(JG 11 × ICC 4958) × 3*JG 11 – 35	RARS, Nandyal	*Fusarium* wilt resistance	2016–2017
5	NBeG 786	(JG 11 × ICC 4958) × 3*JG 11 – 40	RARS, Nandyal	Dual resistance (wilt + root rot)	2016–17

(continued)

Table 10 (continued)

SL No	Name of entry	Pedigree	Source	Key traits	Evaluation year
6	NBeG 1004	(ICCV 10 × ICC 4958) × 3*ICCV 10 – 9	RARS, Nandyal	Heat tolerance	2016–2017
7	RKD 1	(JG 11 × ICC 4958) × 3*JG 11 – 13	ARS, Kalaburagi	Drought tolerance	2016–2017
8	RKD 4	(ICCV 10 × ICC 4958) × 3*ICCV 10 – 2	ARS, Kalaburagi	*Ascochyta* blight resistance	2016–2017
9	RVSS 51	(JG 11 × ICC 4958) × 3*JG 11 – 34	RAKCA, Sehore	High protein (22–24%)	2016–2017
10	RVSSG 52	(ICCV 10 × ICC 4958) × 3*ICCV 10 – 13	RAKCA, Sehore	Salinity tolerance	2016–2017
11	DIBG 305	(JG 11 × ICC 4958) × 4*JG 11 – 5	UAS, Dharwad	Enhanced drought tolerance	2020
12	NBeG 1207	(ICCV 10 × ICC 4958) × 4*ICCV 10 – 18	RARS, Nandyal	Heat + wilt resistance	2021
13	RKD 7	(JG 11 × ICC 4958) × 4*JG 11 – 9	ARS, Kalaburagi	Pod borer tolerance	2022
14	RVSSG 75	(ICCV 10 × ICC 4958) × 4*ICCV 10 – 7	RAKCA, Sehore	Biofortified (high Fe/Zn)	2023
15	Pusa JG 11+	(JG 11 × ICC 4958) × 5*JG 11 – 12	ICAR-IARI	CRISPR-edited for drought genes	2024

6 Conclusion

Significant progress has been made so far in the development of genetic and genomic resources for chickpea during the past few decades. The availability of a large number of molecular markers, dense genetic maps, markers associated with traits, and transcriptomics resources has made it possible to integrate genomics technologies to accelerate the chickpea improvement. The draft genome of chickpea has explored the possibility for breeders and biotechnologists to unlock their genetic potential in order to develop new variety with enhanced resistance and yield. Genomic resources developed in the recent past have enabled both trait dissection and trait improvement through modern molecular breeding approaches. Some of the marker-assisted breeding methods, like MABC and MARS, are currently being integrated into chickpea breeding programs. The integration of genomic technologies in chickpea breeding will greatly improve the efficiency of breeding programs in the development of better cultivars and reduce the time required for cultivar development.

7 Summary

Chickpea is one of the most important food legume crops, and its improvement programs need to be dynamic and efficient in rapidly developing varieties needed for adaptation to existing and evolving cropping systems. Information on genetic progress achieved over time from a breeding program is absolutely essential to develop effective and efficient breeding strategies by assessing the efficiency of past improvement works in genetic yield potential and suggesting future selection directions to facilitate further improvement. Integrating breeding approaches utilizing conventional and genomic resources has been effectively used for fast, precise, and efficient chickpea breeding. Improved lines with enhanced resistance to AB have been developed. *Fusarium* wilt (FW) and *Ascochyta* blight (AB) are two major biotic constraints to chickpea production; therefore, a stepwise MABC approach has been adopted to develop dual-resistant C 214, an elite chickpea cultivar. To develop resistant lines, the *foc1* locus for FW and two quantitative trait loci (QTL) regions were targeted. To improve Annigeri 1 and JG 74, the introgression of QTL conferring resistance against FW race 4 (*foc4)* was carried out through marker-assisted backcrossing using WR 315 as the donor parent. In the present scenario, pod borer continues to remain the most challenging insect-pest of chickpea due to the non-availability of sources with high levels of resistance in germplasm, however, with the advent of modern breeding methods, breeders will also have an opportunity to develop new varieties for complex traits. The integration of genomic technologies in chickpea breeding will greatly improve the efficiency of breeding programs in the development of better cultivars and reduce the time required for their development.

Glossary of Technical Terms Pedigree selection (PS)

Three way cross
Mutational Breeding
Backcross breeding
Marker-Assisted Backcrossing (MABC)
Marker-Assisted Recurrent Selection (MARS)
QTL hotspot
QTL introgression
Genome Wide Selection (GWS)
Next–Generation Sequencing (NGS)
SNPs Single Nucleotide Polymorphisms
Genome wide selection (GWS)

References

Abbo, S., Molina, C., Jungmann, R., Grusak, M. A., Berkovitch, Z., Reifen, R., Kahl, G., Winter, P., & Reifen, R. (2005). Quantitative trait loci governing carotenoid concentration and weight iseeds of chickpea (*Cicer arietinum* L.). *Theoretical and Applied Genetics, 111*, 185–195.

Agarwal, G., Jhanwar, S., Priya, P., Singh, V. K., Saxena, M. S., Parida, S. K., ... & Jain, M. (2012). Comparative analysis of kabuli chickpea transcriptome with desi and wild chickpea provides a rich resource for development of functional markers. *PloS One, 7*(12), e52443.

Ahloowalia, B. S., Maluszynski, M., & Nichterlein, K. (2004). Global impact of mutation-derivedvarieties. *Euphytica, 135*, 187–204.

Allard, R. W. (1960). *Principles of plant breeding*. Wiley.

Alo, F., Rani, A. R., Baum, M., Singh, S., Kehel, Z., Rani, U., Udupa, S., Al-Sham'aa, K., Alsamman, A. M., Istanbuli, T. and Attar, B. (2022). Novel genomic regions linked to Ascochyta blight resistance in two differentially resistant cultivars of chickpea. *Frontiers in Plant Science, 13*, 762002.

Anbessa, Y., Taran, B., Warkentin, T. D., Tullu, A., & Vandenberg, A. (2009). Genetic analyses and conservation of QTL for ascochyta blight resistance in chickpea (*Cicer arietinum* L.). *Theoretical and Applied Genetics, 119*, 757–765.

Argikar, G. P. (1970). Gram. In P. Kachroo (Ed.), *Pulse crops of India* (pp. 54–135). ICAR.

Aryamanesh, N., Nelson, M. N., Yan, G., Clarke, H. J., & Siddique, K. H. M. (2010). Mapping a major gene for growth habit and QTLs for ascochyta blight resistance and flowering time in a population between chickpea and *Cicer reticulatum*. *Euphytica, 173*, 307–319.

Asati, R., Tripathi, M. K., Tiwari, S., Yadav, R. K., & Tripathi, N. (2022). Molecular breeding and drought tolerance in chickpea. *Life., 12*(11), 1846.

Auckland, A. K., & van der Maesen, L. J. G. (1980). Chickpea. In W. R. Fehr & H. H. Hadely (Eds.), *Hybridization of crop plants* (pp. 249–259). American Society of Agronomy.

Bahl, P. N., & Gowda, C. L. L. (1975). Pod setting in crosses of Bengal gram. *Indian Journal of Genetics, 35*, 13–16.

Basu, U., Srivastava, R., Bajaj, D., Thakro, V., Daware, A., Malik, N., Upadhyaya, H. D., & Parida, S. K. (2018). Genome-wide generation and genotypingof informative SNPs to scan molecular signatures for seed yield in chickpea. *Scientific Reports, 8*, 13240.

Bejiga, G., & Tessema, T. (1981). Identification on the optimum time of emasculationand pollination to increase percentageof hybrid seed set in chickpea. *Journal of Agricultural Sciences, 3*, 129–134.

Bernardo, R. (2010a). Genome wide selection with minimal selection in self-pollinated crops. *Crop Science, 50*, 624.

Bernardo, R. (2010b). *Potential and practicality of genome wide selection in plants*. Proceedings of the Plant & Animal Genomes XVIII Conference, 9–13 January 2010.

Bernardo, R., & Charcosset, A. (2006). Usefulness of gene information in marker-assisted recurrent selection: A simulation appraisal. *Crop Science, 46*, 614–662.

Bernardo, R., & Yu, J. (2007). Prospects for genomewide selection for quantitative traits in maize. *Crop Science, 47*(3), 1082–1090.

Bharadwaj, C., Tripathi, S., Soren, K. R., Thudi, M., Singh, R. K., Sheoran, S., Roorkiwal, M., Patil, B. S., Chitikineni, A., Palakurthi, R., & Vemula, A. (2021). Introgression of "QTL-hotspot" region enhances drought tolerance and grain yield in three elite chickpea cultivars. *The Plant Genome., 14*(1), e20076.

Bharathi, C. L., Lavanya, G. R., Reddy, Y. V., & Anirudh, A. (2022). Genetic variability, correlation, direct and indirect effects on seed yield through different quantitative traits in chickpea (*Cicer arietinum* L.) germplasm at central plain zone (Uttar Pradesh). *Journal of Experimental Agriculture International, 44*(2), 1–11.

Bharadwaj, C., Tripathi, S., Soren, K. R., Thudi, M., Singh, R. K., Sheoran, S., Roorkiwal, M., Patil, B. S., Chitikineni, A., Palakurthi, R. and Vemula, A. (2021). Introgression of "QTL-hotspot" region enhances drought tolerance and grain yield in three elite chickpea cultivars. *The Plant Genome, 14*(1), p.e20076.

Bhatia, C. R., Maluszynski, M., Nichterlein, K., & Van Zanten, L. (2001a). Grain legume cultivars derived from induced mutations, and mutation affecting nodulation. *Mutation Breeding, 13*, 1–44.

Bhatia, C. R., Nichterlein, K., & Maluszynski, M. (2001b). Mutation affecting nodulation in grain legumes and their potential in sustainable cropping systems. In M. Maluszynski & K. J. Kasha

(Eds.), *Mutations. In: Vitro and molecular techniques for enviornmentally sustainable crop improvement* (pp. 201–226). Kluwer Academic.
Bhatia, C. R., Nichterlein, K., & Maluszynski, M. (2001c). Mutation affecting nodulationin grain legumes and their potential in sustainable cropping systems. *Euphytica, 120*, 415–432.
Bhutia, K. L., Kisku, A., Lap, B., Sahni, S., Arya, M., Bhutia, N. D., Ahmad, M., Chaturvedi, R., Sudhan, R. A. & Sharma, V. K. (2024). Insight into a region of chickpea (Cicer arietinum L.) Chromosome 2 revealed potential candidate genes linked to Foc4 Fusarium wilt resistance. *Functional Plant Biology, 51*(9).
Bisen, M. S., Singh, S. P., & Rao, S. K. (1985). Effectiveness of selection methods inchickpea *Cicer arietinum* L. under different environments. *Theoretical and Applied Genetics, 70*(6), 661–666.
Brock, R. D. (1977). Prospects and perspectivesin mutation breeding. In A. Muhammed, R. Aksel, & R. C. Von Borstel (Eds.), *Genetic diversity in plants* (pp. 117–132). Plenum Press.
Byth, D. E., Green, J. M., & Hawtin, G. C. (1980). ICRISAT/ICARDA chickpea breeding strategies. In *Proceedings of the International Workshop on Chickpea Improvement* (pp. 11–27). 28 February–2 March 1979. ICRISAT.
Cagirgan, M. I., & Toker, C. (2004). Selection of chickpea mutants for cold tolerance and *ascochyta* blight resistance. *International Chickpea and Pigeonpea Newsletter, 11*, 14–16.
Charcosset, A., & Moreau, L. (2004). Use of molecular markers for the development of new cultivars and the evaluation of genetic diversity. *Euphytica, 137*, 81–94.
Chamarthi, S. K., Kumar, A., Vuong, T. D., Blair, M. W., Gaur, P. M., Nguyen, H. T., & Varshney, R. K. (2011). Trait mapping and molecular breeding. In *Biology and breeding of food legumes* (pp. 296–313). Wallingford UK: CABI.
Chaturvedi, S. K., Kumar, S., Dua, R. P. C., & Breeding. (2003). Chapter 4. In A. Masood, K. Shiv, & N. B. Singh (Eds.), *Chickpea research in India* (pp. 69–98). Indian Institute of Pulse Research.
Cho, S. H., Chen, W. D., & Muehlbauer, F. J. (2004). Pathotype-specific genetic factors in chickpea (*Cicerarietinum* L.) for quantitative resistance to Aschochyta blight. *Theoretical and Applied Genetics, 109*, 733–739.
Cobos, M. J., Fernández, M. J., Rubio, J., Kharrat, M., Moreno, M. T., Gil, J., & Millan, T. (2005). A linkage map of chickpea (Cicer arietinum L.) based on populations from Kabuli× Desi crosses: location of genes for resistance to fusarium wilt race 0. *Theoretical and Applied Genetics, 110*, 1347–1353.
Cobos, M. J., Rubio, J., Fernández-Romero, M. D., Garza, R., Moreno, M. T., Millán, T., & Gil, J. (2007). Genetic analysis of seed size, yield and days to flowering in a chickpea recombinant inbred line population derived from a Kabuli×Desi cross. *Annals of Applied Biology, 151*(1), 33–42.
Cobos, M. J., Winter, P., Kharrat, M., Cubero, J. I., Gil, J., Millan, T., & Rubio, J. (2009). Genetic analysis of Agronomic traits in a wide cross of chickpea. *Field Crops Research, 1*, 130–136.
Collard, B.C.Y., & Mackill, D.J. (2008). Marker-assisted selection: An approach for precision plant
Collard, B. C. Y., Pang, E. C. K., Ades, P. K., & Taylor, P. W. J. (2003). Preliminary investigations of QTL associated with seedlings resistance to *Ascochyta* blight from *Cicer echinospermum*, a wild relative of chick pea. *Theoretical and Applied Genetics, 107*, 719–729.
Concibido, V., La Vallee, B., Mclaird, P., Pineda, N., Meyer, J., Hummel, L., ... & Delannay, X. (2003). Introgression of a quantitative trait locus for yield from Glycine soja into commercial soybean cultivars. *Theoretical and Applied Genetics, 106*, 575–582.
Conger, B. V., Konzak, C. F., & Nilan, R. A. (1977). Radiation sensitivity and modifying factors. In *Manual on mutation breeding* (2nd ed., pp. 40–50). Technical reports series No. 119. International Atomic Energy Agency.
Coyne, C. J., Kumar, S., von Wettberg, E. J., Marques, E., Berger, J. D., Redden, R. J., Ellis, T. N., Brus, J., Zablatzká, L., & Smýkal, P. (2020). Potential and limits of exploitation of crop wild relatives for pea, lentil, and chickpea improvement. *Legume Science, 2*(2), e36.
Croser, J. S., Ahmad, F., Clarke, H. J., & Siddique, K. H. M. (2003). Utilization of wild *Cicer* in chickpea improvement—Progress, constraints, and prospects. *Australian Journal of Agricultural Research, 54*, 429–444.

Dahiya, B. S., Waldia, R. S., Kaushik, L. S., & Solanki, I. S. (1984). Early generation yield testing versus visual selection in chickpea (*Cicer arietinum* L.). *Theoretical and Applied Genetics, 68*(6), 525–529.

Dariva, F. D., Arman, A., Morales, M., Navasca, H., Shah, R., Atanda, S. A., Piche, L., Worral, H., Raymon, G., McPhee, K. & Coyne, C. (2024). Identification of novel candidate genes for Ascochyta blight resistance in chickpea. *Scientific Reports, 14*(1), 31415.

Das, S., Upadhyaya, H. D., Bajaj, D., Kujur, A., Badoni, S., Laxmi, K. V., Tripathi, S., Gowda, C. L. L., Sharma, S., Singh, S., Tyagi, A. K., & Parida, S. K. (2015). Deploying QTL-seq for rapid delineation of a potential candidate gene underlying major trait-associated QTL in chickpea. *DNA Research, 22*, 193–203.

Deokar, A., Sagi, M., & Tar'an, B. (2019). Genome-wide SNP discovery for development of high-density genetic map and QTL mapping of ascochyta blight resistance in chickpea (*Cicer arietinum* L.). *Theoretical and Applied Genetics, 132*, 1861–1872.

De Vries, H. (1901). *Die mutationstheorie* (Vol. 2, pp. 166–166). Leipzig: Veit.

Dua, R. P., Chaturvedi, S. K., & Sewak, S. (2001). *Reference varieties of chickpea for IPR regime*. IIPR.

Eshed, Y., & Zamir, D. (1995). An introgression line population of *Lycopersicon pennellii* in the cultivated tomato enables the identification and fine mapping of yield-associated QTL. *Genetics, 141*, 1147–1162.

Faramarzpour, A., Dezhsetan, S., Hassaneian Khoshro, H., Mirdar Mansuri, R., Pouralibaba, H. R., & Shobbar, Z. S. (2025). The transcriptional response to yellow and wilt disease, caused by race 6 of Fusarium oxysporum f. sp. Ciceris in two contrasting chickpea cultivars. *BMC Genomics, 26*(1), 1–17.

Fayaz, Humara, Sandhya Tyagi, Aijaz A. Wani, Renu Pandey, Sabina Akhtar, Mohd Ashraf Bhat, Annapurna Chitikineni (2022). "Genome-wide association analysis to delineate high-quality SNPs for seed micronutrient density in chickpea (Cicer arietinum L.)." *Scientific Reports, 12*(1), 11357.

FAOSTAT (2024). *Food and agriculture organization corporate statistical database*. http://www.fao.org/faostat/en/#data/QC

Flandez-Galvez, H., Ades, P. K., Ford, R., Pang, E. C. K., & Taylor, P. W. J. (2003). QTL analysis for ascochyta blight resistance in an intraspecific population of chickpea (*Cicer arietinum* L.). *Theoretical and Applied Genetics, 107*, 1257–1265.

Foncéka, D., Hodo-Abalo, T., Rivallan, R., Faye, I., Sall, M. N., Ndoye, O., Fávero, A. P., Bertioli, D. J., Glaszmann, J. C., & Courtois, B. (2009). Genetic mapping of wild introgressions into cultivated peanut: A way toward enlarging the genetic basis of a recent allotetraploid. *BMC Plant Biology, 9*, 103.

Frisch, M., Bohn, M., & Melchinger, A. E. (1999). Minimum sample size and optimal positioning offlanking markers in marker assisted backcrossing for transfer of a target gene. *Crop Science, 39*, 967–975.

Garg, R., Patel, R. K., Tyagi, A. K., & Jain, M. (2011a). De novo assembly of chickpea transcriptome using short reads for gene discovery and marker identification. *DNA Research, 18*(1), 53–63.

Garg, R., Patel, R. K., Jhanwar, S., Priya, P., Bhattacharjee, A., Yadav, G., ... & Jain, M. (2011b). Gene discovery and tissue-specific transcriptome analysis in chickpea with massively parallel pyrosequencing and web resource development. *Plant Physiology, 156*(4), 1661–1678.

Garg, V., Dudchenko, O., Wang, J., Khan, A. W., Gupta, S., Kaur, P., Han, K., Saxena, R. K., Kale, S. M., Pham, M. & Yu, J. (2022). Chromosome-length genome assemblies of six legume species provide insights into genome organization, evolution, and agronomic traits for crop improvement. *Journal of Advanced Research, 42*, 315–329.

Gaur, P. M., Jukanti, A. K., & Varshney, R. K. (2012). Impact of genomic technologies on chickpea breeding strategies. *Agronomy, 2*(3), 199–221.

Gaur, P. M., & Gour, V. K. (2002). A gene producing one to nine flowers per flowering node in chickpea. *Euphytica, 128*, 231–235.

Gujaria, N., Kumar, A., Dauthal, P., Dubey, A., Hiremath, P., Bhanu Prakash, A., ... & Varshney, R. K. (2011). Development and use of genic molecular markers (GMMs) for construction of

a transcript map of chickpea (Cicer arietinum L.). *Theoretical and Applied Genetics, 122*, 1577–1589.
Gupta, P. K., Kumar, J., Mir, R. R., & Kumar, A. (2010). Marker-assisted selection as a component of conventional plant breeding. *Plant Breeding Reviews, 33*(4), 145–217.
Gutiérrez, A. G., Carabalí, S. J., Giraldo, O. X., Martínez, C. P., Correa, F., Prado, G., ... & Lorieux, M. (2010). Identification of a rice stripe necrosis virus resistance locus and yield component QTLs using Oryza sativa× O. glaberrima introgression lines. *BMC Plant Biology, 10*, 1–11.
Haq, M. A., & Singh, K. B. (1994). Induction of cold tolerance in kabuli chickpea (Cicer arietinum L.) through induced mutations. *Mutat Breed Newsletter, 41*, 6.
Haq, M. A., Sadiq, M., & Hassan, M. (1999). CM 98′ (CM311/85) a very high yielding disease resistant mutant variety of chickpea. *Int Chickpea Pigeonpea Newsletter, 6*, 910.
Haq, M. A., Sadiq, M., & Hassan, M. (2001). CM 88 a multiple disease resistant chickpea mutant variety. *Mutat Breed Newsletter, 45*, 56.
Heffner, E. L., Sorrells, M. E., & Jannink, J. L. (2009). Genomic selection for crop improvement. *Crop Science, 49*, 1. –12.
Hiremath, P. J., Farmer, A., Cannon, S. B., Woodward, J., Kudapa, H., Tuteja, R., ... & Varshney, R. K. (2011). Large-scale transcriptome analysis in chickpea (Cicer arietinum L.), an orphan legume crop of the semi-arid tropics of Asia and Africa. *Plant Biotechnology Journal, 9*(8), 922–931.
Hiremath, P. J., Kumar, A., Penmetsa, R. V., Farmer, A., Schlueter, J. A., Chamarthi, S. K., ... & Varshney, R. K. (2012). Large-scale development of cost-effective SNP marker assays for diversity assessment and genetic mapping in chickpea and comparative mapping in legumes. *Plant Biotechnology Journal, 10*(6), 716–732.
Hospital, F. (2005). Selection in backcross programmes. *Philosophical Transactions of the Royal Society B, 360*, 1503–1511.
Hospital, F., & Charcosset, A. (1997). Marker assisted introgression of quantitative trait loci. *Genetics, 147*, 1469–1485.
Hossain, S., Ford, R., McNeil, D., Pittock, C., & Panozzo, J. F. (2010). Inheritance of seed size in chickpea (*Cicer arietinum* L.) and identification of QTL based on 100-seed weight and seed size index. *Australian Journal of Crop Science, 4*, 126–135.
Iruela, M., Rubio, J., Barro, F., Cubero, J. I., Millán, T., & Gil, J. (2006). Detection of two quantitative trait loci for resistance to ascochyta blight in an intra-specific cross of chickpea (Cicer arietinum L.): development of SCAR markers associated with resistance. *Theoretical and Applied Genetics, 112*, 278–287.
Iruela, M., Castro, P., Rubio, J., Cubero, J. I., Jacinto, C., Millan, T., & Gil, J. (2007). Validation of a QTL for resistance to Ascochyta blight linked to resistance to *Fusarium* wilt race 5 in chickpea (*Cicerarietinum* L.). *European Journal of Plant Pathology, 199*, 29–37.
Jaganathan, D., Thudi, M., Kale, S., Azam, S., Roorkiwal, M., Gaur, P. M., ... & Varshney, R. K. (2015). Genotyping-by-sequencing based intra-specific genetic map refines a "QTL-hotspot" region for drought tolerance in chickpea. *Molecular Genetics and Genomics, 290*, 559–571.
Jain, M., Bansal, J., Rajkumar, M. S., & Garg, R. (2022). An integrated transcriptome mapping the regulatory network of coding and long non-coding RNAs provides a genomics resource in chickpea. *Communications Biology, 5*(1), 1106.
Jannink, J. L., Lorenz, A. J., & Iwata, H. (2010). Genomic selection in plant breeding: From theory topractice. *Briefings in Functional Genomics and Proteomics, 9*, 166–177.
Kale, S. M., Jaganathan, D., Ruperao, P., Chen, C., Punna, R., Kudapa, H., & Garg, V. (2015). Prioritization of candidate genes in "QTL-hotspot" region for drought tolerance in chickpea (*Cicer arietinum* L.). *Scientific Reports, 5*, 15296.
Kamenetzky, L., Asís, R., Bassi, S., de Godoy, F., Bermudez, L., Fernie, A. R., ... & Carrari, F. (2010). Genomic analysis of wild tomato introgressions determining metabolism-and yield-associated traits. *Plant Physiology, 152*(4), 1772–1786.

Kaushal, K., Rajani, K., Kumar, R. R., Ranjan, T., Kumar, A., Ahmad, M. F., Kumar, V., Kumar, V., & Kumar, A. (2024). Physio-biochemical responses and crop performance analysis in chickpea upon botanical priming. *Scientific Reports, 14*(1), 9342.

Kharkwal, M. C. (1998). Induced mutations in chickpea (Cicer arietinum L.) I. Comparative mutagenic effectiveness and efficiency of physical and chemical mutagens. *Indian Journal of Genetics and Plant Breeding, 58*(02), 159–167.

Kharkwal, M. C. (1999). Induced mutations in chickpea (Cicer arietinum L.) III. Frequency and spectrum of viable mutations. *Indian Journal of Genetics and Plant Breeding, 59*(04), 451–464.

Kharkwal, M. C., Jain, H. K., & Sharma, B. (1988). Induced mutations for improvement of chickpea, lentil, pea and cowpea. In *Proc. FAO/IAEA workshop on Improvement of Grain Legume Production using Induced Mutations* (pp. 89–109). IAEA.

Kharkwal, M. C., Nagar, J. P., & Kala, Y. K. (2005). BGM 547—A high yielding chickpea (*Cicer arietinum* L.) mutant variety for late sown conditions of North Western Plains Zone of India. *Journal of Genetics, 65*, 229–230.

Kharkwal, M. C., Gopalakrishna, T., & Pawar, S. E. (2008). Mutation breeding for improvement of food legumes. In M. C. Kharkwal (Ed.), *Food legumes for nutritional security and sustainable agriculture* (pp. 194–221). Vol. 1. Proc. Fourth International Food Legumes Research Conference (IFLRC–IV), Oct. 18–22, 2008, New Delhi. Indian Society of Genetics and Plant Breeding.

Khosh-Khui, M., & Niknejad, M. (1972). Plant height and width inheritance and their correlation with some of the yield components in chickpeas (Cicer arietinum L.). *The Journal of Agricultural Science, 78*(1), 37–38.

Konzak, C. F. (1977). Genetic control of the content, amino acid composition, and processing properties of proteins in wheat. *Advances in Genetics, 19*, 407–582.

Kottapalli, P., Gaur, P. M., Katiyar, S. K., Crouch, J. H., Buhariwalla, H. K., Pande, S., & Gali, K. K. (2009). Mapping and validation of QTLs for resistance to an Indian isolate of Ascochyta blight pathogen in chickpea. *Euphytica, 165*(1), 79–88.

Kudapa, H., Ghatak, A., Barmukh, R., Chaturvedi, P., Khan, A., Kale, S., Fragner, L., Chitikineni, A., Weckwerth, W., & Varshney, R. K. (2024). Integrated multi-omics analysis reveals drought stress response mechanism in chickpea (Cicer arietinum L.). *The Plant Genome, 17*(1), e20337.

Kumar, J., & Bahl, P. N. (1992). Direct and indirect selection for yield in chickpea. *Euphytica, 60*, 197–199.

Kushwah, A., Bhatia, D., Barmukh, R., Singh, I., Singh, G., Bindra, S., Vij, S., Chellapilla, B., Pratap, A., Roorkiwal, M. & Kumar, S. (2022). Genetic mapping of QTLs for drought tolerance in chickpea (Cicer arietinum L.). Frontiers in Genetics, 13, p.953898.

Lakmes, A., Jhar, A., Sadanandom, A., Brennan, A. C., & Kahriman, A. (2024). Inheritance of Resistance to Chickpea Fusarium Wilt Disease (Fusarium oxysporum f. sp. ciceris Race 2) in a Wide-Cross Cicer arietinum× Cicer reticulatum Mapping Family. *Genes, 15*(6), 819.

Lal, S., Singh, S. N., Lal, S. B., Tripathi, B. D. H. (1973). Improvement in gram through pedigree method of breeding. In: Proceedings of the 10th Workshop on Rabi Pulses AICRP (ICAR), Bose Institute. 10–14 September. ICAR, New Delhi, India, pp. 70–80.

Li, B., Zhang, N., Wang, Y. G., George, A. W., Reverter, A., & Li, Y. (2018). Genomic prediction of breeding values using a subset of SNPs identified by three machine learning methods. *Frontiers in Genetics, 9*, 237.

Lichtenzveig, J., Bonfil, D. J., Zhang, H. B., Shtienberg, D., & Abbo, S. (2006). Mapping quantitative trait loci in chickpea associated with time to flowering and resistance to *Didymella rabiei* the causal agent of ascochyta blight. *Theoretical and Applied Genetics, 113*, 1357–1369.

Madrid, E., Rubiales, D., Moral, A., Moreno, M. T., Millan, T., Gil, J., & Rubio, J. (2008). Mechanism and molecular markers associated with rust resistance in a chickpea interspecific cross (*Cicer arietinum* × *Cicer reticulatum*). *European Journal of Plant Pathology, 121*, 45–53.

Maluszynski, M., Nichterlein, K., Van Zanten, L., & Ahloowalia, B. S. (2000). Officially released mutant varieties—The FAO/IAEA. *Database, 12*, 1–85.

Mannur, D. M., Babbar, A., Thudi, M., Sabbavarapu, M. M., Roorkiwal, M., Yeri, S. B., ... & Varshney, R. K. (2019). Super Annigeri 1 and improved JG 74: two Fusarium wilt-resistant introgression lines developed using marker-assisted backcrossing approach in chickpea (Cicer arietinum L.). *Molecular Breeding, 39*, 1–13.

Mayor, P. J., & Bernardo, R. (2009). Genome wide selection and marker-assisted recurrent selection in doubled haploid versus F populations. *Crop Science, 49*, 1719–1725.

Micke, A. (1988a). Genetic improvement of food legumes in developing countries by mutation induction. In *World crops: Cool season food legumes: A global perspective of the problems and prospects for crop improvement in pea, lentil, faba bean and chickpea* (pp. 1031–1047). Dordrecht: Springer Netherlands.

Micke, A. (1988b). Genetic improvement of grain legumes using induced mutations. An overview (pp. 1–51).

Micke, A., & Donini, B. (1993). Induced mutations. In M. D. Hayward, N. O. Bosemark, & I. Romagosa (Eds.), *Plant breeding: Principles and prospects* (pp. 53–62). Chapman & Hall.

Millán, T., Rubio, J., Iruela, M., Daly, K., Cubero, J. I., & Gil, J. (2003). Markers associated with Ascochyta blight resistance in chickpea and their potential in marker-assisted selection. *Field Crops Research, 84*, 373–384.

Mohanty, J. K., Jha, U. C., Dixit, G. P., & Parida, S. K. (2022). Harnessing the hidden allelic diversity of wild Cicer to accelerate genomics-assisted chickpea crop improvement. *Molecular Biology Reports, 16*, 1–9.

Moose, S., & Mumm, R. H. (2008). Molecular plant breeding as the foundation for 21st century crop improvement. *Plant Physiology, 147*, 969–977.

Moreno, M. T., & Cubero, J. I. (1978). Variation in *Cicer arietinum* L. *Euphytica, 27*, 465–485.

Muller, H. J. (1927). Artificial transmutation of the gene. *Science LXVI, 66*, 84–87.

Nayak, S. N., Zhu, H., Varghese, N., Datta, S., Choi, H. K., Horres, R., ... & Varshney, R. K. (2010). Integration of novel SSR and gene-based SNP marker loci in the chickpea genetic map and establishment of new anchor points with Medicago truncatula genome. *Theoretical and Applied Genetics, 120*, 1415–1441.

Nguyen, D. T., Hayes, J. E., Atieno, J., Li, Y., Baumann, U., Pattison, A., Bramley, H., Hobson, K., Roorkiwal, M., Varshney, R. K., & Colmer, T. D. (2022a). The genetics of vigour-related traits in chickpea (Cicer arietinum L.): Insights from genomic data. *Theoretical and Applied Genetics, 135*(1), 107–124.

Nguyen, D. T., Hayes, J. E., Harris, J. and Sutton, T. (2022b). Fine mapping of a vigor QTL in chickpea (Cicer arietinum L.) reveals a potential role for Ca4_TIFY4B in regulating leaf and seed size. *Frontiers in Plant Science, 13*, 829566.

Oram, R. N., Shaikh, M. A. Q., & Kman, K. M. S. (1987). Isozyme similarity and genetic difference in morphology between Hyprosala, a high yielding, high protein mutant of chickpea (*Cicer arietinum* L.) and its parental cultivar. *Environmental and Experimental Botany, 27*, 455–462.

Parween, S., Nawaz, K., Roy, R., Pole, A. K., Venkata Suresh, B., & Misra, G. (2015). An advanced draft genome assembly of a desi type chickpea (*Cicer arietinum* L.). *Scientific Reports, 5*, 12806.

Paul, P. J., Samineni, S., Thudi, M., Sajja, S. B., Rathore, A., Das, R. R., & Gaur, P. M. (2018). Molecular mapping of QTLs for heat tolerance in chickpea. *International Journal of Molecular Sciences, 19*(8), 2166.

Pundir, R. P. S., & Reddy, K. N. (1984). Comparison of cross-pollination methods in chickpea. *International Chickpea Newsletter, 11*, 9–11.

Pushpavalli, R., Quealy, J., Colmer, T. D., Turner, N. C., Siddique, K. H., Rao, M. V., & Vadez, V. (2016). Salt stress delayed flowering and reduced reproductive success of chickpea (Cicer arietinum L.), a response associated with Na+ accumulation in leaves. *Journal of Agronomy and Crop Science, 202*(2), 125–138.

Rahman, M. A., & Bahl, P. N. (1985). Comparison of single seed descent, mass selection and random bulk methods in chickpea. *Indian Journal of Genetics, 45*(2), 186–193.

Rajesh, P. N., Tullu, A., Gil, J., Gupta, V. S., Ranjekar, P. K., & Muehlbauer, F. J. (2002). Identification of STMS marker for the double-podding gene in chickpea. *Theoretical and Applied Genetics, 105*, 604–607.

Rajput, S., Jain, S., Tiwari, S., Sikarwar, R. S., Tripathi, N., Barela, A., & Tripathi, M. K. (2023). Evaluation of gene-based markers against Fusarium wilt disease in chickpea (*Cicer arietinum* L.). Current. *Journal of Applied Science and Technology, 42*(6), 26–32.

Rakshit, S., Winter, P., Tekeoglu, M., Juarez Muñoz, J., Pfaff, T., Benko-Iseppon, A. M., Muehlbauer, F. J., & Kahl, G. (2003). DAF marker tightly linked to a major locus for ascochyta blightresistance in chickpea (*Cicer arietinum* L.). *Euphytica, 132*, 23–30.

Rangel, P. N., Brondani, R. P., Rangel, P. H., & Brondani, C. (2008). Agronomic and molecula rcharacterization of introgression lines from the interspecific cross *Oryza sativa* (BG90-2) ×*Oryza glumaepatula* (RS-16). *Genetics and Molecular Research, 7*, 184–195.

Redden, R. J., & Berger, J. D. (2007). History and origin of chickpea. *Chickpea Breeding and Management., 1*, 1–3.

Ribaut, J. M., & Hoisington, D. (1998). Marker-assisted selection: New tools and strategies. *Trends in Plant Science, 3*, 236–239.

Ribaut, J. M., de Vicente, M. C., & Delannay, X. (2010). Molecular breeding in developing countries: Challenges and perspectives. *Current Opinion in Plant Biology, 13*, 1–6.

Robinson, G. K. (1991). That BLUP is a good thing: The estimation of random effects. *Statistical Science, 6*, 15–32.

Roorkiwal, M., Von Wettberg, E. J., Upadhyaya, H. D., Warschefsky, E., Rathore, A., & Varshney, R. K. (2014). Exploring germplasm diversity to understand the domestication process in Cicer spp. using SNP and DArT markers. *PLoS One, 9*(7), e102016.

Roorkiwal, M., Rathore, A., Das, R. R., Singh, M. K., Jain, A., Srinivasan, S., ... & Varshney, R. K. (2016). Genome-enabled prediction models for yield related traits in chickpea. *Frontiers in Plant Science, 7*, 1666.

Roorkiwal, M., Jain, A., Kale, S. M., Doddamani, D., Chitikineni, A., Thudi, M., & Varshney, R. K. (2018). Development and evaluation of high-density Axiom® Cicer SNP Array for high-resolution genetic mapping and breeding applications in chickpea. *Plant Biotechnology Journal, 16*(4), 890–901.

Roorkiwal, M., Bharadwaj, C., Barmukh, R., Dixit, G. P., Thudi, M., Gaur, P. M., ... & Varshney, R. K. (2020). Integrating genomics for chickpea improvement: achievements and opportunities. *Theoretical and Applied Genetics, 133*, 1703–1720.

Rupela, O. P. (1992). Natural occurrence and salient characters of non-nodulating chickpea plants. *Crop Science, 32*, 349–352.

Sabbavarapu, M. M., Sharma, M., Chamarthi, S. K., Swapna, N., Rathore, A., Thudi, M., ... & Varshney, R. K. (2013). Molecular mapping of QTLs for resistance to Fusarium wilt (race 1) and Ascochyta blight in chickpea (Cicer arietinum L.). *Euphytica, 193*, 121–133.

Sagel, Z., Tutluer, M. I., & Peskircioglu, H. (2009). The improvement of TAEK–Sagel Chickpea (*Cicer arietinum* L.) mutant variety in Turkey. In Q. Y. Shu (Ed.), *Induced plant mutations in the genomics era* (pp. 319–321). Food and Agriculture Organization of the United Nations.

Salimath, P. M., & Bahl, P. N. (1985). Early generation selection in chickpea. III predicted and realized gains. *Experimental Genetics, 1*(2), 59–62.

Salina, E., Dobrovolskaya, O., Efremova, T., Leonova, I., & Roder, M. S. (2003). Microsatellite monitoring of recombination around the *Vrn-B1* locus of wheat during early backcross breeding. *Plant Breeding, 122*, 116–119.

Saraf, C. S., Rupela, O. P., & Hegde, D. M. (1998). *Biological nitrogen fixation and residual effects of winter grain legumes in rice and wheat cropping systems of the Indo-Gangetic plain.*

Saxena, H. P., & Raina, A. K. (1970). A bruchid resistant strain of Bengal gram. *Current Science, 39*, 189–190.

Shaikh, M. A. Q. (1977). Grain legumes in Bangladesh. In *Induced mutations for the improvement of grain legumes in South East Asia (1975)* (pp. 61–70). IAEA-203. International Atomic Energy Agency.

Shaikh, M. A. Q., Bhuiya, A. D., Saha, C. S., & Rahman, L. (1983a). Performance and agronomic potential of radiation induced and collected genotypes of chickpea (*Cicer arietinum* L.). In *Induced mutations for improvement of grain legumes production III, IAEA-TECDOC-299* (pp. 47–54). International Atomic Energy Agency.

Shaikh, M. A. Q., Khanum, S., Begum, S., Ahmed, Z. U., Majid, M. A., & Zaman, K. M. S. (1983b). Effects of chemical mutagens on four species of grain legumes. In *Induced mutations for improvement of grain legumes production III, IAEATECDOC-299* (pp. 77–85). International Atomic Energy Agency.

Shamsuzzaman, K. M., Islam, M. M., & Jalaluddin, M. (2004). Development of new chickpea varieties: Binasola–3 and Binasol–4. *Bangladesh Journal of Nuclear Agriculture*, 18.

Sharma, B., & Kharkwal, M. C. (1982). Induced mutations in grain legumes. In *Induced mutations for the improvement of grain legumes production II, IAEA-TECDOC-260* (pp. 59–64). International Atomic Energy Agency.

Sharma, B., & Kharkwal, M. C. (1983). Mutation studies and mutation breeding in grain legumes. In *Induced mutations for improvement of grain legumes production III, IAEA-TECDOC-299* (pp. 65–75). International Atomic Energy Agency.

Sharma, P. C., & Varshney, R. K. (1999). Induced mutations in chickpea improvement. In B. A. Siddiqui & S. Khan (Eds.), *Breeding in crop plants, mutations and in vitro mutation breeding* (pp. 57–67). Kalyani Publishers.

Shrivastava, A. K., Dixit, G. P., & Kumar, N. (2017). Milling properties of desi and kabuli chickpea (*Cicer arietinum*) varieties released in India. *Indian Journal of Agricultural Sciences, 87*(10), 1350–1357.

Sigurbjönsson, B. (1983). Induced mutations. In D. R. Wood (Ed.), *Crop breeding* (pp. 153–176). American Society of Agronomy and Crop Science Society of America.

Sindhu, J. S., Lal, S. B., & Singh, R. P. (1981). Studies on the factors determining crossing success in chickpea (*Cicer arietinum*). *Pulse Crops Newsletter, 1*, 21–22.

Singh, K. B. (1987). Chickpea breeding. In M. C. Saxena & K. B. Singh (Eds.), *The chickpea* (pp. 127–162). CAB International.

Singh, K. B. (1997). Chickpea (*Cicer arietinum* L.). *Field Crops Research, 53*, 161–170.

Singh, B. D. (2005). Mutations in crop improvement. In B. D. Singh (Ed.), *Plant breeding, principles and methods* (pp. 698–731). Kalyani Publishers.

Singh, K. B., & Auckland, A. K. (1975). Chickpea breeding at ICRISAT. In *Proceedings of the international workshop on grain legumes* (pp. 3–17). ICRISAT.

Singh, I., & Waldia, R. S. (1994). Comparison of selection methods in chickpea under late sown condition. *Crop Research, Hissar, 8*(3), 508–511.

Singh, K. B., Ocampo, B., & Robertson, L. D. (1998). Diversity for abiotic and biotic stress resistance in the wild annual *Cicer* species. *Genetic Resources and Crop Evolution, 45*, 9–17.

Singh, V. K., Khan, A. W., Jaganathan, D., Thudi, M., Roorkiwal, M., Takagi, H., Garg, V., Kumar, V., Chitikineni, A., Gaur, P. M., Sutton, T., Terauchi, R., & Varshney, R. K. (2016). QTL-seq for rapid identification of candidate genes for 100-seed weight and root/total plant dry weight ratio under rainfed conditions in chickpea. *Plant Biotechnology Journal, 14*, 2110–2119.

Singh, C., Kumar, R., Sehgal, H., Bhati, S., Singhal, T., Gayacharan, Nimmy, M. S., Yadav, R., Gupta, S. K., Abdallah, N. A., & Hamwieh, A. (2023). Unclasping potentials of genomics and gene editing in chickpea to fight climate change and global hunger threat. *Frontiers in Genetics, 18*(14), 1085024.

Sivasakthi, K., Thudi, M., Tharanya, M., Kale, S. M., Kholová, J., Halime, M. H., ... & Vadez, V. (2018). Plant vigour QTLs co-map with an earlier reported QTL hotspot for drought tolerance while water saving QTLs map in other regions of the chickpea genome. *BMC Plant Biology, 18*, 1–18.

Smartt, J. (1976). Comparative evolution of pulse crops. *Euphytica, 25*, 139–143.

Stadler, L. J. (1928). Mutations in barley induced by x-rays and radium. *Science, 68*, 186–187.

Stam, P. (1994). Marker-assisted breeding. In *Proceedings of the IX Meeting of EUCARPIA Section on Biometrics in Plant Breeding* (pp. 32–44). Wageningen.

Streeck, T. S., & Piepho, H. P. (2010). Genome-wide selection by mixed model ridge regression and extensions based on geo statistical models. *BMC Proceedings, 4*, S1–S8.

Sudheesh, S., Kahrood, H. V., Braich, S., Dron, N., Hobson, K., Cogan, N. O., & Kaur, S. (2021). Application of genomics approaches for the improvement in ascochyta blight resistance in chickpea. *Agronomy, 11*(10), 1937.

Tanksley, S. D., & Nelson, J. C. (1996). Advanced backcross QTL analysis: A method of the simultaneous discovery and transfer of valuable QTLs from unadapted germplasm into elite breeding lines. *Theoretical and Applied Genetics, 92*, 191–203.

Tekeoglu, M., Rajesh, P. N., & Muehlbauer, F. J. (2002). Integration of sequence tagged microsatellite sites to the chickpea genetic map. *Theoretical and Applied Genetics, 105*, 847–854.

Tester, M., & Langridge, P. (2010). Breeding technologies to increase crop production in a changing world. *Science, 327*, 818–822.

Thudi, M., Bohra, A., Nayak, S. N., Varghese, N., Shah, T. M., Penmetsa, R. V., Thirunavukkarasu, N., Gudipati, S., Gaur, P. M., Kulwal, P. L., & Upadhyaya, H. D. (2011). Novel SSR markers from BAC-end sequences, DArT arrays and a comprehensive genetic map with 1, 291 marker loci for chickpea (Cicer arietinum L.). *PLoS One, 6*(11), e27275.

Thudi, M., Roorkiwal, M., Kudapa, H., Chaturvedi, S. K., Singh, N. P., & Varshney, R. K. (2017). An overview of chickpea research: From discovery to delivery. *Pulse India, 2*(5), 22–25.

Thudi, M., Samineni, S., Li, W., Boer, M. P., Roorkiwal, M., Yang, Z., Ladejobi, F., Zheng, C., Chitikineni, A., Nayak, S. and He, Z. (2023). Whole genome resequencing and phenotyping of MAGIC population for high resolution mapping of drought tolerance in chickpea. *The Plant Genome*.

Thudi, M., Samineni, S., Li, W., Boer, M. P., Roorkiwal, M., Yang, Z., Ladejobi, F., Zheng, C., Chitikineni, A., Nayak, S., & He, Z. (2024). Whole genome resequencing and phenotyping of MAGIC population for high resolution mapping of drought tolerance in chickpea. *The Plant Genome, 17*(1), e20333.

Toker, C. (2005). Preliminary screening and selection for cold tolerance in annual wild *Cicer* species. *Genetic Resources and Crop Evolution, 52*, 1–5.

Toker, C., & Cagirgan, M. I. (2004). Spectrum and frequency of induced mutations in chickpea. *International Chickpea and Pigeonpea Newsletter, 11*, 8–10.

Toker, C., Uzun, B., Canci, H., & Ceylan, F. O. (2005). Effects of gamma irradiation on the shoot length of *Cicer* seeds. *Radiation Physics and Chemistry, 73*, 265–367.

Turano, M. J., Baird, L. M., & Webster, B. D. (1983). Characteristics of the Stigma of Chickpea 1. *Crop Science, 23*(6), 1033–1036.

Udupa, S. M., & Baum, M. (2003). Genetic dissection of pathotype-specific resistance to ascochyta blight disease in chickpea (*Cicer arietinum* L.) using microsatellite markers. *Theoretical and Applied Genetics, 106*, 1196–1202.

Upadhyaya, H. D., Bajaj, D., Das, S., Kumar, V., Gowda, C. L. L., Sharma, S., Tyagi, A. K., & Parida, S. K. (2016). Genetic dissection of seed-iron and zinc concentrations in chickpea. *Scientific Reports, 6*, 24050.

Vadez, V., Krishnamurthy, L., Thudi, M., Anuradha, C., Colmer, T., Turner, N., Siddique, K., Gaur, P., & Varshney, R. K. (2012). Assessment of ICCV 2 × JG 62 chickpea progenies shows sensitivity of reproduction to salt stress and reveals QTL for seed yield and yield components. *Molecular Breeding, 30*, 9–21.

Van der Maesen, L. J. G., & Cicer, L. (1972). *A monograph of the genes, with special reference to the chickpea (Cicer arietinum), its ecology and cultivation*. Mededlingen Landbouwhoge school (Communication Agricultural University), Wageningen.

Van Harten, A. M. (1998). *Mutation breeding: Theory and practical applications*. Cambridge University Press.

Varshney, R. K. (2016). Exciting journey of 10 years from genomes to fields and markets: some success stories of genomics-assisted breeding in chickpea, pigeonpea and groundnut. *Plant Science, 242*, 98–107.

Varshney, R. K., Hiremath, P. J., Lekha, P., Kashiwagi, J., Balaji, J., Deokar, A. A., Vadez, V., Xiao, Y., Srinivasan, R., & Gaur, P. M. (2009). A comprehensive resource of drought- and salinity-responsive ESTs for gene discovery and marker development in chickpea (*Cicer arietinum* L.). *BMC Genomics, 10*, 523.

Varshney, R. K., Gaur, P. M., Chamarthi, S. K., Krishnamurthy, L., Tripathi, S., Kashiwagi, J., Samineni, S., Singh, V. K., Ihudi, M., & Jaganathan, D. (2013). Fast-track introgression of

"QTL-hotspot" for root traits and other drought tolerance traits in JG 11, an elite and leading variety of chickpea. *Plant Genome, 6*, 1–9.

Varshney, R. K., Terauchi, R., & McCouch, S. R. (2014). Harvesting the promising fruits of genomics: applying genome sequencing technologies to crop breeding. *PLoS Biology, 12*(6), e1001883.

Varshney, R. K., Thudi, M., Nayak, S. N., Gaur, P. M., Kashiwagi, J., Krishnamurthy, L., & Rathore, A. (2014a). Genetic dissection of drought tolerance in chickpea (*Cicer arietinum* L.). *Theoretical and Applied Genetics, 127*(2), 445–462.

Varshney, R.K.; Mohan, S.M.; Gaur, P.M.; Chamarthi, S.K.; Singh, V.K; Srinivasan, S.; Swapna, N.;, Sharma M.; Singh, S.; Kaur, L. And Pande, S. Marker-assisted backcrossing to introgress resistance to Fusarium wilt (FW) race 1 and Ascochyta blight (AB) in C 214, an elite cultivar of chickpea Plant Genome, 2014b, 7 (1): 1–11.

Varshney, R. K., Terauchi, R., & McCouch, S. R. (2014c). Harvesting the promising fruits of genomics applying genome sequencing technologies to crop breeding. *PLoS Biology, 12*, e1001883.

Varshney, R. K., Thudi, M., Roorkiwal, M., He, W., Upadhyaya, H. D., Yang, W., Bajaj, P., Cubry, P., Rathore, A., Jian, J. & Doddamani, D. (2019). Resequencing of 429 chickpea accessions from 45 countries provides insights into genome diversity, domestication and agronomic traits. *Nature Genetics, 51*(5), 857–864.

Wallace, T. C., Murray, R., & Zelman, K. M. (2016). The nutritional value and health benefits of chickpeas and hummus. *Nutrients, 8*(12), 766.

Winter, P., Benko-Iseppon, A. M., Hüttel, B., Ratnaparkhe, M., Tullu, A., Sonnante, G., Pfaff, T., Tekeoglu, M., Santra, D., Sant, V. J., Rajesh, P. N., Kahl, G., & Muehlbauer, F. J. (2000). A linkage map ofthe chickpea (*Cicer arietinum* L.) genome based on the recombinant inbred lines from *C. arietinum* × *C. reticulatum* cross: Localization of resistance genes for *Fusarium* races 4 and 5. *Theoretical and Applied Genetics, 101*, 1155–1163.

Wood, J. A., & Grusak, M. A. (2007). Nutritional value of chickpea. In *Chickpea breeding and management* (pp. 101–142). Ed. S S Yadav.

Yadav, R. K., Tripathi, M. K., Tiwari, S., Tripathi, N., Asati, R., Patel, V., Sikarwar, R. S., & Payasi, D. K. (2023). Breeding and genomic approaches towards development of fusarium wilt resistance in chickpea. *Life, 13*(4), 988.

Yadava, Y. K., Chaudhary, P., Yadav, S., Rizvi, A. H., Kumar, T., Srivastava, R., Soren, K. R., Bharadwaj, C., Srinivasan, R., Singh, N. K., & Jain, P. K. (2023). Genetic mapping of quantitative trait loci associated with drought tolerance in chickpea (Cicer arietinum L.). *Scientific Reports, 13*(1), 17623.

Zhang, X., Zhou, S., Fu, Y., Su, Z., Wang, X., & Sun, C. (2006). Identification of a drought tolerant introgression line derived from Dongxiang common wild rice (*O. rufipogon* Griff.). *Plant Molecular Biology, 62*, 247–259.

Cross-Species *Vigna* Gene Pool for Improving Biotic Stress Tolerance in Mung Bean

Chirag, Poornima Singh, Ruchi, Priyanka Kaundal, Mukul Kumar, Brijesh Pandey, Aditya Pratap, Awdhesh Kumar Mishra, Narendra Pratap Singh, and Chandra Mohan Singh

Abstract Mung bean (*Vigna radiata* L. Wilczek) is a short-duration and widely adaptable leguminous crop grown across different agroclimatic zones and kinds of soils. The development of preferred climate-resilient mung bean cultivars has advanced significantly over the past two decades mainly through introgression breeding, which has made extensive use of novel genetic resources. However, as new competitors or biotypes of biotic stress agents emerge, the utilization of already existing genetic resources to counteract these stresses may be insufficient. Thus, employing wild and exotic plant genetic resources to disseminate genetics will pay dividends in the long run. Through the utilization of linkages between various research groups, many agriculturally significant genes have been uncovered in recent years, and the output of mung bean improvement has been improved by harnessing these genes. Utilization of the genetic and genomic resources toward the creation of new genetic diversity by tagging, mapping, and use of genes of interest has been ameliorated. For the purpose of effectively using foreign genes/QTL to reduce sexual stress in mung bean, this chapter offers the recent information on donor and genetic, genomic advancements, functional genomics and transcriptomics, and molecular marker technology.

Keywords Introgression breeding · Transcriptomics · Marker-assisted breeding · Phytonutrients · Whole-genome sequencing · Phylogenetic analysis · SNPs · InDels analysis

---

Chirag · Ruchi · P. Kaundal · M. Kumar · N. P. Singh · C. M. Singh (✉)
Banda University of Agriculture and Technology, Banda, Uttar Pradesh, India

P. Singh · B. Pandey
Mahatma Gandhi Central University, Marihari, Bihar, India

A. Pratap
ICAR-Indian Institute of Pulses Research, Kanpur, India

A. K. Mishra
Yeungnam University, Gyeongsan, South Korea

A. Pratap et al. (eds.), *Integrated Improvement of Food Legumes*,
https://doi.org/10.1007/978-3-031-82629-0_7

Being a short-duration and widely adaptable crop, mung bean is one of the most important grain legumes among the pulses, suited to tropical and sub-tropical climates. Globally it is cultivated over 7.2 million ha. with the productivity of about 750 kg/ha (Nair & Schreinemachers, 2020). It is mainly grown in Asia, Africa, and Latin America (Pratap et al., 2021; Sriphadet et al., 2007) and consumed as a food, feed, and green manuring crop (Karuppanapandian et al., 2006; Keatinge et al., 2011; Pratap et al., 2021). India is the largest producer, processor, importer, and consumer of mung bean globally, with a production of 3.68 million tons from an area of 5.55 million hectares during 2022–2023 (Project Cordinator's Report). It has a wider adaptability to an array of environmental and soil conditions, including extreme environments (Pratap et al., 2021; Gupta & Pratap, 2016; Basu et al., 2019). The short life cycle (55–70 days), low input requirements, and nitrogen-fixing ability (Goyal et al., 2021) make it a valuable sustenance crop. Mung bean seed consists of high proteins (~24% easily digestible protein), fiber, antioxidants, and phytonutrients (Itoh et al., 2006). Owing to these features, there is an ample scope for vertical as well as horizontal expansion of mung bean worldwide. Nonetheless, the productivity of mung bean is still compromised by poor crop management, changing environments, and various biotic and abiotic factors.

1 Major Production Constraints

Among the various factors affecting yield potential, the major ones include environmental fluctuations and high genotype x environment interactions, subsequently leading to various kinds of biotic and abiotic stresses (Singh et al., 2014, Pratap et al., 2021; Douglas et al., 2020). The environment plays an important role in developing proper phenotype and successful crop growth to harvest its actual potential yield. Likewise, diseases and insect pests together can cause significant yield losses ranging from 10% to 100% (Nair et al., 2019; Pratap et al., 2020; Singh et al., 2019, 2022a, b, Sahu et al., 2023). Yellow mosaic disease (YMD), *Cercospora* leaf spot (CLS), powdery mildew (PM), and web blight (WB) are the chief causes of concern among the diseases, whereas the major pests include *Ophiomyia phaseoli* (Tryon), *Bemisia tabaci* (Gennadius), *Empoasca* spp., *Polyphagotarsonemus latus* (Banks), *Aphis craccivora* Koch, *Spodoptera litura* (Fabricius), *Maruca* spp., *Helicoverpa armigera* (Hubner), *Lampides boeticus* Linnaeus, *Megalurothrips distalis* (Karny), and *Callosobruchus* spp. (Pratap et al., 2022). Four different species of Begomoviruses, namely, *mung bean yellow mosaic India virus* (*MYMIV*), *mung bean yellow mosaic virus* (*MYMV*), *horsegram yellow mosaic virus* (*HgYMV*), and *Dolichos yellow mosaic virus* (*DoYMV*), are found to closely share their genomic content in most of the regions, in which *MYMV*, *MYMIV*, and *HgYMV* are recognized as the causal agents of YMD in mung bean in different parts of the world (Akram et al., 2021). The *DoYMV* is still limited to the *Dolichos* bean only (Fig. 1).

Among the viral diseases, yellow mosaic disease (YMD) caused by bipartite begomoviruses is the most important constraint on mung bean worldwide,

Fig. 1 Major diseases of mung bean: (**a**) Leaf infected by yellow mosaic disease; (**b**) YMD infection on pods; (**c**) Anthracnose; (**d**) Powdery mildew; (**e**) Cercospora leaf spot; (**f**) Web blight; (**g**) Leaf curl; (**h**) Mixed infection of YMD and LC

particularly in the Asian and Southeast Asian countries. Yield losses due to YMD vary widely, ranging from 10% to 100%, depending on the stage of the crop at the time of infection and disease intensity. It has been observed that the mung bean plants that get affected at an early growth stage tend to suffer more than those that get infected later in the growth cycle (Akram et al. 2021). Additionally, YMD reduces the height and fresh weight of mung bean and urdbean plants by up to 38.2% and 28.5%, respectively (Chand & Varma, 1983). The major fungal diseases of mung bean include powdery mildew (*Erysiphe* spp.) (Savaliya et al., 2018), anthracnose (*Colletotrichum* spp.) (Shen et al., 2010), *Cercospora* leaf spot (*Cercospora* spp.) (Joshi et al., 2006), and web blight (*Rhizoctonia solani*) (Iqbal et al. 2014). Yield losses due to *Cercospora* leaf spot were reported from 23% to 96% in field trials conducted in different states of India (Kaur et al., 2006). The foliar diseases cause significant yield reduction, especially when incidence takes place at an early crop growth stage (Reddy et al., 1994; Naimuddin et al., 2023). *Erysiphe polygoni* is an obligate biotroph on leaves and fruits that infects cells of the epidermal layer from which nutrients are extracted by the pathogen through a haustorium (Kumar & Singh, 1981). The anthracnose disease of mung bean was first reported in India from Assam (Majid, 1953). It causes considerable damage by reducing seed quality and yield (Sharma et al., 1971). The average seed yield loss of 40.18% and stalk yield loss of 46.90% was noticed due to anthracnose of mung bean (Kulkarni & Rao, 2009). The process of infection/penetration by *Colletotrichum* spp. is thought to be triggered by cutin-degrading enzymes such as cutinase that have been suggested to participate in carbon acquisition for saprophytic growth

(Koller & Parker, 1989). They were also presumed to have a role in fungal penetration (Dickman et al., 1989; Skamnioti et al., 2007) and in surface adhesion (Deising et al., 1992). About 30–40% of yield loss due to web blight was reported from Rajasthan, and 20–40% seedling mortality due to *Rhizoctonia* infection was reported from central India (Tiwari, 1993). Among the minor fungal diseases, 20% yield loss was reported due to Fusarium wilt (Anderson, 1985) and 10% due to Alternaria leaf spot (Maheshwari & Krishna, 2013). Toward control of these diseases, the efficacy of bactericides, fungicides, bio-fungicides, and botanicals has been studied in seed treatment and foliar spray. Besides, the impact of cultural practices has also been seen to reduce mung bean diseases (Pandey et al., 2018). Nevertheless, the deployment of varieties with genetic resistance is the most effective and durable method for integrated disease management (Pratap & Gupta, 2009). Therefore, efforts should be made toward the development of multiple stress-resistant varieties in mung bean.

2 Inheritance of Ket Stress Tolerance Traits

Understanding the inheritance of key traits is crucial for breeding improved cultivars and also enhancing the resilience of crops. While traits with simple inheritance patterns can be easily transferred using morphological markers or classical breeding techniques, many diseases are controlled by genes with quantitative inheritance, posing challenges for breeders to effectively stack resistance. Researchers have identified numerous sources of resistance in different *Vigna* gene pools (Pratap et al., 2014, 2021) (Table 1). While genes from cultivated backgrounds within the crop species are readily transferable, resistance breakdown is common due to pathogen evolution and a limited genetic variability. To bolster resistance durability, breeders must stack target genes from diverse backgrounds, even drawing desirable genes from wild relatives and cross-specific gene pools. However, the inheritance patterns of traits from cross-species and crop wild relatives remain understudied, even though these hold promise for improving mung bean breeding programs.

3 Cross-Species *Vigna* Diversity and Gene Flow

Cross-species gene flow in *Vigna* has been a successful approach toward the deployment of desirable genes in mung bean improvement. Mung bean × urdbean hybridization has been routinely practiced in breeding programs, and the derivatives from these hybridizations exhibit many desirable features, namely, resistance to climate vagaries, synchronous podding, and non-shattering pods (Pratap et al., 2019a, b). Similarly, mung bean × rice bean and mung bean × *V. radiata* var. *sublobata* hybridization have also been practiced by many breeders, and progenies were derived that were resistant to MYMV (Verma & Brar, 1996). Mahalingam and Manivannan (2023)

Table 1 Inheritance of disease/insect resistance in *Vigna* and its derivatives (Modified from Singh et al. 2022b)

Stress factor	Crop	Population	Approach	Genetics	Marker used	Reference
MYMIV	Mung bean	RILs of KPS 2 × NM 10–12-1	Linkage mapping	QTL	SSR	Kitsanachandee et al. (2013)
		BM6 × BM1	Linkage mapping	QTL	SSR	Alam et al. (2014)
		A panel of 130 genotypes	Association mapping	QTL	SSR	Singh et al. (2020)
	Blackgram	AKU 9904 × DPU 88–31	BSA	Single dominant gene	SSR	Gupta et al. (2013)
	Blackgram × Rice bean	KUG253 × Mash114	QTL-seq	Major QTL	SNP	Dhaliwal et al. (2022)
MYMV	Mung bean	KMG 189 × VBN (Gg) 2	Segregation analysis	Single recessive gene	–	Sudha et al. (2013)
		KMG 189 × VBN (Gg) 2	Segregation analysis	Single recessive gene/ linkage mapping	SCAR	Sai et al. (2017)
		NM92 × VC2272, 6601 × VC2272, 6601 × Pusa Baisakhi, VC3902A × NM92, VC3902A × ML-5, NM92 × Pusa Baisakhi, VC 1560D × 6601, VC 1560D × NM92	F_2 segregation analysis	–	–	Khan et al. (2007)
		NM 92 × NM 98	F_2 segregation analysis	–	–	Khattak et al. (2008)
		VBN(Gg)2 × KMG189	F_2 segregation analysis	Single recessive gene	–	Sai et al. (2017)
	Blackgram	MDU1 × Mash 1008	Linkage mapping	QTLs	SSR	Vadivel et al. (2021)
		MDU 1 × TU 68	Linkage mapping	Major QTL	SSR	Subramaniyan et al. (2022)
	Rice bean × Mungbean	TNAU red × VRM (Gg) 1	Segregation analysis	Single recessive gene	–	Sudha et al. (2013)
	Mung bean × Rice bean	VRM (Gg) 1 × TNAU RED	GBS approach	One major QTL and three minor QTLs	SNP	Mathivathana et al. (2019)
YMV (Cuasel virus not identified)	Blackgram	IC436656 × KKB14045	Segregation analysis and linkage mapping	Single recessive gene	SSR	Sathees et al. (2022)

(continued)

Table 1 (continued)

Stress factor	Crop	Population	Approach	Genetics	Marker used	Reference
Cercospora leaf spot	Mung bean	KPS1 × V4718	Linkage mapping	QTL	SSR	Chankaew et al. (2011)
		Kamphaeng Saen 1 × V4718	Fine mapping	QTL	SNP	Choudhary et al. (2021)
		Kopargaon × HUM12; Kopargaon × ML 1720	GMA	Quantitative inheritance	–	Choudhary et al. (2021)
		KPS1 × V4718	GBS	QTL	SNP	Yundaeng et al. (2021)
Powdery mildew	Mung bean	Kamphaeng Saen 1 × VC6468-11-1A (RILs)	Linkage mapping	Two QTLs	SSR	Kasettranan et al. (2010)
		Berken × ATF 3640	Linkage mapping	Major QTL	RFLP	Humphry et al. (2003)
		Chai Nat 72 × V4758	Segregation analysis and BSA	QTL	ISSR-RGA	Tantasawat et al. (2022)
	Cowpea	ZN 016 × Zhijiang282	Linkage mapping	Major QTL	SSR and SNP	Wu et al. (2014)
Bruchid (*C. chinensis*)	Mung bean	V2709	–	Single dominant gene	RAPD/SSR/STS	Sun et al. (2008)
Callosobruchus spp.	Mung bean × *V. sublobata*	VC2778A × TC 1966	Comparative genomics	Major QTL	SNP	Chen et al. (2022)
C. chinensis; C. maculatus		KPS1″ and ″V2709	Segregation analysis and linkage mapping	Major QTL	SNP	Kaewwongwal et al. (2017)
	V. vexillata	TVNu 240 × TVNu 1623	SLAF sequencing	QTL	SNP	Amkul et al. (2019)
Bruchid (*C. maculatus*)	Urdbean × *Vigna mungo* var. *silvestris*	TU 94–2 × *Vigna mungo* var. *silvestris*	Segregation analysis	QTLs	RAPD, ISSR, SSR	Souframanien & Gopalakrishna, 2007
	Urdbean × *Vigna mungo* var. *silvestris*	BC48 × TC2210	Linkage mapping	Two QTLs	SNP	Somta et al. (2019)
	Urdbean	MDU1 × TU 68	Linkage mapping	QTL	SSR	Subramaniyan et al. (2022)

could produce successful hybrids between *V. radiata* and *V. umbellata* with intermediate morphology and MYMV resistance. An investigation of the crossability barriers in interspecific crosses between *V. radiata* and *V. umbellata* was carried out by Basavaraja et al. (2019). Five MYMV-resistant genotypes of rice beans and five MYMV-susceptible genotypes of mung beans were hybridized in an attempt to create interspecific crosses; mung bean genotypes were used as the female parents and rice bean genotypes as the male parents. A total of 12 of the 25 interspecific crosses exhibited a successful pod set. The crosses that had the highest crossability percentages were Yellow Mung × KBR-1 (17.30%), DGGV-2 × RBL-35 (16.0%), Selection-4 × KBR-1 (11.80%), Chinamung × KBR-1 (11.0%), and BGS-9 × RBL-35 (10.20%). These crossings were regarded as effective, suggesting that there is a chance to spread the genetic variety of interspecific hybrids and transfer beneficial genes between the two species. Additionally, the cross Chinamung × KBR-1 showed considerable hybrid pollen fertility (78.24%), which was followed by the Selection-4 × KBR-1 cross. These results were combined with decreased hybrid lethality (21.43%) and lower hybrid breakdown (7.14%). These results point to intriguing directions for future investigation into gene transfer and interspecific hybridization between *V. radiata* and *V. umbellata*. Pandiyan et al. (2012) reported a crossability percentage of 10.25 while examining the crossability relationship between *V. radiata* and *V. trilobata*. 34.21% of hybrid seeds were found to germinate, while 75.56% of hybrid pollen samples were fertile. A hybrid breakdown of 46.15% and a hybrid lethality rate of 53.84% were also discovered by the investigation. Through interspecific crosses with *V. trilobata*, these results point to possible ways to improve the qualities of *Vigna radiata*, especially with regard to softness and suitability for calf feed. Important insights into crop improvement techniques and genetic enhancement may result from more studies in this field.

Bhanu et al. (2018) studied the crossability relationship between 3 *Vigna* species, namely, *V. radiata*, *V. mungo*, and *V.* umbellata, through a 80 interspecific crosses. In total, 8 crosses of *V. radiata* × *V. mungo* (with *V. radiata* as the female), 36 crosses of *V. mungo* × *V. umbellata* (with *V. mungo* as the female), and 36 crosses of *V. radiata* × *V. mungo* (with *V. radiata* as the female) were among the several combinations. The cross HUM 12 × RBL 9 (16.27%) had the highest crossability among the *V. radiata* × *V. umbellata* crosses, followed by HUM 12 × RBL 9 (15.78%). Mash 338 × RBL 9 showed the highest crossability of 11.36% in the *V. mungo* × *V. umbellata* instance. The hybrid ML 1464 × Mash 338 showed the highest crossability at 37.5% among the *V. radiata* × *V. mungo* crosses.

4 Potential Donors for Specific Traits

The *Vigna* genus is highly diverse, with more than 200 species. Since mung bean is near the bottom of the taxonomic range, it has freely absorbed genes from other *Vigna* species. Nevertheless, biases against foreign alleles have hindered the success of gene transfer, even though it has mostly come from wild species inside the basic gene pool. A few deviations exist, such as creole beans, but most crops in the *Vigna* genus are diploid and self-pollinating. Significant variation can be observed in yield-related and adaptive attributes, such as resistance to biotic and abiotic stresses, among wild relatives. The ploidy level influences the probability that foreign genes will successfully transmit through distant hybridization. Mung bean biotic stress resistance may be improved by using possible donors from wild relatives in breeding programs.

Mung bean improvement within the *Vigna* species sometimes involves hybridization between urdbean and mung bean. Such crossings frequently produce offspring with desirable features like synchronous maturity, resilience to different stressors, and excellent seed qualities. Being the first interspecific mung bean variety that carried MYMIV resistance from *V. mungo*, the variety "Meha" represented a notable milestone. More recently, interspecific hybridization between urdbean and mung bean has produced varieties such as "Tripura Mung-1" and "IPM 312–20." In addition to urdbean, additional wild species offer promising opportunities for breeding initiatives targeted at improving stress tolerance and creating new varieties of mung bean (Table 2).

5 Status of Genomics Resources

Genome sequencing efforts have been initiated across various *Vigna* crops, with mung bean serving as the model species. Prior to decoding the genome sequence, researchers utilized transferable markers from other *Vigna* species for analyzing mung bean and other legumes. Kim et al. (2021) conducted sequencing on a mung bean pure line, VC1973A, along with its relatives *V. reflexo-pilosa* var. *glabra* and *V. radiata* var. *sublobata*, resulting in the construction of a draft genome sequence. The estimated genome size ranged from 579 Mb to 968 Mb (*Vigna reflexo-pilosa* var. *glabra*). Additionally, they sequenced a wild relative, TC1966, a *Vigna radiata* var. *sublobata* accession, with an estimated genome size of 501 Mb. In addition, two mung bean accessions, namely, Salu and AL127, were re-sequenced to map the lma locus, identifying 236,998 single nucleotide polymorphisms and 8896 insertion/deletions (InDels). The availability of the mung bean whole-genome sequence is expected to bolster genomics research in *Vigna* species and expedite mung bean breeding programs.

Following the completion of mung bean's whole-genome sequencing, the genomic analysis of other *Vigna* crops has been expedited, with several species sequenced to date. Adzuki bean's genome size (*V. angularis* var. angularis) was estimated by Kaul et al. (2019), and approximately 75% of its genome assembly was drafted. This assembly comprised 3883 scaffolds with proper read coverage statistics, including the pseudo-library from Newbler assembly, with an N50 scaffold length of 703 kb totaling around 443 Mb. Analysis identified 4524 segregating SNP sites and predicted 26,857 high-confidence genes using the MAKER pipeline, with 15,976 located on 11 pseudo chromosomes. Moreover, 9196 orthologs between

Table 2 Potential sources of disease/insect resistance in *Vigna* species (Modified from Singh et al. 2022)

Disease/pest	Donor genotypes	Species	Method of screening	References
MYMIV	IC277021	*V. silvestris*	Field screening	Kumari et al. (2021)
	IC248326, IC248326, IC248343	*V. vexillata*	Field screening	
	LRM/13-43, LRM/13-32, LRM/13-34, IC276983, IC331436, IC331454, IC331456, Trichy local, Kumur local, IIPRW17-3	*V. trilobata*	Field screening	
	RBL-50, IC251445, PRR 2007-2, PRR 2008-2, RB-5-1, IC251439, IC251442, IC251446, IC251447, IC528878, IC197812, IIPRW 17-1	*V. umbellata*	Field screening	
	PRR 2008-2	*V. umbellata*	Field screening	
	LRM/13-11, LRM/13-33, TMV-1, LRM/13-26, LRM/13-37, LRM/13-38, LRM/13-36	*V. aconitifolia*	Field screening	
	IC331450	*V. hainiana*	Field screening	
	Trichy Local-1, Trichy Local-2	*V. stipulaceae*	Field screening	
	TCR-20	*V. glaberesense*	Field screening	
	TCR-7	*V. umbellata*	Field screening	
	TU-68	*V. mungo*	Bioassay and GC-MS analysis	Ragul et al. (2022)
	TCR-79, TCR-82, TCR-239	*V. radiata*	Field screening	Singh et al. (2021)
	TCR-7, TCR-238, JAP/10-36, TCR-110, JAP/10-47, TCR-160, TCR-88, JAP/10-51, NSB 007	*V. sublobata*	Field screening	
	TCR-64, LRM/13-43, LRM/ 13-34, LRM/13-32, LRM/13-24, LRM/13-30, ZAP/10-5, ZAP/10-7, ZAP/10-9, TCR-192, TCR-305, TCR-319, TCR-320, TCR-513, LRM/13-44, LRM/13-33, LRM/13-26, LRM/13-38, LRM/13-36, LRM/13-37	*V. trilobata*	Field screening	
	TCR-254, TCR-390	*V. silvestris*	Field screening	
	TCR-314, TCR-315, TCR-24, TCR-29	*V. haniana*	Field screening	
	TCR-20	*V. glabrascens*	Field screening	
	TLC1, TLC2	*V. stipulaceae*	Field screening	
	RBL-1, TCR-93, PRR-2007-2, PRR-2008-2, RB-5-1, TCR-91, TCR-87, TCR-90, TCR-94, TCR-95, TCR-279	*V. umbellata*	Field screening	

(continued)

Table 2 (continued)

Disease/pest	Donor genotypes	Species	Method of screening	References
	IC 0611670, UH 87-7, UH 198, SYNP-2, IC 56048, IC 472020, IPU 99-243, 35-15, PLU 59, IPU 99-168, IPU 99-190, UH 86-4, IPU 99-211, IPU 99-18, PU 1541, BUC 18-2, KPU 128-105, PU 14-28, PU 14-19, LBG 623, PU 31, BUC 18-3, BIG 0067, BUC 17-1, BUS 20-2	*Vigna mungo*	Field screening	Tripathi et al. 2024
MYMIV	IC-546453, IPU 11-02, IC-548278, IC-43647, COBG-653, pant-Urd-19, UR-218, Shekhar-2, STY-2289, VBG-04-008, IPU 31-1, PDU-19, PDU-3, IPU 99-211, PLU-110, DPU88-31	*V. mungo*	Field screening	Kumar et al. (2022)
	UPU 8335, IPU 99-205, PGRU 95004, SPS 43		Field screening	Shamim & Pandey (2014)
	PLU-710, UH 86-4, TCR-20, PRR 2008-2	*V. mungo*	Field screening	Reddy et al. (1994)
Powdery mildew	LBG 645, LBG 17, IC-281977	*V. mungo*	Field and artificial screening	Rangaiah, (2015)
Callosobruchus maculatus	V2802BG, V2709, BSR-GG-1-49-3-1, BSRGG-1-56-2-2, BSR-GG-1-160-5-3, BSR-GG-1-170-2-4, BSR-GG-1-198-1-4	*V. radiata*	Bioassay	Samyuktha et al. (2022)
Bruchid (*C. analis*)	IC251439, IC251442, PRR 2007-2, IC251440, TCR 279	*V. umbellata*	Bioassay	Aidbhavi et al. (2021)
	JAP/10-51	*V. trinervia*		
	IC251435, IC553527, IC553526, JAP/10-7, IC349701	*V. trilobata*		
	TMV 1, LRM 13-44	*V. aconitifolia*		
	Trichy local 1	*V. stipulacea*		
	Kumur local	*V. khandalensis*		
	Mung seed 1, IC571775, IC251434	*V. radiata*		
	IC251390, IC251387	*V. mungo*		
	IC210580	*V. pilosa*		

Bruchid (*C. maculatus*)	Mung Seed-1, IC251426A, IC251426B	*V. radiata*	Bioassay	Aidbhavi et al. (2021)
	IC247408	*V. dalzelliana*		
	IC248326, IC248343	*V. vexillata*		
	Kumur local	*V. khandalensis*		
	JAP/10–51	*V. trinervia*		
	IC210575	*V. pilosa*		
	IC251394, IC251390, IC251385, IC251387	*V. mungo*		
	JAP/10-5, IC251435, IC553527	*V. trilobata*		
	IC251439, IC251442, PRR 2007-2, IC251440	*V. umbellata*		
Bruchid (*C. chinensis*)	Kumur local	*V. khandalensis*	Bioassay	Aidbhavi et al. (2021)
	IC251397, IC251390, IC251385, IC251387	*V. mungo*		
	IC247408	*V. dalzelliana*		
	IC248326, IC248343	*V. vexillata*		
	IC251439, IC251442, PRR 2007-2, IC251440	*V. umbellata*		
	JAP/10-51	*V. trinervia*		
	IC247407	*V. trinervia* var. *bournei*		
	IC210575	*V. pilosa*		
	Mung seed-1	*V. radiata*		

V. angularis var. *angularis* and *V. radiata* var. *radiata* exhibited persistent tissue specificity, suggesting extensive retention of gene functions.

Lonardi et al. (2019) presented the draft genome of cowpea cv. IT97K-499-35, totaling approximately 519 Mbp, with 29,773 protein-coding loci annotated, including alternatively spliced transcripts. Singh et al. (2022) sequenced the whole genome of *V. stipulacea* cv. JP245503, predicting 26,038 protein-coding genes based on gene models and transcript and protein alignments. These efforts collectively enhance understanding of *Vigna* crop genomics and the underlying mechanism controlling key traits, which can aid in the mung bean breeding program. Singh et al. (2022) conducted the genome sequencing of rice bean (*Vigna umbellata*), estimating around 31,276 highly confidential genes with 96.08% functional coverage. The genome assembly was found to be quite similar to adzuki bean, followed by mung bean and cowpea. Additionally, a draft genome sequence of *Vigna marina* cv. ANBp-14-03 was generated, totaling 23.7 Gb of sequence data, producing 68,731 scaffolds with an N50 length of 10,272 bp and a total assembly size of 365.6 Mb. Genome analysis revealed 50,670 genes with a mean coding sequence length of 1042 bp. Phylogenetic analysis indicated the highest sequence similarity with *V. angularis*, followed by *V. radiata*. Comparison with *V. angularis* genome revealed 16,699 SNPs and 2253 InDels, while comparison with *V. radiata* genome revealed 17,538 SNPs and 2300 InDels.

Souframanien & Gopalakrishna (2007) constructed the draft genome sequence of blackgram (*Vigna mungo* L. Hepper) cv. PU-31 using the hybrid genome assembly method, identifying 42,115 genes with 80.60% annotation coverage. About 50% of the assembled sequence consisted of repetitive elements. Singh et al. (2022) performed long-read-based draft genome sequencing of blackgram cv. Uttara, identifying 119 NBS-encoding genes associated with various biotic stress resistances. The close resemblance of many *Vigna* species with mung bean offers insights into stress tolerance mechanisms, aiding mung bean breeding efforts (Fig. 2).

6 Omics-Assisted Breeding

High yield, stress tolerance, stability, and sustainability are among the agricultural attributes that plant breeding attempts to enhance. This is combining genes from the genetic resources to improve traits that are desired, such as resilience to diseases and stress tolerance. A key factor in boosting breeding efforts is the use of genomic resources and molecular markers. DNA markers help identify allelic variants because of genetic linkage; this process is known as marker-assisted selection (MAS), which increases breeding efficiency. For gene mapping and tagging, a number of plant populations are used, such as F2 segregating populations, backcross populations, and fixed populations like double haploids (DH) and recombinant inbred lines (RILs). However, the efficiency of linkage mapping is limited by the fact that these approaches mainly comprise bi-parental segregation. Target loci may now be fine-mapped quickly and accurately, thanks to the development of

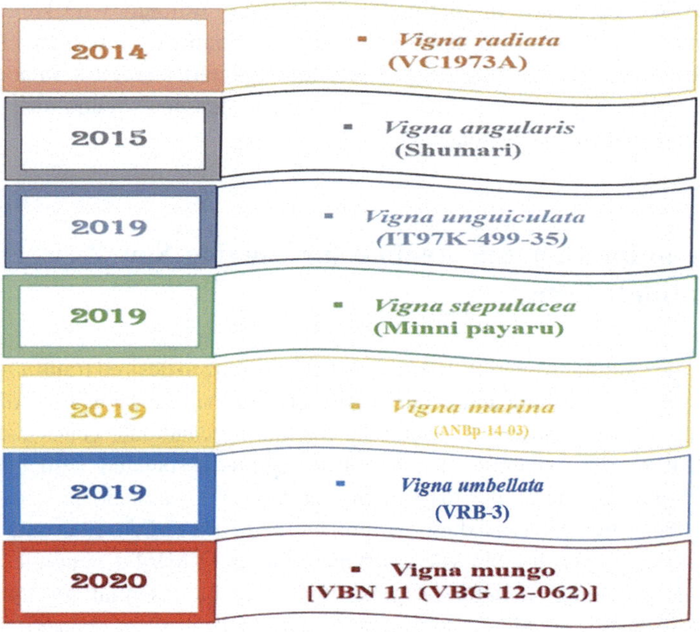

Fig. 2 The timeline of availability of the genome sequence of *Vigna* species

association analysis as an effective method for locating quantitative trait loci (QTLs). For instance, Noble et al. (2018) created a mapping panel for mung bean that maps complicated characteristics like heat and drought tolerance up to the F_5 generation using 30 crosses, including interspecific crosses. This is a perfect example of how novel techniques are used in plant breeding to improve agricultural productivity and resilience.

When association mapping is used instead of bi-parental populations, it can identify more QTLs because it uses a wider range of germplasm, which leads to more allelic diversity and more random events during parental evolutionary history. Important features in a variety of crops, such as cotton fiber quality, MYMIV resistance in soybean, agronomic traits in lentil and mung bean, seed coat color, and resistance to powdery mildew, have been mapped by numerous researchers using marker-trait association studies and association mapping. Previously, there was a belief that markers identified from preliminary mapping studies could be directly used in MAS, but recent consensus emphasizes the need for QTL confirmation and validation. The choice of donor parents in developing mapping populations and the background of molecular markers significantly impact QTL detection, with cross-specific molecular markers showing efficiency in identifying QTLs. Various studies have successfully identified QTLs using different approaches, such as genotyping by sequencing (GBS) and specific marker panels tailored to particular crosses. Furthermore, investigations have uncovered potential candidate genes associated with QTLs, offering insights into the genetic basis of traits like resistance to pests and diseases. Studies have highlighted the potential of utilizing wild *Vigna* species like *V. umbellata*, *V. sylvistris*, and *V. sublobata* as valuable resources for enhancing biotic stress tolerance in mung bean. These findings underscore the importance of leveraging diverse genetic resources and advanced genomics techniques for crop improvement and stress resilience.

7 Expanding Genomic Regions for Tagging New Candidate Genes

In current agricultural research, identifying QTLs linked to desired traits has become common, although cloning and detailed characterization of these QTLs still remain challenging. However, there is a potential for finding candidate genes causing the desired features by expanding the genomic regions associated with QTLs. For instance, Mathivathana et al. (2019) identified five QTLs associated with MYMV resistance using rice bean as a donor parent. Of them, qMYMV4-1 became a prominent and reliable QTL for MYMV resistance. The qMYMV4-1 genomic area was further enlarged, and 16 potential genes that may be essential for transferring MYMV resistance were identified. Likewise, a mung bean x rice bean RILs population's bruchid-resistant QTLs were mapped to chromosome 5 by Mariyammal et al. (2019). They found 16 potential genes linked to bruchid resistance through their analysis. Further investigation into the genetic pathways driving resistance against MYMV and bruchid pests could potentially focus on these candidate genes. Enhancing crop resistance and advancing agricultural practices require deepening our knowledge of the genomic regions associated with QTLs and locating potential genes within these regions (Table 3).

8 Comprehensive RNA-Seq Approach

RNA-seq, a technique that examines the complete set of RNA transcripts produced by an individual's genome under specific cellular or tissue conditions, is emerging as a powerful method for analyzing gene expression patterns and understanding gene function. Earlier methods such as cDNAs-AFLP, differential display-PCR (DD-PCR), and SSH provided limited resolution, but advanced techniques like microarrays, digital gene expression profiling, NGS, RNA-seq, and SAGE have greatly enhanced our ability to understand candidate genes. Mapping tagged candidate genes against decoded whole-genome sequences of crops and their relatives offers insights into structural and functional diversity. Wu et al. (2014) highlighted single molecule sequencing (SMS) as a cutting-edge technique for gene discovery and annotation. Baruah et al. (2017) used qPCR to analyze the expression of 20 defense-related genes, revealing insights into gene regulation upon bruchid

Table 3 The candidate genes associated with major disease/insect-pest resistance

Gene name	LG	Stress	Crop	Functional characterization	Reference
Vradi04g06770	04	*MYMV*	Mung bean × rice bean	Protein kinase superfamily protein (serine/threonine kinase activity)	Mathivathana et al. (2019)
Vradi04g06840	04			Small GTP-binding protein (disease resistance protein)/leucine-rich repeat/P-loop-containing nucleoside triphosphate hydrolase	
Vradi04g06900	04			Zinc finger, RING/FYVE/PHD-type (RING finger protein 165-like)	
Vradi04g06950	04			Receptor-like kinase/leucine-rich repeats	
Vradi04g06960	04			Zinc finger, RING/FYVE/PHD-type (U-box domain-containing protein 15-like)	
Vradi04g07000	04			Protein kinase superfamily protein (serine/threonine kinase activity)	
Vradi04g07100	04			WRKY family transcription factor	
Vradi04g07130	04			WRKY family transcription factor	
Vradi04g07220	04			MYB transcription factor MYB64	
Vradi04g07240	04			Transcription factor bHLH79-like (basic helix-loop-helix (bHLH) domain)	
Vradi04g0727	04			MYB transcription factor MYB183	
Vradi04g07290	04			DNA/RNA helicase, DEAD/DEAH box type, N-terminal/P-loop-containing nucleoside triphosphate hydrolase	
Vradi04g07440	04			Zinc finger, RING/FYVE/PHD-type (U-box domain-containing protein 38-like)	
Vradi04g07450	04			Jasmonic acid carboxyl methyltransferase (SAM-dependent carboxyl methyltransferase)/methyltransferase activity	
Vradi04g07490	04			Zinc finger, RING/FYVE/PHD-type (RING-H2 finger protein 2B)	
Vradi04g07540	04			Cytochrome P450 (oxidation-reduction process)	

Table 3 (continued)

Gene name	LG	Stress	Crop	Functional characterization	Reference
Vradi05g09450	05	*Bruchid*		WRKY family transcription factor	Mariyammal et al. (2019)
Vradi05g09480	05			Kelch repeat F-box protein	
Vradi05g09650	05			Aminoacyl-tRNA synthetase	
Vradi05g09830	05			Flavin-binding monooxygenase family protein	
Vradi05g09990	05			Cellulose synthase family protein	
Vradi05g10080	05			Ethylene-responsive transcription factor (ERF)	
Vradi05g10110	05			F-box family protein (leucine-rich repeat)	
Vradi05g10130	05			Ascorbate peroxidase (peroxidase activity)	
Vradi05g10140	05			Receptor-like serine/threonine protein kinase	
Vradi05g10200	05			Chloroplastic ATP synthase	
Vradi05g10210	05			Protein kinase superfamily protein/protein kinase activity/ATP binding/protein phosphorylation	
Vradi05g10410	05			Cellulose synthase family protein	
Vradi05g10460	05			Protein kinase superfamily protein/concanavalin A-like lectin/glucanase/protein phosphorylation	
Vradi05g10480	05			Ethylene-responsive transcription factor (ERF)	
Vradi05g10500	05			Zinc finger protein 1-like (zinc finger, C2H2)	
Vradi05g10560	05			Calmodulin-binding transcription activator 4-like isoform X5	
Vradi05g10580	05			Calmodulin-binding transcription activator 4 isoform X3 protein	

Vigan.05G027700	05	*C. maculatus*	Black gram	Zinc finger RING/FYVE/PHD-type, CTLH/CRA C-terminal to LisH motif	Subramaniyan et al. (2022)
Vigan.05G028300	05			Leucine-rich repeat—N-terminal, protein kinase family	
Vigan.05G029200	05			Pathogenesis-related protein 1-like/cysteine-rich secretory protein allergen V5/Tpx-1 family	
Vigan.05G030000	05			Myc-type basic helix–loop–helix (bHLH) type	
Vigan.05G030500	05			Zinc finger proteins (C2H2 type)	
Vigan.05G031900	05			Protein kinase family (serine–threonine/tyrosine-protein kinase catalytic), Concanavalin A-like lectin/glucanase subgroup	
Vigan.05G035600	05			F-box family protein	
Vigan.05G036000	05			Diacylglycerol kinase catalytic protein, ATP-NAD kinase-like	
Vigan.05G036200	05			Target SNARE site (coiled-coil structure), syntaxin N-terminal	
Vigan.05G038400	05			Toll/interleukin-1 receptor homology (TIR)	
Vigan.05G042200	05			Ubiquitin-conjugating enzyme E2/RWD-like	
Vigan.05G044400	05			Leucine-rich repeat (malectin-like carbohydrate binding)	
Vigan.05G046400	05			Glutaredoxin-like protein/thioredoxin-like fold	
Vigan.05G046700	05			NB-LRR family proteins	
Vigan.05G048300	05			Pathogenesis-related genes transcriptional activator (PTI5)	
Vigan.05G049800	05			Chloramphenicol acetyltransferase-like	
Vigan.05G056900	05			Ankyrin repeat-containing protein	
Vigan.05G066300	05			F-box proteins/Kelch repeat type 1	
Vigan.05G075700	05			F-box/kelch-repeat protein At3g23880-like isoform X1	
Vigan.08G002100	08			Bi-functional inhibitor/seed storage helical protein	
Vigan.08G002200	08			Lipid-transfer protein DIR1 family	
Vigan.08G002600	08			Myc-type basic helix–loop–helix (bHLH) type	
Vigan.08G003600	08			Zinc finger CCCH-type K homology protein	
Vigan.08G003900	08			Cytochrome P450 conserved protein	
Vigan.08G004400	08			Protein kinase superfamily (serine–threonine-dual specificity protein)/concanavalin A-like lectin/glucanase subgroup	
Vigan.08G004700	08			Basic-leucine zipper protein/transcription factor TGA-like	

Table 3 (continued)

Gene name	LG	Stress	Crop	Functional characterization	Reference
VrPGIP1, *VrPGIP2*	05	*Callosobruchus chinensis* and *Callosobruchus maculatus*	Green gram	Polygalacturonase-inhibiting protein	Kaewwongwal et al. (2017) and Chotechung et al. (2016)
Vradi04g00919	04	*C. chinensis*		Polygalacturonase inhibitor	Chen et al. (2021)
Vradi05g03810	05	*Callosobruchus* spp.	*V. radiata* × *V. sublobata*	Resistant specific protein-1	Chen et al. (2022)
Vradi05g03830	05			Resistant specific protein-2	
Vradi05g03840	05			Resistant specific protein-2	
Vradi05g03860	05			Resistant specific protein-2	
Vradi05g03870	05			Resistant specific protein-1	
Vradi05g03880	05			Resistant specific protein-1	
Vradi05g03930	05			Resistant specific protein-2	
Vradi05g03940	05			Polygalacturonase inhibitor	
Vradi05g03950	05			Polygalacturonase inhibitor	

oviposition. Lin et al. (2019) conducted transcriptome and proteome analysis, identifying differentially expressed genes and proteins related to resistance mechanisms.

Liu et al. (2016) compared genomic and transcriptomic data in mung bean related to bruchid resistance, identifying nucleotide variations and sequence-changed-protein genes associated with resistance. Raizada and Jegadeesan (2020) performed comparative transcriptome analysis of wild and cultivated blackgram, annotating differentially expressed genes related to bruchid infestation, YMD, and seed size.

Das et al. (2021) performed transcriptomic analysis, indicating the induction of defense responses in black gram upon bruchid oviposition, potentially mediated by SA signaling pathways. These studies underscore the role of transcriptomics in understanding stress responses and development in crops. RNA-seq has also facilitated the development of genic SSR, or functional markers linked to phenotypic traits. Kumar et al. (2021) analyzed lentils under heat stress, identifying differentially expressed genes and developing genic SSR markers. Comparative transcriptomic approaches across different crop species offer insights into differential expression profiles in response to specific stresses. Given the close resemblance of *Vigna* species and their gene transferability, comparative transcriptomics can elucidate target gene functions and aid in pyramiding candidate genes into mung bean for improvement. Overall, transcriptomic techniques offer valuable tools for enhancing mung bean and other crop species' resilience and productivity.

9 Gene-Based Functional Markers

Functional markers can be developed through various approaches such as transcriptomics, NGS, TILLING, homologous recombination (HR), association mapping, allele mining, and more. Utilizing transcriptome sequences available in databases provides a cost-effective and valuable resource for developing new molecular markers. For example, Gupta and Gopalkrishna (2010) developed 1071 SSRs in cowpea unigene sequences, successfully designing primer pairs for 803 SSR motifs and ultimately characterizing and validating 102 SSR markers. Similarly, Gupta et al. (2014) identified 12,596 EST sequences from mung bean and developed 1848 in-silico EST-SSRs. They further used 100 randomly selected primer pairs for characterization. These EST-SSRs hold potential for QTL mapping of target traits and can be used for functional characterization and the development of functional markers, proving more effective than random markers in crop improvement.

Baruah et al. (2017) also developed EST-SSRs associated with MYMV in blackgram. However, the development of gene-based functional markers (FM) remains limited in *Vigna* species, warranting accelerated efforts in this area. The advantage of FMs over other molecular markers lies in their close genomic resemblance to phenotypic traits. FMs enable the direct selection of genes associated with a phenotype, thus increasing selection efficiency for crop improvement. Additionally, there's no need to use flanking markers during gene introgression, making FMs pivotal in marker-assisted selection strategies. Therefore, the development of gene-based functional markers holds significant promise for enhancing crop breeding programs.

10 Developing Potential SCARs

A diverse array of molecular markers has been developed and employed in crop improvement, with PCR-based markers being considered cost-effective and requiring minimal DNA quantities. Pratap et al., Kumari et al., and numerous other researchers have evaluated molecular diversity using cross-specific molecular markers, demonstrating their efficiency and effectiveness in mung bean. For instance, Singh et al. utilized molecular markers from adzuki bean to map loci for MYMIV and agronomic traits in mung bean populations.

The development of sequence characterized amplified region (SCAR) markers is one strategy for identifying tightly linked markers to species-specific traits. Souframanien and Gopalakrishna used RAPD markers, sequencing marker ISSR8111357, to design SCAR-YMV1 linked to MYMV. Dhole and Reddy developed another marker, OPB-07600, closely linked (6.8 cM) to a MYMV resistance gene. Similarly, Zhang and Panthee developed a codominant SCAR marker linked to resistance genes in tomatoes for gene pyramiding. Certain wild species closely related to mung bean and also each other require robust taxonomy-based phenotyping for identification. Gore et al. identified two genotypes of the section Aconitifoliae (*V. trilobata* and *V. stipulacea*) using 47 descriptive traits. SCAR markers could facilitate the easy identification of these species, enhancing their utilization. Zheng et al. developed SCAR markers for discriminating toxic genotypes from non-toxic ones in Dendrobium officinale, highlighting the potential of SCAR markers in *Vigna* improvement. These developments underscore the importance of molecular markers in crop enhancement and germplasm characterization.

11 Hormonal Regulations and Nanotechnology for Improving Resistance

Plant defense mechanisms against diverse biotic and abiotic challenges depend significantly on phytohormones such as jasmonic acid (JA), salicylic acid (SA), and its derivatives. These hormones serve important roles in a number of plant signaling pathways, enabling plants to respond and adjust to harsh environmental conditions. They have a comparable number of advantageous and negative effects. While SA is essential in the fight against biotrophic diseases, JA primarily serves as defense against pathogenic organisms and herbivores. Furthermore, research has shown that by controlling particular gene expression, the application of JA and SA can strengthen pulses' resilience to abiotic challenges such as drought, temperature extremes, metal toxicity, and salt stress.

Under stressful conditions, the amount of JA and SA is controlled by an effective signaling system that includes coordinated activities of enzyme regulation, both post-transcriptional and transcriptional, and modification of essential proteins by external substances. The transcription factor WRKY70 (TF) is involved in the control and modulation of genes such as *NPR1*, *VSP1*, *VSP2*, *PR2*, and *PR10* at the post-transcriptional level. These genes are linked to the stimulation of the plant defense response in pulse crops. However, pulse crops' modulation of the JA and SA signaling pathways under stress is complex and requires an in-depth understanding of the underlying mechanisms. In order to mitigate the effects of these stresses, it is essential to create comprehensive plans for the exogenous application of JA/SA and its derivatives in pulses. This requires understanding the complex interaction of these hormones and their subsequent effects on the control of protein and gene expression. Sultana et al. (2024) observed that under waterlogging (WL) stress conditions, the growth and yield output of two elite mung bean genotypes, BU Mung bean-4 and BU Mung bean-6, were dramatically increased by pretreatment with salicylic acid (SA). Both genotypes showed significant increases in shoot dry weight, individual leaf area, and photosynthetic pigment contents after SA pretreatment. Interestingly, BU Mung bean-6 showed a greater degree of improvement than BU Mung bean-4. Moreover, SA pretreatment increased total flavonoid levels and improved the activities of peroxidase, glutathione S-transferase, catalase, and ascorbate peroxidase, thereby reducing the oxidative burden resulting from reactive oxygen species in both genotypes. Proline and free amino acid accumulation was also improved by SA pretreatment in both genotypes, indicating that the use of these osmo-protectants improved osmotic balance. Elevated levels of leaf water content and leaf succulence in BU Mung bean-6 were particularly indicative of these improvements. In BU Mung bean-6 under WL circumstances, the improvement of physiological and biochemical mechanisms mediated by SA resulted in larger yield-associated traits. Limited work has been done regarding the influence of nanoparticles on the biotic stresses of *Vigna* species (Table 4). However, certain studies in various legume crops have demonstrated the application of nanoparticles against biotic stresses and reported them to enhance plant defense response by improving nutrition and suppressing bacterial, fungal, and viral infections. In legumes, they possess antimicrobial activity against important phytopathogens, such as *Cercospora canescens*, *Fusarium solani*, *F. oxysporum*, *Botrytis cinerea*, *Alternaria alternata*, *Rhizoctonia solani*, *Aspergillus niger*, *Sclerotium rolfsii*, and *Macrophomina phaseolina*, and of course, yellow mosaic virus (Table 4). Application of ZnO NPs at 1200 ppm considerably inhibited the mycelial growth of *Cercospora canescens* by 89.86% and controlled the disease by 30% in *Vigna radiata* (Rafiq et al., 2022). Few studies have investigated the impact of metal and metal oxide nanoparticles in reducing seed-borne pathogens. Chatterjee et al. (2016) synthesized titanium dioxide (TiO_2) nanoparticles using *Vigna radiata* seed extract and accessed their antimicrobial activity against different pathogens. The nanoparticles exhibited significant inhibitory effects on all tested pathogens by dissolving their outer membranes

Table 4 Different nanoparticles against various biotic stresses in *Vigna* crops

Crop species	Nanoparticles	Size (nm)	Concentration	Biotic stress/Causal organism	References
Mung bean, *Vigna radiata* L.	ZnO	15–20	1200 ppm	*Cercospora canescens*	Rafiq et al. (2022)
	MgO	–	25, 50, 100 μg/ml	*Fusarium solani* and *F. oxysporum*	Abdallah et al. (2022)
	TiO_2	–	50, 75, 100 μl	*Escherichia coli*, *Staphylococcus aureus*, *Serratia marcescens*, *Salmonella* sp., *Pseudomonas aeruginosa*, *Klebsiella pneumoniae*, *Enterobacter* sp., *Proteus mirabilis*, and *Shigella* sp.	Chatterjee et al. (2016)
Cowpea, *Vigna unguiculata* Walp	TiO_2	–	50, 75, 100 μl	*E. coli*, *Salmonella*, *Enterobacter*, *Pseudomonas aeruginosa*, and *Serratia marcescens*	Chatterjee et al. (2017)
	TiO_2	30–60	125 cc/ha	Cercospora leaf spot, brown blotch, seed-borne fungal pathogens	Owolade and Ogunleti (2008), Maity et al. (2018)
	SiO_2 and Ag	20–60	1, 1.5, 2, 2.5 g/kg	*Callosobruchus maculatus* F.	Rouhani et al. (2013)
	ZnO CuO Ag TiO_2	20–50 50–80 5–15 30–60	1000–1250 mg/kg seed	Seed-borne fungal pathogens	Maity et al. (2018)

through the presence of hydroxyl groups, leading to the organisms' deaths. MgONPs significantly decreased the severity of disease caused by *F. solani* and *F. oxysporum* in *Vigna radiata* plants (Abdallah et al., 2022). The treatment of nanoparticles (ZnO, CuO, Ag, and TiO_2) at 1000–1250 mg/kg seed in *Vigna unguiculata* inhibited the seed-borne fungal pathogens. Out of these, AgNP inhibited the fungal attack on seed at the highest degree, and TiO_2-treated seeds showed maximum attack of pathogens (Maity et al., 2018). Biosynthesized silver (Ag) nanoparticles were tested against pathogens responsible for root rot disease of groundnut, namely, *Sclerotium rolfsii*, *Macrophomina phaseolina*, and *Aspergillus* niger, exhibited up to 60–65% effective mycelial growth control (Raja et al., 2021). NPs employ a variety of antimicrobial mechanisms against pathogens. They have the ability to damage DNA, rupture cell membranes, and stop electron transport by internalizing and releasing ions. Additionally, they have the ability to produce ROS, which can damage DNA, denaturize proteins, and interfere with the function of enzymes (Dizaj et al., 2014).

Nanopesticides and nanoformulations show an enhanced range of pesticide efficacy as compared to the conventional pesticides, thus having a positive impact on the control of plant diseases and pests (Chhipa, 2017). Nanopesticides such as SiO_2NPs are believed to impair the physiological processes of insects by causing cuticular damage, desiccation, respiratory obstruction, and damage to the midgut epithelium (Caceres et al., 2019). Studies have clearly shown that SiNPs are the effective pesticides that may be used alone or in combination with other commercially available pesticides. Rouhani et al. (2013) employed SiNPs as nanopesticides, which showed an entomotoxic effect against cowpea seed beetle, *Callosobruchus maculatus*. Recently, the entomotoxic effects of silica nanoparticles were also tested at various concentrations (75, 150, 225, 300, 375, and 425 mg/l) on the pests of soybean (*Spodoptera littoralis*) and faba bean (*Liriomyza trifolii* and *Aphis craccivora*), and 50–75 mg/l was found to be the best for their management (Thabet et al., 2021). Application of silica and zinc as nanoparticles at 1000 ppm and 750 ppm, respectively, showed the highest mortality of the bruchid, *Callosobruchus chinensis*, affecting chickpea seeds (Khandaitaray et al., 2023). Seed priming of SiO_2 NPs together with *R. leguminosarum* in peas has shown a reduction in the severity of blight disease (*Pseudomonas syringae* pv. *pisi*), galling and multiplication of *Meloidogyne incognita* (Kashyap & Siddiqui, 2020). Chitosan NPs were evaluated for the control of bean yellow mosaic virus in faba bean (*Vicia faba*). At 300 and 400 mg/l concentrations, ChiNPs completely inhibited the disease's infectivity and viral accumulation content when applied curatively 48 hours after virus inoculation.

12 Marker-Assisted Breeding

Marker-assisted selection (MAS) involves the use of DNA markers to aid in phenotypic selection for crop improvement. Various markers such as RAPD, SCoT, ISSR, SSR, and SNPs have been effectively utilized in breeding programs. Identifying and exploiting tightly linked markers to the trait of interest has resulted in practical achievements, enhancing varietal development with increased efficiency and accuracy in shorter timeframes. Marker-assisted backcrossing has been successfully deployed in breeding disease-resistant chickpea cultivars among pulses (Pratap et al., 2017; Bharadwaj et al., 2022). Next-generation sequencing (NGS) technologies have brought remarkable advancements, providing ultra-throughput sequences for plant genotyping. To extend the application of sequencing technologies to large crop genomes or in cases where reference sequences are unavailable, genotyping-by-sequencing (GBS) has been developed for marker discovery.

In recent times, bioinformatic pipelines have been deployed to identify candidate genes, which could be valuable for developing functional markers for MAS programs aimed at mung bean improvement. While many QTLs/genes for biotic stress resistance have been identified from cross-specific donors using recombinant inbred lines (RILs) populations, their cloning, characterization, and introgression into high-yielding varieties for enhancing biotic stress tolerance remain limited.

Wu et al. (2014) successfully introgressed the bruchid-resistant gene VrPGIP-2 from V2802 into KPS-1 using VrBR-SSR013 and DMB-SSR-158 as the foreground markers. Mariyammal et al. (2019) identified 15 candidate genes for bruchid resistance in mung bean x rice bean populations. Chen et al. (2022) identified the bruchid-resistant gene Vradi05g03810 encoding a probable resistant-specific protein from TC1966 in the background of mung bean. Mathivathana et al. (2019) identified 19 candidate genes on chromosome 04 in mung bean x rice bean populations for MYMV resistance. These identified genes can be tagged and introgressed into mung bean to improve bruchid resistance. The identification of resistant genes from different sources will also aid in gene pyramiding, enhancing the durability of resistance (Fig. 3).

13 Conclusion and Future Prospects

In general, there are numerous biotic and abiotic factors affecting mung bean productivity, many of which lead to the crop growing overly large. A great deal of research has been done in the last few years to improve resistance to diseases, particularly yellow mosaic and leaf disease. Nonetheless, the complexity of the stress response has hindered progress. Biologic stressors also frequently result in immune dysfunction. Thus, the construction of QTLs/genes from related species utilizing genetic markers, along with the discovery and characterization of genes, would aid in the resolution of this issue. It can be utilized for this purpose because many of the wild relatives of *Vigna* are resistant to numerous pests and diseases. With the advent of genome-wide data, transcripts, and replication data, evaluation of disease variety, genetic interaction, and genome-wide research are now possible, leading to the development of specific indicators and symptoms that can alleviate depressive symptoms. The full genome sequence of the mung bean and similar plants has also made it possible to study strain variations that cause anxiety to: stress and stress-encoding gene families across the genome. Studies on the expression of genes in different tissues under stress and throughout normal development provided knowledge that could be useful in understanding individuals who encounter stress.

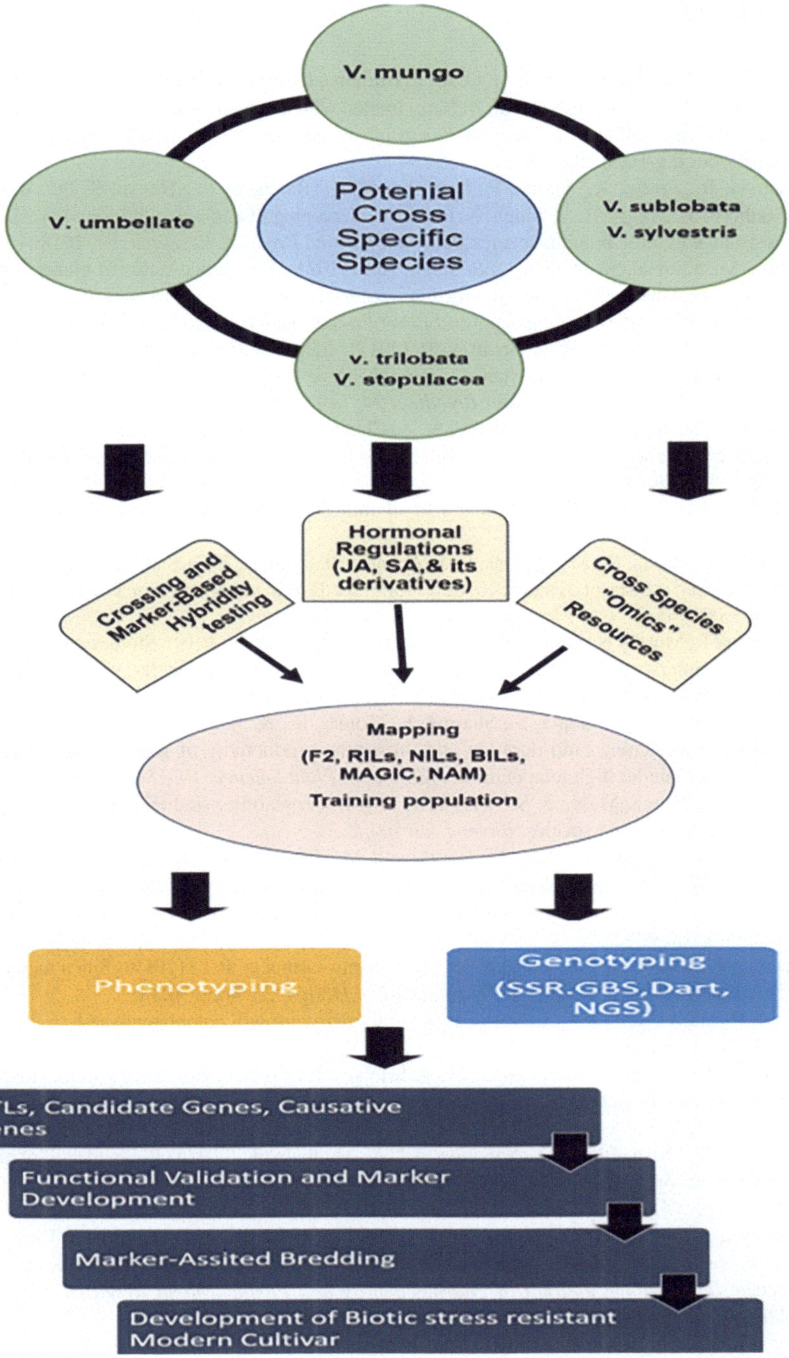

Fig. 3 Ways to utilize alien genetic and omics resources for mung bean improvement

References

Abdallah, Y., Hussien, M., Omar, M. O. A., Elashmony, R. M. S., Alkhalifah, D. H. M., & Hozzein, W. N. (2022). Mung bean (*Vigna radiata*) treated with magnesium nanoparticles and its impact on soilborne *Fusarium solani* and *Fusarium oxysporum* in clay soil. *Plants., 11*(11), 1514. https://doi.org/10.3390/plants11111514

Aidbhavi, R., Pratap, A., Verma, P., Lamichaney, A., Bandi, S. M., Nitesh, S. D., Akram, M., Rathore, M., Singh, B., & Singh, N. P. (2021). Screening of endemic wild *Vigna* accessions for resistance to three bruchid species. *Journal of Stored Products Research, 93*, 101864.

Akram, M., Kamaal, N., Pratap, A., & Singh, N. P. (2021). Resistance status of mungbean (*Vigna radiata (L.) Wilczek) advanced breeding materials against mungbean yellow mosaic India virus. Archives of Phytopathology and Plant Protection, 54(19–20), 2533–2546.*

Alam, A. K. M., Somta, P., & Srinives, P. (2014). Identification and confirmation of quantitative trait loci controlling resistance to mungbean yellow mosaic disease in mungbean [*Vigna radiata* (L.) Wilczek]. *Molecular Breeding, 34*, 1497–1506.

Amkul, K., Wang, L., Somta, P., Wang, S., & Cheng, X. (2019). Construction of a high density linkage map and genome dissection of bruchid resistance in zombi pea (*Vigna* vexillata (L.) A Rich). *Scientific Reports, 9*, 11719.

Anderson, T. R. (1985). Root rot and wilt of mungbean in Ontario. *Canadian Plant Disease Survey, 65*, 3–6.

Baruah, I. K., Panda, D., MV, J., Das, D. J., Acharjee, S., Sen, P., & Sarmah, B. K. (2017). Bruchid egg induced transcript dynamics in developing seeds of black gram (Vigna mungo). *PLoS One, 12*(4), e0176337.

Basavaraja T., Murthy Niranjana, Kumar L. Vijay, Mallikarjun K. (2019). Studies on cross compatibility in interspecific crosses of vigna radiata × vigna umbellata species. www.indianjournals.com/ijor.aspx?target=ijor:lr&volume=42&issue=5&article=021

Basu, P. S., Pratap, A., Gupta, S., Sharma, K., Tomar, R., & Singh, N. P. (2019). Physiological traits for shortening crop duration and improving productivity of greengram (Vigna radiata L. Wilczek) under high temperature. *Frontiers in Plant Science, 10*, 1508.

Bhanu, A. N., Singh, M. N., & Srivastava, K. (2018). Crossability studies of interspecific hybridization among *Vigna* species. *Biomed Journal, 2*, 7.

Bharadwaj, C., Jorben, J., Rao, A., Roorkiwal, M., Patil, B. S., Jayalakshmi, ... & Varshney, R. K. (2022). Development of high yielding Fusarium wilt resistant cultivar by pyramiding of "genes" through marker-assisted backcrossing in chickpea (Cicer arietinum L.). *Frontiers in Genetics, 13, 924287.*

Caceres, M., Vassena, C. V., Garcera, M. D., & Santo-Orihuela, P. L. (2019). Silica nanoparticles for insect pest control. *Current Pharmaceutical Design, 25*, 4030–4038.

Chand, P., Varma, J., (1983). Effect of yellow mosaic on growth components and yield of mungbean and urdbean.

Chankaew, S., Somta, P., Sorajjapinun, W., & Srinives, P. (2011). Quantitative trait loci mapping of Cercospora leaf spot resistance in mungbean, *Vigna radiata* (L.) Wilczek. *Molecular Breeding, 28*, 255–264.

Chatterjee, A., Nishanthini, D., Sandhiya, N., & Abraham, J. (2016). Biosynthesis of titanium dioxide nanoparticles using *Vigna radiata. Asian Journal of Pharmaceutical and Clinical Research, 9*(4), 85–88.

Chatterjee, A., Ajantha, M., Talekar, A., Revathy, N., Abraham, J., et al. (2017). Biosynthesis, antimicrobial and cytotoxic effects of titanium dioxide nanoparticles using *Vigna unguiculata* seeds. *International Journal of Pharmacognosy and Phytochemical Research, 9*(1), 95–99. https://doi.org/10.25258/ijpapr.v9i1.8048

Chen, H., Hu, L., Wang, L., Wang, S., & Cheng, X. (2021). *Identification of a candidate gene for bruchid resistance by combining fine mapping and transcriptome profiling in mung bean (Vigna radiata L.).* Chinese Academy of Agricultural Sciences Institute of Crop Sciences.

Chen, T., Hu, L., Wang, S., Wang, L., Cheng, X., & Chen, H. (2022). Construction of high-density genetic map and identification of a Bruchid resistance locus in mung bean (*Vigna radiata* L.). *Frontiers in Genetics, 13*, 903267.

Chhipa, H. (2017). Nanofertilizers and nanopesticides for agriculture. *Environmental Chemistry Letters, 15*, 15–22.

Chotechung, S., Somta, P., Chen, J., Yimram, T., Chen, X., & Srinives, P. (2016). A gene encoding a polygalacturonase-inhibiting protein (PGIP) is a candidate gene for bruchid (Coleoptera: Bruchidae) resistance in mungbean (*Vigna radiata*). *Theoretical and Applied Genetics, 129*, 1673–1683.

Choudhary, S., Kaurav, H., & Chaudhary, G. (2021). Vaibidang (Embelia ribes): A potential herbal drug in ayurveda with anthelmintic property. *International Journal for Research in Applied Sciences and Biotechnology, 8*, 237–243.

Das, D., Baruah, I. K., Panda, D., Paswan, R. R., Acharjee, S., & Sarmah, B. K. (2021). Bruchid beetle ovipositioning mediated defense responses in black gram pods. *BMC Plant Biology, 21*, 1–22.

Deising, H., Nicholson, R. L., Haug, M., Howard, R. J., & Mendgen, K. (1992). Adhesion pad formation and the involvement of cutinase and esterases in the attachment of uredospores to the host cuticle. *The Plant Cell, 4(9), 1101–1111.*

Dhaliwal, S. S., Sharma, V., & Shukla, A. K. (2022). Impact of micronutrients in mitigation of abiotic stresses in soils and plants—A progressive step toward crop security and nutritional quality. *In Plant Nutrients and Abiotic Stress Tolerance; Sparks, D.L.B.T.-A. in A., Ed.; Academic Press, 173*, 1–78. ISBN 0065-2113.

Dickman, M. B., Podila, G. K., & Kolattukudy, P. E. (1989). Insertion of cutinase gene into a wound pathogen enables it to infect intact host. *Nature, 342*, 446–448.

Dizaj, S. M., Lotfipour, F., Barzegar-Jalali, M., Zarrintan, M. H., & Adibkia, K. (2014). Antimicrobial activity of the metals and metal oxide nanoparticles. *Materials Science and Engineering: C, 44*, 278–228.

Douglas, C., Pratap, A., Rao, B. H., Manu, B., Dubey, S., Singh, P., & Tomar, R. (2020). Breeding progress and future challenges: abiotic stresses. *The Mungbean Genome, 81–96.*

Goyal, L., et al. (2021). Association and diversity analysis for yield attributing traits in advance generation of green gram (Vigna radiata (L.) Wilczek). *International Journal of Chemical Studies.*

Gupta, S. K., & Gopalakrishna, T. (2010). Development of unigene-derived SSR markers in cowpea (Vigna unguiculata) and their transferability to other Vigna species. *Genome, 53*(7), 508–523.

Gupta, S. K., & Pratap, A. (2016). Breeding Methods. *Biology and Breeding of Crucifers, 79.*

Gupta, S., Gupta, D., Sen, Anjum, T. K., Pratap, A., & Kumar, J. (2013). Inheritance and molecular tagging of MYMIV resistance gene in blackgram (*Vigna* mungo L. Hepper). *Euphytica, 193*, 27–37.

Gupta, S. K., Bansal, R., & Gopalakrishna, T. (2014). Development and characterization of genic SSR markers for mungbean (Vigna radiata (L.) Wilczek). *Euphytica, 195*, 245–258.

Humphry, M. E., Magner, T., McIntyre, C. L., Aitken, E. A. B., & Liu, C. J. (2003). Identification of a major locus conferring resistance to powdery mildew (Erysiphe polygoni DC) in mungbean (*Vigna radiata* L. Wilczek) by QTL analysis. *Genome, 46*, 738–744.

Iqbal, U., & Mukhtar, T. (2014). Morphological and pathogenic variability among macrophomina phaseolina isolates associated with mungbean (Vigna radiata L.) Wilczek from Pakistan. *Scientific World Journal, 2014*, 950175.

Itoh, T., Garcia, R. N., Adachi, M., Maruyama, Y., Tecson-Mendoza, E. M., Mikami, B., & Utsumi, S. (2006). Structure of 8Sα globulin, the major seed storage protein of mung bean. *Acta Crystallographica. Section D, Biological Crystallography, 62*(7), 824–832. https://doi.org/10.1107/s090744490601804x

Joshi, A., Souframanien, J., Chand, R., & Pawar, S. E. (2006). Genetic diversity study of Cercospora canescens (Ellis & Martin) isolates, the pathogen of Cercospora leaf spot in legumes. *Current Science, 90*, 564–568.

Kaewwongwal, A., Chen, J., Somta, P., Kongjaimun, A., Yimram, T., Chen, X., & Srinives, P. (2017). Novel alleles of two tightly linked genes encoding polygalacturonase-inhibiting proteins (VrPGIP1 and VrPGIP2) associated with the Br locus that confer bruchid (Callosobruchus spp.) resistance to mungbean (*Vigna radiata*) accession V2709. *Frontiers in Plant Science, 8*, 1692.

Karuppanapandian, T., Sinha, P. B., Haniya, A. M. K., & Mamoharan, K. (2006). Differential antioxidative responses of ascorbate-glutathione cycle enzymes and metabolites to chromium stress in green gram (Vigna radiata L. wilczek) leaves. *Journal Of Plant Biology, 49*(6), 440–447. https://doi.org/10.1007/bf03031124

Kasettranan, W., Somta, P., & Srinives, P. (2010). Mapping of quantitative trait loci controlling powdery mildew resistance in mungbean (*Vigna radiata* (L.) Wilczek). *Journal of Crop Science and Biotechnology, 13*, 155–161.

Kashyap, D., & Siddiqui, Z. A. (2020). Effect of silicon dioxide nanoparticles and rhizobium *leguminosarum* alone and in combination on the growth and bacterial blight disease complex of pea caused by Meloidogyne incognita and pseudomonas syringae pv. pisi. *Archives of Phytopathology and Plant Protection*. https://doi.org/10.1080/03235408.2020.1843306

Kaul, T., Eswaran, M., Thangaraj, A., Meyyazhagan, A., Nehra, M., Raman, N. M., et al. (2019). Rice Bean (Vigna umbellata) draft genome sequence: unravelling the late flowering and unpalatability related genomic resources for efficient domestication of this underutilized crop. *bioRxiv*, 816595.

Kaur, L. (2006). Multiple disease resistant sources of mungbean. *Acta Horticulturae, 752*, 423–426.

Keatinge, J. D. H., Easdown, W. J., Yang, R. Y., Chadha, M. L., & Shanmugasundaram, S. (2011). Overcoming chronic malnutrition in a future warming world: The key importance of mungbean and vegetable soybean. *Euphytica, 180*(1), 129–141. https://doi.org/10.1007/s10681-011-0401-6

Khan, M. G., Ahmad, W., Khattak, G. S. S., & Ahmad, H. (2007). Mode of inheritance of resistance to mungbean yellow mosaic virus (MYMV) in mungbean (Vigna radiata (L.) Wilczek). *Sarhad Journal of Agriculture, 23(4), 1071.*

Khandaitaray, T., Mishra, P. R., Satapathy, S. N., Shankar, T., Badjena, T., Tripathy, B., & Paikaray, S. S. (2023). Elevated CO_2 and nanoparticles for the management of pulse beetle, *Callosobruchus chinensis* L. in stored chickpea. *Legume Research*. https://doi.org/10.18805/LR-5232

Khattak, G. S. S., Saeed, I., & Shah, S. A. (2008). Breeding high yielding and disease resistant mungbean (*Vigna radiata* (L.) Wilczek) genotypes. *Pakistan Journal of Botany, 40*, 1411–1417.

Kim, B. C., Lim, I., Jeon, S. Y., Kang, M., & Ha, J. (2021). Effects of irrigation conditions on development of mungbean (Vigna radiata l.) sprouts. *Plant Breeding and Biotechnology, 9*(4), 310–317.

Kitsanachandee, R., Somta, P., Chatchawankanphanich, O., Akhtar, K. P., Shah, T. M., Nair, R. M., Bains, T. S., Sirari, A., Kaur, L., & Srinives, P. (2013). Detection of quantitative trait loci for mungbean yellow mosaic India virus (MYMIV) resistance in mungbean (*Vigna radiata* (L.) Wilczek) in India and Pakistan. *Breeding Science, 63*, 367–373.

Koller, W., & Parker, D. M. (1989). Purification and characterization of cutinase from Venturia inaequalis. *Phyto-pathology, 79, 278–283.*

Kulkarni, A., & Rao, M. (2009). Differential elicitation of an aspartic protease inhibitor: Regulation of endogenous protease and initial events in germination in seeds of Vigna radiata. *Peptides, 30(12), 2118–2126.* https://doi.org/10.1016/j.peptides.2009.08.024

Kumar, H., & Singh, R. B. (1981). Genetic analysis of adult plant resistance to powdery mildew in pea (Pisum sativum L.). *Euphytica, 30, 147–151.*

Kumar, J., Gupta, D. S., Kesari, R., Verma, R., Murugesan, S., Basu, P. S., et al. (2021). Comprehensive RNAseq analysis for identification of genes expressed under heat stress in lentil. *Physiologia Plantarum, 173*(4), 1785–1807.

Kumar, S., Kumar, K., Tewari, K., Sagar, P., Pandey, J., Ps, S., Rathore, M., Kumar, V., Akram, M., Singh, A., et al. (2022). Gene expression and biochemical profiling of contrasting *Vigna* mungo genotypes against Mungbean Yellow Mosaic India Virus (MYMIV). *Journal of food and legumes, 35*, 107–116.

Kumari, G., Pratap, A., Lavanya, R. G., Akram, M., Rathore, M., Madhavan, L., Singh, Y., & Singh, N. P. (2021). Potential resistant donors for yellow mosaic disease identified from endemic wild *Vigna* species. *Journal of Food Legumes, 34*, 10–16.

Lin, Y., Li, W., Zhang, Y., Xia, C., Liu, Y., Wang, C., et al. (2019). Identification of genes/proteins related to submergence tolerance by transcriptome and proteome analyses in soybean. *Scientific Reports, 9*(1), 14688.

Liu, M. S., Kuo, T. C. Y., Ko, C. Y., Wu, D. C., Li, K. Y., Lin, W. J., et al. (2016). Genomic and transcriptomic comparison of nucleotide variations for insights into bruchid resistance of mungbean (Vigna radiata [L.] R. Wilczek). *BMC Plant Biology, 16*, 1–16.

Lonardi, S., Muñoz-Amatriaín, M., Liang, Q., Shu, S., Wanamaker, S. I., Lo, S., et al. (2019). The genome of cowpea (Vigna unguiculata [L.] Walp.). *The Plant Journal, 98*(5), 767–782.

Maheshwari, S. K., & Krishna, H. (2013). Field efficacy of fungicides and bioagents against Alternaria leaf spot of mungbean. *Annals of Plant Protection Sciences, 21*, 364–367.

Maity, A., Natarajan, N., Pastor, M., Vijay, D., Gupta, C. K., & Wasnik, V. K. (2018). Nanoparticles influence seed germination traits and seed pathogen infection rate in forage sorghum (*Sorghum bicolour*) and cowpea (*Vigna unguiculata*). *Indian Journal of Experimental Biology, 56*, 363–372.

Mahalingam, A., & Manivannan, N. (2023). Interspecific hybridization between Vigna radiata and Vigna mungo towards the broadening of genetic base for MYMV disease resistance and generating variability. Legume Research-An International *Journal. 46(4): 395-402.* https://doi.org/10.18805/LR-4495.

Mariyammal, I., Seram, D., Samyuktha, S. M., Karthikeyan, A., Dhasarathan, M., Murukarthick, J., Kennedy, J. S., Malarvizhi, D., Yang, T.-J., & Pandiyan, M. (2019). QTL mapping in *Vigna radiata*× *Vigna* umbellata population uncovers major genomic regions associated with bruchid resistance. *Molecular Breeding, 39*, 1–13.

Mathivathana, M. K., Murukarthick, J., Karthikeyan, A., Jang, W., Dhasarathan, M., Jagadeeshselvam, N., Sudha, M., Vanniarajan, C., Karthikeyan, G., Yang, T.-J., et al. (2019). Detection of QTLs associated with mungbean yellow mosaic virus (MYMV) resistance using the interspecific cross of *Vigna radiata* × *Vigna* umbellata. *Journal of Applied Genetics, 60*, 255–268.

Majid, S. (1953). Annals Report of department of agriculture, Assam for year ending 31st March 1950 II. *The Grow More Food Campaign, 11, 107.*

Naimuddin, K., & Akram, M., Pratap, A., Kumar, D., & Nair, R. (2023). Urdbean Leaf Crinkle Virus: A Mystery Waiting to Be Solved. *Viruses, 15(10), 2120.* https://doi.org/10.3390/v15102120

Nair, R., & Schreinemachers, P. (2020). *Global Status and Economic Importance of Mungbean.* https://doi.org/10.1007/978-3-030-20008-4_1

Nair, R. M., Pandey, A. K., War, A. R., Hanumantharao, B., Shwe, T., Alam, A., Pratap, A., Malik, S. R., Karimi, R., Mbeyagala, E. K., et al. (2019). Biotic and abiotic constraints in mungbean production—Progress in genetic improvement. *Frontiers in Plant Science, 10*, 1340.

Noble, T. J., Tao, Y., Mace, E. S., Williams, B., Jordan, D. R., Douglas, C. A., & Mundree, S. G. (2018). Characterization of linkage disequilibrium and population structure in a mungbean diversity panel. *Frontiers in Plant Science, 8*, 2102. https://doi.org/10.3389/fpls.2017.02102

Owolade, O. F., & Ogunleti, D. O. (2008). Effects of titanium dioxide on the diseases, development and yield of edible cowpea. *Journal of Plant Protection Research, 48*(3), 329–335.

Pandiyan, M., Senthil, N., Suersh, R., Chakravarthy, N., Packiaraj, D., & Jagadeesh, S. (2012). Interspecific hybridization of Vigna radiata× Vigna trilobata. *Euphytica, 35*, 1017–1022.

Pandey, A. K., Burlakoti, R. R., Kenyon, L., & Nair, R. M. (2018). Perspectives and challenges for sustainable management of fungal diseases of mungbean [Vigna radiata (L.) R. Wilczek var. radiata]: a review. *Frontiers in Environmental Science, 6, 53.*

Pratap, A., & Gupta, S. K. (2009). Biotechnxvc ological interventions in host plant resistance. Integrated Pest Management: Innovation-Development Process: *Volume 1, 183–206.*

Pratap, A., Basu, P. S., Gupta, S., Malviya, N., Rajan, N., Tomar, R., ... & Singh, N. P. (2014). Identification and characterization of sources for photo-and thermo-insensitivity in Vigna species. *Plant Breeding, 133(6), 756–764.*

Pratap, A., Chaturvedi, S. K., Tomar, R., Rajan, N., Malviya, N., Thudi, M., ... & Singh, N. P. (2017). Marker-assisted introgression of resistance to fusarium wilt race 2 in Pusa 256, an elite cultivar of desi chickpea. *Molecular Genetics and Genomics, 292(6), 1237–1245.*

Pratap, A., Gupta, S., Basu, P. S., Tomar, R., Dubey, S., Rathore, M., et al. (2019a). Towards development of climate smart mungbean: Challenges and opportunities. In C. Kole (Ed.), *Genomic designing of climate-smart pulse crops* (pp. 235–264). Springer. https://doi.org/10.1007/978-3-319-96932-9_5

Pratap, A., Gupta, S., Nair, R. M., Gupta, S. K., Schafleitner, R., Basu, P. S., Singh, C. M., Prajapati, U., Gupta, A. K., Nayyar, H., et al. (2019b). Using plant phenomics to exploit the gains of genomics. *Agronomy, 9*, 126.

Pratap, A., Douglas, C., Prajapati, U., Kumari, G., War, A. R., Tomar, R., ... & Dubey, S. (2020). Breeding progress and future challenges: biotic stresses. The Mungbean Genome, 55–80.

Pratap, A., Gupta, S., Rathore, M., Basavaraja, T., Singh, C. M., Prajapati, U., Singh, P., Singh, Y., & Kumari, G. (2021). Mungbean. In A. Pratap & S. B. T.-T. B. Gupta (Eds.), *The beans and the peas: From orphan to mainstream crops* (pp. 1–32). Woodhead Publishing. ISBN 978-0-12-821450-3.

Pratap, A., Kumar, S., Polowick, P. L., Blair, M. W., & Baum, M. (2022). Accelerating genetic gains in pulses. *Frontiers in Plant Science, 13, 879377.*

Rafiq, H., Aftab Z-e-H, Anjum, T., Ali, B., Akram, W., Bashir, U., Mirza, F. S., Aftab, M., Ali, M. D., Li, G., et al. (2022). Bio-fabrication of zinc oxide nanoparticles to rescue mung bean against cercospora leaf spot disease. *Frontiers in Plant Science, 13*, 1052984. https://doi.org/10.3389/fpls.2022.1052984

Ragul, S., Manivannan, N., Ganapathy, N., & Karthikeyan, G. (2022). Screening and biochemical analysis on blackgram genotypes for resistance against storage pest bruchine [Callosobruchus maculatus (F.)]. *Legume Research International Journal, 45*, 371–378.

Raizada, A., & Jegadeesan, S. (2020). Comparative transcriptomic analysis revealed complex molecular mechanisms underlying pests, pathogens resistance and seed development in wild and cultivated Blackgram. *bioRxiv, 2020*, 11.

Raja, M., Sharma, R. K., Jambhulkar, P., Sharma, K. R., & Sharma, P. (2021). Biosynthesis of silver nanoparticles from *Trichoderma harzianum* Th3 and its efficacy against root rot complex pathogen in groundnut. *Materials Today Proceedings, 43*(5), 3140–3143.

Rangaiah, S. (2015). Evaluation of blackgram (Vigna mungo L. Hepper) genotypes for root traits as a measure of drought tolerance. *International Journal of Tropical Agriculture, 33, 1463–1468.*

Reddy, K. S., Pawar, S. E., & Bhatia, C. R. (1994). Inheritance of powdery mildew (Erysiphe polygoni DC) resistance in mungbean (*Vigna radiata* L. Wilczek). *Theoretical and Applied Genetics, 88*, 945–948.

Rouhani, M., Samih, M. A., & Kalantari, S. (2013). Insecticidal effect of silica and silver nanoparticles on the cowpea seed beetle, *Callosobruchus maculatus* F. (Col.: Bruchidae). *Journal of the Entomological Research Society, 4*, 297–305.

Sai, C. B., Nagarajan, P., Raveendran, M., Rabindran, R., Senthil N. (2017). Understanding the inheritance of mungbean yellow mosaic virus (MYMV) resistance in mungbean (Vigna radiata L. Wilczek). *Mol Breed 37, 63.*

Sahu, B., Choudhary, V. K., Sahu, M. P., Kumar, K. K., Sujayanand, G. K., Gopi, R., ... & Ghosh, P. K. (2023). Biotic Stress Management 23. Trajectory of 75 years of Indian Agriculture after Independence, 619.

Samyuktha, S. M., Malarvizhi, D., Mariyammal, I., Karthikeyan, A., Seram, D., Dhasarathan, M., Juliet Hepziba, S., Sheela, V., Thanga Hemavathy, A., & Kavithamani, D. (2022). The hunt for Mungbean (*Vigna radiata* (L.) Wilczek) genotypes and breeding lines resistance to south Indian Bruchid strain. *Agriculture, 12*, 1050.

Sathees, N., Shoba, D., Mani, N., Saravanan, S., Kumari, M. P., & Pillai, M. A. (2022). Tagging of SSR markers associated to yellow mosaic virus resistance in black gram (*Vigna* mungo (L.) Hepper). *Euphytica, 218*, 1–11.

Savaliya, A. S., Chopada, G. B., Sekhada, M. R., & Bhimani, A. A. (2018). Survey of the powdery mildew of Mungbean (Vigna radiata (L). Wilczek) in selected districts of South Gujarat. *Trends in Biosciences, 11*, 3004–3009.

Shamim, M. Z., & Pandey, A. (2014). Identification of yellow mosaic virus (YMV) resistant black gram (Vigna mungo L.) genotypes for cultivation in Northern India. *Journal of Agriculture and Natural Resource Management, 1, 48–50.*

Sharma, H. C., Khare, M. N., Joshi, L. K., & Kumar, S. M. (1971). *Efficacy of fungicides in the control of diseases of kharif pulses mung and urid* (p. 2). In Proceedings of the All India Workshop on Kharif Pulses.

Shen, Y. M., Liu, H. L., Chang, S. T., & Chao, C. H. (2010). First report of anthracnose caused by Colletotrichum acutatum on mung bean sprouts in Taiwan. *Plant Disease, 94*, 131.

Singh, A., Dikshit, H. K., Jain, N., & Yadav, R. N. (2014). Efficiency of SSR, ISSR and RAPD markers in molecular characterization of mungbean and other Vigna species. *Indian Journal of Biotechnology, 13*, 81–88.

Singh, C. M., Singh, P., Pratap, A., Pandey, R., Purwar, S., Douglas, C. A., ... & Mishra, A. K. (2019). Breeding for enhancing Legumovirus resistance in mungbean: Current understanding and future directions. *Agronomy, 9(10), 622.*

Singh, C. M., Pratap, A., Gupta, S., Biradar, R. S., & Singh, N. P. (2020). Association mapping for mungbean yellow mosaic India virus resistance in mungbean (*Vigna radiata* L. Wilczek). *3 Biotech, 10*, 1–12.

Singh, P., Mishra, A. K., & Singh, C. M. (2021). Genome-wide identification and characterization of lectin receptor-like kinase (LecRLK) genes in mungbean (*Vigna radiata* L. Wilczek). *Journal of Applied Genetics, 62*, 223–234.

Singh, C. M., Pratap, A., Kumar, H., Singh, S., Singh, B. K., Prasad, D., et al. (2022a). Recent advances in omics approaches for mungbean improvement. *Technologies in Plant Biotechnology and Breeding of Field Crops*, 181–200.

Singh, P., Pandey, B., Pratap, A., Gyaneshwari, U., Nair, R. M., Mishra, A. K., & Singh, C. M. (2022b). Genetic and genomics resources of cross-species vigna gene pools for improving biotic stress resistance in mungbean (Vigna radiata L. Wilczek). *Agronomy, 12*(12), 3000.

Somta, P., Chen, J., Yundaeng, C., Yuan, X., Yimram, T., Tomooka, N., & Chen, X. (2019). Development of an SNP-based high-density linkage map and QTL analysis for bruchid. (Callosobruchus maculatus F.) resistance in black gram (*Vigna* mungo (L.) Hepper). *Scientific Reports, 9*, 1–9.

Souframanien, J., & Gopalakrishna, T. (2007). Source for bruchid resistance and its inheritance in Trombay wild urdbean (*Vigna mungo var. silvestris). Journal of Food Legumes, 20, 19.*

Sriphadet, S. C. J. P. (2007). Inheritance of agronomic traits and their interrelationship in Mungbean (Vigna radiata(L.) Wilczek). https://koreascience.kr/article/JAKO200716637994336.page

Subramaniyan, R., Narayana, M., Krishnamoorthy, I., Natarajan, G., & Gandhi, K. (2022). Novel and stable QTL regions conferring resistance to MYMV disease and its inheritance in blackgram (*Vigna* mungo (L.) Hepper). *Journal of Genetics, 101*, 1–7.

Sudha, M., Anusuya, P., Mahadev, N. G., Karthikeyan, A., Nagarajan, P., Raveendran, M., Senthil, N., Pandiyan, M., Angappan, K., & Balasubramanian, P. (2013). Molecular studies on mungbean (*Vigna radiata* (L.) Wilczek) and ricebean (*Vigna* umbellata (Thunb.)) interspecific hybridisation for Mungbean yellow mosaic virus resistance and development of species-specific SCAR marker for ricebean. *Archives of Phytopathology and Plant Protection, 46*, 503–517.

Sultana, S., Rahman, M. M., Das, A. K., Haque, M. A., Rahman, M. A., Islam, S. M. N., Ghosh, P. K., Keya, S. S., Tran, L.-S. P., & Mostofa, M. G. (2024). Role of salicylic acid in improving the yield of two mung bean genotypes under waterlogging stress through the modulation of antioxidant defense and osmoprotectant levels. *Plant Physiology and Biochemistry, 206*, 108230. https://doi.org/10.1016/j.plaphy.2023.108230

Sun, L., Cheng, X.-Z., Wang, S., Wang, L., Liu, C., Mei, L., Xu, N., Lei, S., Cheng, X.-Z., Wang, S., et al. (2008). Heredity analysis and gene mapping of Bruchid resistance of a Mungbean cultivar V2709. *Agricultural Sciences in China, 7*, 672–677.

Tantasawat, P. A., Poolsawat, O., Kativat, C., Arsakit, K., Papan, P., Chueakhunthod, W., & Pookhamsak, P. (2022). Inheritance and identification of ISSR-RGA markers associated with powdery mildew resistance in mungbean for marker-assisted breeding. *Chilean Journal of Agricultural Research, 82*, 3–9.

Thabet, A. F., Boraei, H. A., Galal, O. A., El-Samahy, M. F. M., Mousa, K. M., Zhang, Y. Z., Tuda, M., Helmy, E. A., Wen, J., & Nozaki, T. (2021). Silica nanoparticles as pesticide against insects of different feeding types and their non-target attraction of predators. *Scientific Reports, 11*(1), 14484. https://doi.org/10.1038/s41598-021-93518-9

Tiwari, A. (1993). Studies on the diseases caused by Rhizoctonia Solani Kuhn in green gram (Phaseolus aureus Roxb), Ph. D. thesis. *Rani Durgavati Vishwavidyalaya (Jabalpur), 338*.

Vadivel, K., Manivannan, N., Mahalingam, A., Satya, V. K., Vanniarajan, C., & Ragul, S. (2021). Identification and validation of quantitative trait loci of mungbean yellow mosaic virus disease resistance in blackgram [*Vigna* mungo (L). Hepper]. Legume Res. *Legume Research International Journal, 1*, 7.

Verma, M. M., & Brar, J. S. (1996). Breeding approaches for increasing yield potential of mungbean. In A. N. Asthana & D. H. Kim (Eds.), *Recent advances in Mungbean research* (pp. 102–123). Indian Society of Pulses Research and Development.

Wu, X., Wang, B., Lu, Z., Wu, X., Li, G., & Xu, P. (2014). Identification and mapping of a powdery mildew resistance gene Vu-Pm1 in the Chinese asparagus bean landrace ZN016. *Legume Research International Journal*, 37.

Yundaeng, C., Somta, P., Chen, J., Yuan, X., Chankaew, S., & Chen, X. (2021). Fine mapping of QTL conferring Cercospora leaf spot disease resistance in mungbean revealed TAF5 as candidate gene for the resistance. *Theoretical and Applied Genetics, 134*, 701–714.

R-Genes for Improving Disease and Insect-Pest Resistance in Legumes

Shalini Purwar, Poornima Singh, Brijesh Pandey, Vishal Chugh, Chirag, Shashi Kashyap, Priyanka Kaundal, and Chandra Mohan Singh

Abstract Plants have evolved intricate defense mechanisms to combat pathogens, deploying a distinct line of protection that involves the activation of genetic responses within their cells through specialized proteins encoded by a class of defense genes known as R genes. In this regard, numerous defense-related genes, such as nucleotide-binding site–leucine-rich repeat (NBS–LRR) genes, play pivotal roles. NBS-LRR proteins, equipped with an amino-terminal variable domain, a central nucleotide-binding site (NBS), and a carboxy-terminal leucine-rich repeats (LRR) domain, exhibit the remarkable ability to recognize pathogens either directly or indirectly. Once activated, these proteins initiate downstream signaling pathways, triggering a robust defense response in plants against various biotic stresses, including bacterial, fungal, viral, nematode, and insect attacks. Notably, the NBS domain serves as a critical hub, binding to and hydrolyzing ATP to facilitate signal transduction upon pathogen recognition. Equally significant, the LRRs possess exceptional adaptability as structural domains, primarily involved in facilitating protein–protein interactions. Furthermore, LRRs can evolve with diverse binding specificities, adding to the plant's repertoire of defense strategies.

Within this chapter, we delve into the comprehensive genomic analysis of the resistance (R) gene family in leguminous plants. Our objectives encompass a wide array of facets, including the precise identification and meticulous annotation of R genes, an exploration of their intricate structural organization, a comparative genomics perspective, an examination of their expression patterns when exposed to biotic stress conditions, and insights into their practical applications in breeding programs. By undertaking this multifaceted approach, we aspire to elevate our understanding of the R gene family's dynamics within the legume realm. Ultimately,

S. Purwar · V. Chugh · Chirag · P. Kundal · C. M. Singh (✉)
Banda University of Agriculture and Technology, Banda, Uttar Pradesh, India

P. Singh · B. Pandey
Mahatma Gandhi Central University, Marihari, Bihar, India

S. Kashyap
Division of Molecular Biology and Biotechnology, CSIR-NBRI, Lucknow, Uttar Pradesh, India

A. Pratap et al. (eds.), *Integrated Improvement of Food Legumes*,
https://doi.org/10.1007/978-3-031-82629-0_8

this knowledge holds the promise of contributing significantly to the development of disease-resistant leguminous varieties, thereby bolstering agricultural sustainability in the face of ever-evolving pathogenic challenges.

Keywords Resistance (R) genes · NBS-LRR (nucleotide-binding site–leucine-rich repeat) · Effector-triggered immunity (ETI) · Hypersensitive response (HR) · Signal transduction pathways · Genome-wide association studies (GWAS) · Bioinformatics tools

Abbreviations

CC	Coiled-coil
Chr	Chromosomes
DAMP	Damage-associated molecular patterns
ETI	Effector-triggered immunity
HR	Hypersensitive response
LRR-TM	Leucine-rich repeat plus transmembrane receptor),
MAMPs	Microbe-associated molecular patterns
NBS-LRR	Nucleotide binding site and leucine-rich repeat domains
PR	Pathogen-related
PAMPs	Pathogen-associated molecular patterns
PCD	Programmed cell death
PRRs	Pattern recognition receptors
PTI	Pattern-triggered immunity
R proteins	Resistance protein
RLK	Receptor-like kinase
SA-CC	Signal anchor plus coiled-coil
STK	Serine-threonine kinase

1 Introduction

The interactions between plants and pathogens hold immense significance in the realm of agriculture, exerting a pivotal influence on crop yields, food availability, and economic resilience. As the plants and pathogens have co-evolved, co-existence with each other has led to developing effective defense mechanisms to avoid pathogen and to protect themselves. So the resistance against parasites/pathogen is a natural and universal heritable trait of plants under which they defend themselves against parasites/pathogens or their activities. Understanding the importance of these interactions is crucial for implementing effective strategies to manage and mitigate plant diseases.

2 Role of R Genes in Plant Defense Against Pathogens

Plants are equipped with an inherent immune system that operates at the cellular level, with each cell contributing to the collective defense effort. When faced with a pathogen incursion, plants exhibit a remarkable ability to release localized and precise signals, culminating in a visually discernible programmed cell death termed the hypersensitive response (HR) at the site of infection. This stands in stark contrast to the defense mechanisms of animals, which rely on specialized defense cells. Plant cells, on the other hand, possess the unique capability to recognize foreign molecules. Within plant cells, an array of immune receptors is produced, enabling them to detect the presence of pathogens and facilitating crucial communication in response to potential invasions (Caplan et al., 2008).

The interaction between plants and pathogens reveals various naturally occurring inherited physical and chemical defense mechanisms. When pathogen propagules come into contact with the host surface, the host rapidly and efficiently initiates signaling cascades that eventually trigger downstream responses. These defense mechanisms can be divided into three distinct layers, with the first layer being triggered by universally recognized foreign molecules referred to as microbe-associated molecular patterns (MAMPs) or pathogen-associated molecular patterns (PAMPs). MAMPs consist of microbial or pathogen structures such as flagellins, lipopolysaccharides, chitins, glucans, and damaged host components (Ali et al., 2018). In response to injury, the host's tissues release molecules known as damage-associated molecular patterns (DAMPs), encompassing various plant signaling molecules. These compounds are detected by a wide-ranging family of plant receptors referred to as pattern recognition receptors (PRRs), ultimately initiating a defense mechanism termed pathogen-associated molecular pattern-triggered immunity (PTI) (Kushalappa et al., 2016).

The secondary line of defense, known as effector-triggered immunity (ETI), is specialized in identifying intracellular effector proteins produced by pathogens. This targeted recognition process relies on intracellular receptors called resistance proteins (R proteins), which are encoded by classical plant disease resistance (R) genes. Once these R proteins detect the presence of pathogen proteins within the host cell, they set off a sequence of signaling events that involve calcium ion signaling, the generation of reactive oxygen species, and the buildup of pathogen-related (PR) proteins (Araújo et al., 2019). R genes have been categorized into at least five diverse classes or families, including NBS-LRR (nucleotide binding site and leucine-rich repeat domains), LRR-TM (leucine-rich repeat plus transmembrane receptor), STK (serine-threonine kinase), RLK (receptor-like kinase), and SA-CC (signal anchor plus coiled-coil) (Wei et al., 2020). The cytoplasmic nucleotide-binding site (NBS) leucine-rich repeat (LRR) proteins constitute the most extensive category within the R proteins. ETI serves as a form of qualitative resistance, often referred to as vertical resistance, and is frequently linked with the gene-for-gene hypothesis (Flor, 1971; Kushalappa et al., 2016).

Finally, the tertiary defense mechanism hinges on the principles of natural selection, which compel pathogens to evade ETI by either shedding or diversifying the recognized effector genes. Alternatively, pathogens may acquire additional effectors capable of suppressing ETI as a survival strategy (Table 1).

A comprehensive exploration of the R gene family on a genome-wide scale in leguminous plants holds profound significance for unraveling the molecular underpinnings of plant defenses against pathogens and advancing crop protection strategies. Leguminous crops, encompassing pivotal ones such as soybean, chickpea, and lentil, confront a multitude of pathogenic threats that can substantially compromise their yield and quality. The R gene family assumes a pivotal role in the plant's defense arsenal, as it identifies and responds to specific pathogen effectors, thereby instigating immune responses to thwart pathogen incursions. Through the systematic genome-wide analysis of the R gene family in legumes, researchers gain the ability to identify and characterize the entirety of R genes within the genome. This analytical approach yields invaluable insights into the breadth, distribution, and evolutionary trajectories of these genes. Furthermore, it paves the way for the development of precise breeding methodologies tailored to enhance disease resistance in legume crops. Such analysis permits the identification of prospective R genes, which can be subject to further investigation and incorporation into breeding programs aimed at nurturing resilient varieties fortified against distinct pathogens.

Additionally, genome-wide scrutiny unveils genetic variations and polymorphism residing within R genes. This revelation contributes significantly to our comprehension of the intricate mechanisms governing susceptibility and resistance to pathogens in legumes. It unravels the intricate signaling pathways and molecular interplay that underlie the defense responses mediated by R genes. This wealth of knowledge can be harnessed to engineer or manipulate R gene-mediated immune responses, ushering in sustainable and effective strategies for managing diseases afflicting legume crops.

Table 1 R gene and R protein list in plants

S. No	R Gene	R Protein	Role	References
1	NBS-LRR	NBS-LRR proteins	Activation of defense signaling pathways upon recognition of pathogen effectors	Collier and Moffett (2009)
2	LRR-TM	PR4b	The inhibition of spore germination and the growth of mycelium in fungal plant pathogens	Muthamilarasan et al. (2014)
3	STK	Serine/threonine protein kinase	Regulation of gene expression and cell division	Macho and Zipfel (2014)
4	RLK	Receptor-like kinases	Activation or inactivation of downstream proteins through phosphorylation events	Shiu and Bleecker (2001), Meng et al. (2020)
5	SA-CC	PR1, PR2, PR5	The synthesis of pathogenesis-related proteins as a part of plant defense mechanisms	van Loon et al. (2006), Ali et al. (2018)

3 NBS-LRR Structure and Its Biological Function

One of the largest and most important gene families involved in disease resistance in plants is the nucleotide-binding (NB) site leucine-rich repeat (LRR) that plants use to detect effector proteins by inducing effector-triggered immunity (ETI). In contrast to their animal counterparts, NB-LRR immune receptors in plants have undergone specialized evolution, endowing them with the capacity to precisely identify a broad spectrum of effector proteins originating from a variety of pathogens, encompassing fungi, oomycetes, bacteria, viruses, and insects (Caplan et al., 2008). NBS-LRR proteins are typically encoded by genes ranging from approximately 860–1900 amino acids in length. These proteins exhibit a characteristic structure composed of at least three discernible domains that are interconnected by linker regions. These domains encompass a variable amino-terminal domain, the NBS domain, and the LRR region (Wei et al., 2020). NBS-LRR genes can be further categorized into two distinct subfamilies: TIR-NBS-LRR (TNL) proteins, distinguished by the presence of the TIR domain (Toll/interleukin 1 receptor), and non-TIR-NBS-LRR (nTNL) proteins, recognized by the absence of the TIR domain. Within the nTNL group, a significant proportion of members exhibit a coiled-coil (CC) domain in their N-terminal region, leading to their classification as CC-NBS-LRR (CNL) proteins.

The NB domain, often referred to as NB-ARC, is noteworthy for its size and composition, encompassing the nucleotide-binding region (NB) along with subdomains ARC1 and ARC2. This nomenclature is attributed to its sequence similarities with the apoptotic protease 1 activation factor (APAF-1) found in humans, which shares homology with APAF-1 from the nematode *Caenorhabditis elegans* (CED-4) (Bezerra-Neto et al., 2020; Van Ooijen et al., 2008). Shao et al., 2016 and Zhang et al., 2016a, b have described a group of nTNL proteins with a particular N-terminal domain called RPW8 (Resistance to Powdery mildeW8), possibly representing a new class of NBS proteins, the RPW8-NBS-LRR (RNL) (Meyers et al., 2002; Nandety et al., 2013). In the plant tobacco (*Nicotiana benthamiana*), researchers have identified two additional protein families containing the TIR domain, namely TIR-NB site (TN) and TIR-unknown/random (TX). These protein families are distinct from the longer TIR-NB-LRR (TNL) R proteins. TN proteins are characterized by the presence of both TIR and NBS domains but lack the leucine-rich repeat (LRR) domains. On the other hand, TX proteins do not possess either NBS or LRR domains, although they often feature a small and variable C-terminal domain. Within the NBS domain of these proteins, researchers have confirmed the existence of eight distinct conserved motifs. These motifs include P-loop, RNBS-A, Kinase2, RNBS-B, RNBS-C, GLPL, RNBS-D, and MHDV (Fig. 1).

The NBS and LRR domains appear as the most important of plant R proteins, and all major classes of R proteins containing at least one NBS and/or LRR domain have a role in pathogen recognition specificity (Dubey & Singh, 2018). N-terminal domain has been identified as similar to residues that can enhance gene expression and protein stability (Wei et al., 2020). NTPase activity is a pivotal function of the

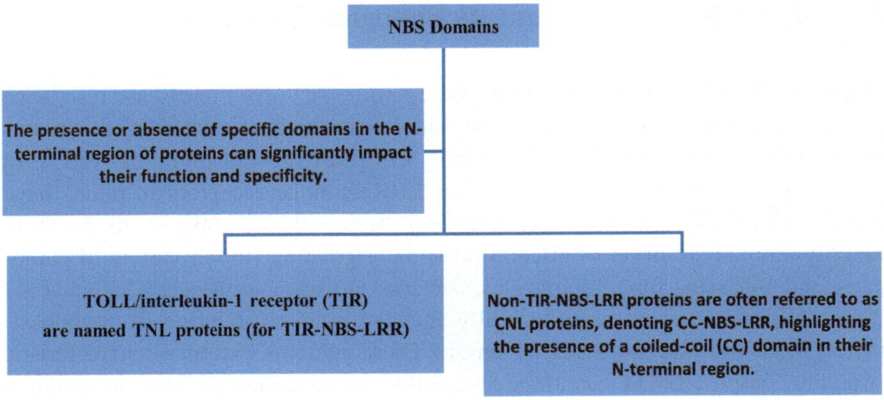

Fig. 1 NBS nucleotide-binding site domain classes

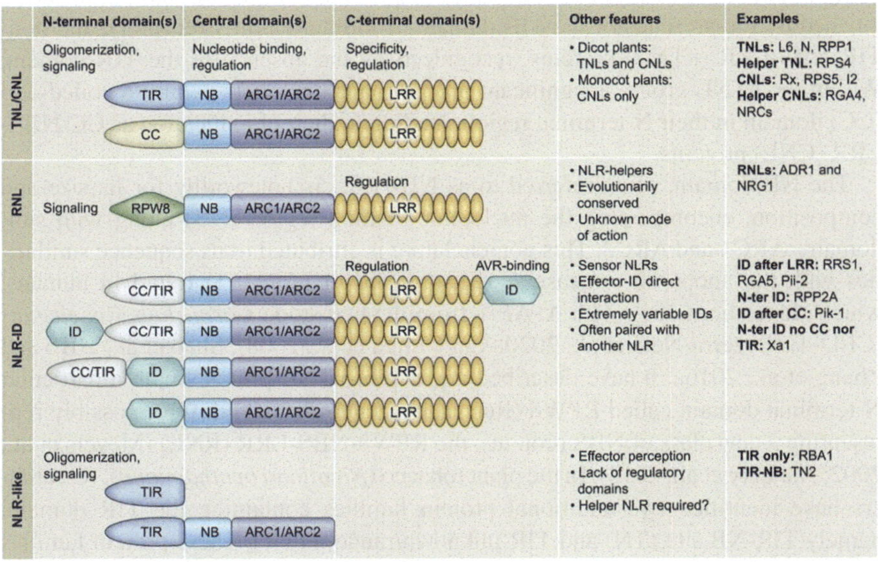

Fig. 2 Rearranging plant NBS-LRRs: Exploring predicted and experimentally verified domain functions—CC (Coiled-Coil), TIR, NB (nucleotide-binding), ARC1/2 (Apaf-1, R-gene products, and CED4), LRR (leucine-rich repeat), and ID (integrated domain). Understanding these domains' roles enhances our grasp of plant immunity mechanisms (Source: Cesari, 2018)

NBS domain, characterized by highly conserved motifs as noted by Dubey and Singh (2018). This activity plays a critical role in catalyzing the conversion of molecules, thereby initiating signal transduction pathways. In stark contrast, the LRR domain exhibits a sequence that varies significantly and is frequently associated with facilitating protein–protein interactions. This LRR-mediated interaction profoundly influences the specificity of pathogen effector molecule recognition (Wei et al., 2020) (Fig. 2).

4 Molecular Mechanism of NB-LRR Recognizes Pathogen Effectors

In the gene-for-gene model of R gene-mediated resistance, initially introduced by Flor (1942) as a binary system involving a single plant resistance gene and a corresponding pathogen virulence gene, R proteins play a pivotal role in recognizing pathogen effectors. Pathogen effectors can be identified through two distinct approaches: a direct method, involving direct protein–protein interactions with NB-LRRs, and an indirect method, where effectors target and modify host proteins.

During infection, pathogen effectors target and alter host proteins to promote susceptibility (S). In the resistant plant, by the direct effector recognition method, NLR interacts with the effector and induces resistance (R) (Fig. 3) (Cesari, 2018). Interestingly, direct interactions between effectors and host target proteins do not typically display signatures of balancing selection. This is likely because many of these target proteins are central components targeted by multiple effectors, and they cannot tolerate extensive variations without compromising their function (Weßling et al., 2014). Various models have been proposed to explain indirect interactions. The "Guarde Model" suggests the involvement of a second host protein, referred to as a "guardee," in the defense process. R proteins are thought to effectively monitor and safeguard the integrity of other essential host molecules, which serve as targets for pathogen effector proteins. This, in turn, triggers defense responses (Lozano-Torres et al., 2012; Van Es et al., 2008). The Guard Model provides insights into why the protein kinase Pto necessitates the NLR protein Prf for defense activation

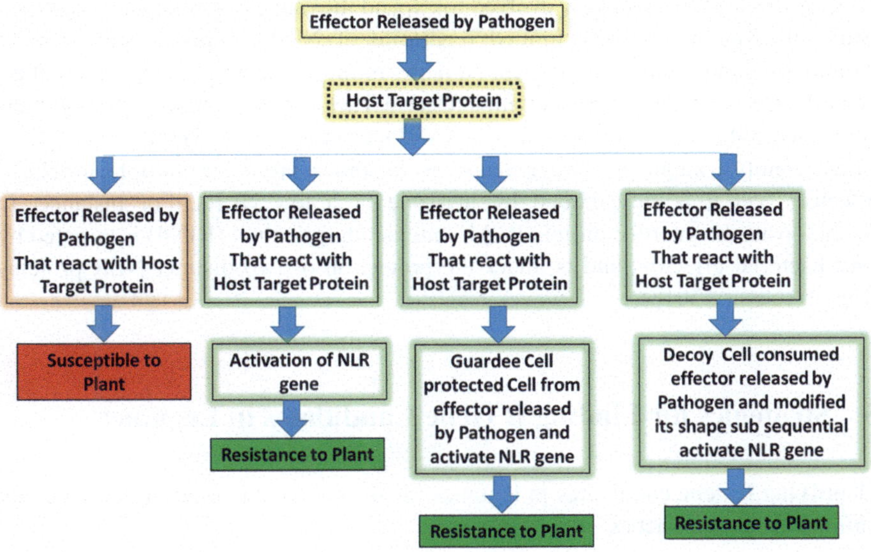

Fig. 3 Different way of NBS-LLRs detect specific pathogen effectors through diverse mechanisms

in the context of R proteins. In this model, a guardee protein is implicated in the defense process. When an effector protein interacts with this guardee protein, it leads to the activation of defense responses (Dangl & Jones, 2001). The central premise of the Guard Model is that both the pathogen effector and the plant NLR co-evolve to bind the plant guardee protein. This co-evolutionary process ensures that the plant's defense mechanisms are activated only when there's a specific interaction between the pathogen effector and the guardee, thus preventing unnecessary or indiscriminate immune responses (Mukhtar et al., 2011). In the Decoy Model, an additional host protein assumes a distinct role as a decoy, which sets it apart from the Guard Model. Unlike the guardee in the Guard Model, the decoy does not actively participate in plant defense. Instead, its primary function is to imitate a genuine defense component, essentially masquerading as a true effector target. By doing so, the decoy lures the pathogen effector and diverts its attention away from actual defense components, helping to thwart the pathogen's attack (Cesari, 2018). In the Decoy Model, effector-induced modifications to the decoy serve as a trigger for an R protein, leading to the activation of the plant's immune response. This modified version of the Guard Model helps elucidate the role of proteins like Pto, which function as decoys for specific effector proteins like AvrPto in tomato. Decoy proteins are dedicated to facilitating indirect NLR detection of effectors, serving as a means to activate plant immunity (see Fig. 1a). These decoy proteins have no purpose other than "trapping" effectors. Consequently, decoys with enhanced effector-binding capabilities may be favored by natural selection (see Fig. 3). In comparison to direct NLR-ligand recognition, indirect recognition offers more versatile options for the plant immune system, as pathogen effectors must either lose their activity or adapt to avoid recognition by these decoys (Cesari, 2018). Indirect effector recognition has likely evolved to expand the range of pathogen recognition while utilizing a relatively limited repertoire of NLRs, approximately 150 in Arabidopsis and around 460 in rice. Notably, intraspecific and interspecific pathogen effectors frequently interact with a restricted set of host proteins, forming interconnected hubs that facilitate infection (Mukhtar et al., 2011; Singh et al., 2023). Consequently, a single guardee or decoy can become a target for multiple, unrelated effectors, yet it is safeguarded by just one or a few NLRs. For instance, in Arabidopsis, the guardee protein RPM1 interacting protein 4 (RIN4) is targeted by four bacterial effectors and is under the protection of two distinct NLR proteins (Fig. 3) (Cesari, 2018).

5 Strategies for Finding R Gene Candidates in Legumes

Identifying R gene candidates in legumes involves a combination of genomic and bioinformatic approaches.

5.1 Reference Genome Analysis

The availability of reference genome sequences for legume species, such as common bean, soybean, pea, and chickpea, enables the identification of R gene candidates. Researchers can search the genomes for sequences encoding conserved domains like the NBS and LRR domains, which are characteristic of R genes. Comparative genomics approaches offer a valuable means to identify potential R genes. This involves comparing the genomes of legume species with established R genes in other plant species.

5.2 Transcriptome Analysis

Transcriptomic data provides valuable insights into gene expression patterns and can be used to identify R gene candidates. RNA sequencing (RNA-seq) technology can be employed to generate transcriptome data from legume tissues infected with pathogens or treated with pathogen-associated molecular patterns (PAMPs). Differential gene expression analysis can identify upregulated genes encoding NBS-LRR-like proteins, which can serve as candidate R genes (Garg et al., 2011).

5.3 Candidate Gene Approaches

Prior knowledge of R genes from other plant species can be used as a basis for identifying homologous R gene candidates in legumes. This strategy entails the exploration of legume genomes or transcriptomes in search of genes that exhibit sequence similarities to established R genes. Moreover, the process may involve cloning and conducting functional analyses of R gene orthologs from closely related plant species to gain insights into potential R gene candidates within legumes (Kwon & Goate, 2000).

5.4 Resistance Gene Enrichment Sequencing (RenSeq)

RenSeq is a targeted sequencing approach specifically designed to capture and sequence R genes. This technique involves the enrichment of genomic DNA using baits designed based on conserved R gene domains. RGESeq has been successfully applied in several legume species, including common bean and soybean, to identify R gene candidates (Fernandez-Gutierrez & Gutierrez-Gonzalez, 2021; Jupe et al., 2013).

5.5 Genome-Wide Association Studies (GWAS)

GWAS leverages natural genetic variation within legume populations to identify genomic regions associated with disease resistance. By correlating the presence or absence of resistance traits with genetic markers, R gene candidates can be identified. GWAS studies have been conducted in legume crops like common bean and soybean, leading to the discovery of R gene candidates (Chen et al., 2021).

5.6 Functional Validation

Once potential R gene candidates are identified, functional validation is crucial to confirm their role in disease resistance. This process can entail the heterologous expression of these candidates in model plants like *Arabidopsis thaliana*, followed by pathogen inoculation assays to evaluate their resistance phenotypes. Additionally, genome editing techniques such as CRISPR-Cas9 can be utilized to validate the functionality of these candidate R genes through the creation of knockout or knockdown mutants.

By employing these strategies, researchers can effectively pinpoint and elucidate R gene candidates in legume species. This not only enhances our comprehension of disease resistance mechanisms but also paves the way for the development of improved legume varieties endowed with heightened resistance to pathogens.

5.7 Bioinformatics Tools for R Gene Prediction

Several bioinformatics tools and databases are available for predicting and analyzing R genes. These tools utilize various algorithms and sequence analysis methods to identify potential R gene candidates in genomic or transcriptomic data. Some commonly used bioinformatics tools for R gene prediction and more information are presented in Table 2.

5.7.1 NLR-Annotator

NLR-Annotator is a web-based tool designed for the annotation and prediction of nucleotide-binding leucine-rich repeat (NLR) genes, including NBS-LRR genes. It utilizes Hidden Markov Models (HMMs) to identify conserved domains within protein sequences and predict NLR genes.

Table 2 Common software identify for RGA domain and motif

Software	Description	URL
HMMER	HMMER: An application for identifying sequence homologs in protein and DNA databases	hmmer.janelia.org
MEME	The MEME suite: A toolbox for sequence motif discovery and analysis	Meme-suite.org
mCUDA-MEME	Ultrafast scalable motif discovery program utilizing GPU acceleration, employing a Hybrid approach of CUDA, MPI, and OpenMPI parallel programming models, based on MEME algorithm	bit.ly/18X8LmA
BLAST+	Advanced bioinformatics data mining toolkit with enhanced CPU speed and RAM efficiency for long queries in the latest version	blast.ncbi.nlm.nih.gov
pfam_scan.pl	Perl script for PFAM database search using "hmmscan" from the HMMER toolkit package to query known HMM models	bit.ly/1M411KRu
InterproScan	Combining protein signature recognition methods from InterPro member databases into a unified tool with InterPro and GO annotation lookup	
Phobius	HMM-based tool for predicting transmembrane topology and signal peptides in proteins with embedded pre-training HMM model	phobius.sbc.su..se
GSDS	Gene structure display server program	gsds.gao-lab.org
nCOILS	A software for detecting CC domains via sequence comparison and scoring against a coiled-coils database using the MTK or MTIDK calculation matrix to determine the probability of coiled-coil conformation	embnet.vital-it.ch

Modified from Sekhwal et al. (2015)

5.7.2 NB-ARC Finder

NB-ARC Finder is a software tool that identifies and annotates NBS-LRR encoding genes in genomic sequences. It employs a combination of HMMs and sequence similarity searches to detect NBS domains and classify the genes into different NBS-LRR subgroups.

5.7.3 Plant Resistance Gene Database (PRGdb)

PRGdb is a comprehensive online resource that provides curated information and tools for R gene prediction and analysis. It offers various prediction tools, including NLR-parser for NLR gene identification and classification, as well as additional resources for functional annotation and comparative genomics.

5.7.4 RGAugury

RGAugury is a web-based tool for predicting resistance gene analogs (RGAs) in plant genomes. It employs profile Hidden Markov Models (HMMs) trained on known R genes to identify potential RGAs, including NBS-LRR genes.

5.7.5 HMMER

HMMER is a widely used software suite for sequence analysis based on profile Hidden Markov Models (HMMs). It can be employed for the identification of conserved domains, including NBS and LRR domains, in protein sequences.

5.7.6 BLAST and PSI-BLAST

Basic Local Alignment Search Tool (BLAST) and Position-Specific Iterated BLAST (PSI-BLAST) are sequence similarity search tools that can be used to identify homologous R genes based on known R gene sequences or conserved domains.

5.7.7 InterProScan

InterProScan is a tool that scans protein sequences against various protein domain databases, including Pfam and SMART. It can be used to identify conserved domains, such as NBS and LRR domains, in protein sequences.

These tools provide valuable resources for the identification and annotation of R genes, aiding researchers in understanding the genetic basis of disease resistance in plants. It is recommended to consider using a combination of these tools to maximize the accuracy and coverage of R gene prediction.

5.8 NBS-LRR Domain Architecture and Motif Analysis

In plants, NBS-LRR proteins are commonly categorized into two primary groups based on their N-terminus: TIR-NBS-LRR (TNL) and non-TIR-NBS-LRR (nTNL). However, this classification can sometimes lack precision. nTNL proteins encompass NBS-LRR proteins that feature diverse N-terminal domains instead of the TIR domain. In addition to the NBS-LRR region, NBS-LRR genes typically incorporate supplementary domains like CC or RPW8 at the N-terminus, along with a variable number of LRR domains at the carboxy-terminus (Shao et al., 2014; Zhang et al., 2016a, b).

The domain architecture of NBS-LRR proteins and motif analysis plays a crucial role in understanding their structure and function. NBS-LRR proteins typically consist of several conserved domains, each contributing to their overall architecture.

5.8.1 Domain Architecture

N-Terminal Domains: The categorization of NBS-LRR proteins into TIR-NBS-LRR (TNL) and non-TIR-NBS-LRR (nTNL) hinges on the presence or absence of a Toll/interleukin-1 receptor (TIR) domain at the protein's N-terminus.

TIR Domain (TNL): TIR domains are involved in protein–protein interactions and are found in TNL proteins. The TIR domain (toll/interleukin-1 receptor domain) is a conserved domain found in TIR-NBS-LRR (TNL) proteins. TNL proteins represent a subset of NBS-LRR proteins characterized by the presence of the TIR domain at their N-terminus. These proteins play a pivotal role in bolstering plant defense mechanisms against pathogens through the recognition of Avr proteins (Balint-Kurti, 2019). This recognition event initiates robust immune responses against pathogenic attacks, notably triggering a hypersensitive response (HR) as a defense mechanism. It's worth emphasizing that the involvement of NRIP1, an additional host protein, is essential for facilitating this interaction (Nandety et al., 2013). Recent studies have indicated that the TIR domain also plays a role in pathogen recognition and signaling cascades (Williams et al., 2014).

In legumes, including plants such as soybean, pea, and lentil, TNL proteins with the TIR domain have been identified. The TIR domain serves as a pivotal mediator of protein–protein interactions and signaling cascades, culminating in the activation of downstream genes associated with defense responses. This domain, spanning approximately 200 amino acids, encompasses three remarkably conserved motifs in Arabidopsis (Verma & Sowdhamini, 2022). There are two well-conserved TIR-1 and TIR-2 domains present at N terminus site of TNL protein and both the domains overlapped each other. Immediately after the translation, initiation codon presence of either alanine or polyserine sequences (SSSSSYD) was found in TIR (Purwar et al., 2023). Several conserved motifs were confirmed that had been identified previously in the TIR domain of plant NBS-LRRs and related proteins (motifs TIR-1, TIR-2, TIR-3, and TIR-4) in Arabidopsis (Meyers et al., 1999, 2002), in Common bean (Wu et al., 2017), and (TIR-1 and TIR-2) in *Vigna radiata* (Purwar et al., 2023). The order of these motifs was well conserved (Table 3), previous findings had noted duplications of the TIR motifs in some Arabidopsis proteins (Meyers et al., 1999). In soyabean Xun et al., 2019 report the over-expression of GmKR3, a soybean TIR–NBS–LRR type R gene on virus resistance in soybean. When over-expressed in soybean, GmKR3 enhanced the plant's resistance to several strains of SMV, the closely related potyviruses bean common mosaic virus (BCMV) and watermelon mosaic virus (WMV), and the secovirus bean pod mottle virus (BPMV) (Xun et al., 2019). Legume TNL proteins with the TIR domain are known to be involved in various defense mechanisms, including resistance against bacterial, fungal, and viral pathogens. They recognize pathogen-derived molecules, known as effectors, through the TIR domain and trigger immune responses to prevent pathogen colonization and infection. Research on legume TNL proteins and their TIR domain-mediated signaling pathways provides insights into the mechanisms of plant defense and can contribute to the development of improved disease-resistant legume varieties through genetic engineering or breeding strategies.

5.8.2 nTNL Domains

nTNL proteins possess various domains other than TIR at their N-terminus, such as CC (coiled-coil) or RPW8 domains. NON TIR-NBS-LRR class was further classified into four subgroups, i.e., subgroup 1, which showed the presence of coiled-coil domain thus referred as CC-NBS-LRR group, subgroup 2, which include RPW8 conserved domain in their sequences thus named RPW8 group, subgroup 3, which showed the presence of both coiled-coil and RPW8 domain and thus named as CC+RPW8 group and subgroup 4, denoted as NBS-LRR group, which lack any of these conserved domains (Meyers et al., 2002).

5.8.3 Nucleotide-Binding Site (NBS) Domain

The NBS domain is a highly conserved region responsible for nucleotide binding and hydrolysis, which is crucial for the activation of the immune response. The NBS domain hydrolyzes ATP or GTP to obtain energy and thus it confers switching of R protein from active to inactive state (Dubey and Singh). The NBS domain is involved in signaling and includes several highly conserved and strictly ordered motifs such as P-loop, kinase-2, and GLPL motifs. Overexpression of CC or TIR domains can reduce hypersensitive response in plants. Song et al., (2019) found three sequences that simultaneously contained an NBS domain and WRKY domain in *A. hypogaea* cv. Tifrunner. In other legumes, NBS–WRKY fusion proteins have only been identified in *G. max*, *A. duranensis*, *A. ipaensis*, and *Vigna radiata* (Mota et al., 2018; Purwar et al., 2023). The bacterial effectors AvrRps4 or PopP2 can trigger WRKY transcription factors that are involved in active *NBS–LRR* gene responses to pathogens. NBS–WRKY fusion proteins can play a crucial role in response to biotic stress in *A. hypogaea* cv. Tifrunner (Song et al., 2019).

5.8.4 Leucine-Rich Repeat (LRR) Domain

LRR domains are found at the C-terminus of NBS-LRR proteins and are involved in pathogen recognition. They form a horseshoe-shaped structure with repetitive units of leucine-rich motifs. The LRR domain exhibits a higher degree of genetic diversity, possibly due to more relaxed or positive selection, primarily because this domain engages with pathogenic effectors (Song et al., 2019). This diversity sets it apart from the NBS, TIR, and CC domains. LRR domains play pivotal roles in mediating protein–ligand and protein–protein interactions, particularly in the context of plant immune responses. In the case of *A. hypogaea* cv. Tifrunner, four distinct types of LRR domains, namely LRR1, LRR3, LRR4, and LRR8, have been identified. Among these sequences, LRR8-type domains are the most prevalent, followed by LRR3, LRR4, and LRR1. Conversely, both *A. duranensis* and *A. ipaensis* display five types of LRR domains: LRR1, LRR3, LRR4, LRR5, and LRR8 (Mota et al., 2018). In *A. duranensis*, LRR8-type domains are the most abundant, followed by LRR4, LRR3, and LRR5. Meanwhile, in *A. ipaensis*, LRR8-type domains are also the most numerous, followed by LRR4, LRR3, LRR5, and LRR1. Notably, the LRR5 domain is exclusively observed in CNL proteins within *A. duranensis* and *A. ipaensis* (Mota et al., 2018). It's worth noting that Song et al. (2019) proposed that *A. hypogaea* cv. Tifrunner may have lost the LRR5 domain, potentially as a result of genetic exchange or gene loss following tetraploidization or whole genome duplication (WGD). This genetic alteration could have contributed to the development of disease resistance in *A. hypogaea* cv. Tifrunner.

5.8.5 Motif Analysis

The ATP-Binding Motif

It is a crucial element within the NBS (nucleotide-binding site) domain of NBS-LRR proteins. This motif is responsible for binding and hydrolyzing ATP (adenosine triphosphate), which is essential for the activation of the immune response. The ATP-binding motif is highly conserved among NBS-LRR proteins and plays a fundamental role in their function. The ATP-binding motif within the NBS domain typically consists of two conserved sequence motifs.

P-loop (Walker A Motif)

The P-loop motif, also known as the Walker A motif, is named after the late Jack L. Walker, a pioneer in ATP-binding research. It is characterized by the presence of the sequence motif GxxxxGK(T/S), where "x" denotes any amino acid. The P-loop motif is responsible for binding the phosphate groups of ATP.

Kinase-2 Motif (Walker B Motif)

The kinase-2 motif, also referred to as the Walker B motif, is another conserved sequence motif found in the ATP-binding region. It is characterized by the sequence motif DxxG, where "x" denotes any amino acid. The kinase-2 motif is involved in coordinating the magnesium ion (Mg^{2+}) required for ATP hydrolysis.

These ATP-binding motifs facilitate the binding and hydrolysis of ATP, providing the energy necessary for conformational changes and downstream signaling events in NBS-LRR proteins. The ATPase activity associated with the ATP-binding motif is crucial for the activation of immune responses and the regulation of plant defense against pathogens.

LRR Motifs

The LRR domain contains characteristic motifs, including the LxxLxLxxNxL motif, which is critical for pathogen recognition and ligand binding.

Coiled-Coil Motifs

In nTNL proteins, coiled-coil motifs within the CC domain facilitate protein–protein interactions and signal transduction. The coiled-coil (CC) domain plays a role similar to the TIR domain by mediating interactions within the signaling cascade. The EDVID motif is a conserved motif within the coiled-coil domain of CC-NBS-LRR proteins, except for certain exceptions like RPS2, RPS5, and Dm3 proteins. Mutations in this motif disrupt intramolecular interactions between the NBS and LRR domains, leading to decreased resistance response against pathogen attack (Zhang et al., 2016a, b).

Signature Motifs

NBS-LRR proteins may have specific signature motifs that distinguish them within different subclasses or clades, providing insights into their evolutionary relationships like RPW8, WRKy, etc. In addition to CNL genes, a subset of nTNL genes has been identified to possess a unique N-terminal domain called the RPW8 (resistance to powdery mildew 8) domain. These nTNL genes, known as RPW8-NBS-LRR or RNL genes, constitute a distinct class of NBS-LRR genes, characterized by the presence of the RPW8 domain. RPW8 domain also reported in legume plat like pigeon pea and *Vigna radiata* (Table 3).

The analysis of domain architecture and motif patterns in NBS-LRR proteins helps in understanding their functional diversity, interactions with other proteins, and specific roles in plant defense against pathogens.

Table 3 Motif analysis in different legumes along with *Vigna radiata*

Domain	Subgroup	Motif	Sequence
TIR	TNL	TIR1	AIEGSRISIVVFSKNYASSTWCLDELAKI
		TIR2	YDVFJSFRGEDTRNSFTDHLYSALLRKGIFVFKDDESLQKGESISPELLQ
		TIR3	KTSGQIVLPVFYDVDPSEVRHQTGDYEEAFVEHEEKF
CC		CC	ELKSYIEKLEHNKERLQHQVD
NBS	CNL	P-loop	MIGVYGVGGVGKTTLVKEVAKKA
		Kinase-2	EKKLFEVVVMANITRNPBIKK
		RNBS-A	IQGQIAGMLGMKLEEESEIARADRIRKRL
		RNBS-B	NTLIILDDLWEGLDFNKLGIPLNDDASGS
		RNBS-C	ITTRSKEVCRSMQCQSIIEVNLLSDEEAWTLF
		GLPL	IVKKCKGLPLAIVTIGSLLK
		RNBS-D	DKTIANFELALSRLENSKPLD
		MHD-linker	MHDIVRD
NBS	TNL	P-loop	VKVJGIWGMGGIGKTTJAKALYBK
		RNBS-A	SHRFEASCFJEBVREVYSKH
		Kinase-2	GLVSLQEQJLSDLLG
		RNBS-B	GSSJIKNRLRRKKVLIVL
		RNBS-C	DDVDELEQLEALAGNREWFGAGSRIIITT
		RNBS-D	RDKHJLKVYEVDVVYEVKELN
		GLPL	ELFSRHAFKEEEPKEGFEELSKZVVKYAGGLPLALEVLGSYLFGRSVEEW
		MHD-Linker	MHELI
LRR	CNL		PRLEVLHIDNKNESFQIPDEFFKSMVRLRVLVLTG
	TNL		FTAENLVELBLPHSKIKKLWEGVKNLPN
CNR		RxN	
		RPW8	

(continued)

Table 3 (continued)

Domain	Subgroup	Motif	Sequence
	NBS	P-loop	VJSITGMGGLGKTTLAKKVYNDEQVKKHF
		Kinase-2	NIFFVTVSKTPKLKTIVERLFEHCGYEVPKFQSDEDAVN
		RNBS-A	QKKRYLLVLDDVWN
		RNBS-B	LVVQIPGSKILVTTR
		RNBS-C	GKPYELKPLSEDDAWELFCHKAFLGDSSP
		GLPL	ELEKVGKEIVKKCGGLPLAIKVIGGLLAT
		RNBS-D	EWRMVCQNLGSEJEGNNKLE
		MHD-linker	–

5.9 *Comparative Genomic Analysis of NBS-LRR Gene in Legumes*

Comparative analysis of the NBS-LRR gene family in legumes provides valuable insights into the evolutionary dynamics, functional diversity, and genomic organization of these disease resistance genes in legume species.

5.9.1 Gene Family Size and Distribution

Comparative analysis allows the examination of the NBS-LRR gene family size and distribution across different legume species. By comparing the genomes of legume crops like common bean, soybean, pea, and chickpea, researchers can identify variations in the number of NBS-LRR genes among species and their distribution patterns, such as clustering or dispersion (Table 4).

In many surveyed genomes, the majority of NBS-LRR genes are found in a clustered organization. Most clusters are homogeneous, with the only duplicated member occupying the same phylogenetic lineage; whereas heterogeneous clusters are composed of members from distantly related clades (Meyer et al., 2010). With more genomic data available in certain angiosperm families, NBS-LRR genes should be further investigated among phylogenetically distant species to fill the gaps in the understanding of their long-term evolutionary patterns. The legume family contains many economically important crop species, such as clover, soybean, peanut, and common bean. Although these legumes are constantly threatened by various pathogens, only a few functional legume R genes have been characterized, and all of them belong to the NBS-LRR class (Meyer et al., 2010).

Table 4 Genome size and distribution of NBS-LRR gene

S. No	Legume crop	Genome size	Reference	Full length NBS-LRR	References
1.	Common bean	587-Mb	Schmutz et al., (2014)	324	Wu et al. (2017)
2.	Soybean, pea	1.1–1.15 Gb	Li et al. (2021)	319	Kang et al. (2012)
3.	Chickpea	740 Mbp	Arumuganathan and Earle (1991)	121	Sagi et al. (2017)
4.	Mung bean	479.35 Mb	Liu et al. (2022)	87	Purwar et al. (2023)
5.	Pea nut	2.7 Gb	Bertioli et al. (2019)	713	Song et al. (2019)
6.	*Arachis duransis*	1.05 Gb	Bertioli et al. (2016)	278	Mota et al. (2018)
7.	*A. ipaensis*	1.56 Gb	Bertioli et al. (2016)	303	Mota et al. (2018)

5.9.2 Comparative Gene Structure Analysis of NBS-LRR

Recent advancements in sequencing technologies, such as Solexa, Roche, and Illumina, have revolutionized the field by enabling high-throughput sequencing, which has opened new avenues for studying plant–microbe interactions. These technologies have facilitated genome-wide analyses of NBS-LRR genes, which are characterized by the NB-ARC domain. Several studies have identified hundreds of NBS-LRR genes in various plant species, including *Arabidopsis thaliana*, and many legumes plant like (McHale et al., 2006). The first NBS-LRR gene sequence identified in plants was in *Arabidopsis thaliana*, which possesses a total of 165 NBS-LRR genes, consisting of 52 CNLs, 106 TNLs, and 7 RNLs, as reported by (Zhang et al., 2016a, b). A total of 2451 NBS-encoding genes were identified from the eight major legume genomes like 571 from *M. truncatula*, 289 from pigeon pea, 337 from common bean, 465 from soybean, 121 from chickpea, 278 from *Arachis duransis*, 303 from *Arachis ipaensis*, 303 and 87 in moogbean (Purwar et al., 2023). The number of NBS-LRR genes varies among plant species, with legume family plants having more genes than Arabidopsis (Table 5). TNLs (non-TIR-NBS-LRR) are believed to have evolved after the divergence of monocots and dicots (Zhang et al., 2016a). The rapidly evolving NBS-LRR gene family exhibits differences in composition even at the genus level (Guo et al., 2022). Novel NBS-LRR genes have been identified using the R gene enrichment and sequencing method (RNASeq), and their overexpression has revealed their involvement in basal defense responses (Mota et al., 2018). By conducting a genome-wide analysis of NBS-LRR genes, a better understanding of their role in plant disease resistance can be achieved. These genes can be classified into TIR-NBS-LRR (TNL) and non-TIR-NBS-LRR (nTNL) categories based on their N-terminus. The nTNL proteins comprise NBS-LRR proteins with N-terminus domains other than the TIR domain, such as CC or RPW8. The functionality of NBS-LRR proteins depends on various domains, including CC, RPW8, and LRR domains.

5.10 Genomic Distribution of NBS-Encoding Genes in Legume Genomes: A Comprehensive Analysis

The genomic distribution of NBS-encoding genes in legume genomes has been the subject of comprehensive analysis. These genes, which encode nucleotide-binding site (NBS) domains associated with resistance (R) genes, play a crucial role in plant defense against pathogens. Through genome-wide studies, researchers have examined the distribution patterns of NBS-encoding genes across different legume species. These analyses have revealed variations in the number, arrangement, and clustering of NBS-encoding genes, indicating their dynamic evolution and adaptation in response to diverse pathogenic threats. Understanding the genomic distribution of NBS-encoding genes in legume genomes provides valuable insights into the

Table 5 Number of identified NBS-encoding genes in the different legume genomes

Predicted protein	Letter code	Chickpea	*A. duranensis*	*A. ipaensis*	*Vigna radiata*	*M. truncatula*	Pigeon Pea	Common bean	Soybean
TNL- subclass									
TIR-NBS-LRR	TNL	39	83	90	4	121	78	76	124
TIR-NBS	TN	3	19	16	6	49	6	13	24
NBS-LRR	NL_{TR}	–	–	–	1	44	15	12	17
NBS	N_{TR}	–	–	–	6	45	11	0	12
Other						16	2	2	1
nTNL subclass									
CNL and homologs									
CC-NBS-LRR	CNL	34	30	38	4	94	63	128	109
CC-NBS	CN	3	7	12	25	16	7	9	8
NBS-LRR	NL_{CC}	21	123	135	1	95	53	84	120
NBS	N_{cc}	16	16	12	17	63	43	2	30
Other					8	11	1	1	2
RNL and homologs									
RPW8-NBS-LRR	RNL	4	–	–	2	8	4	5	9
RPW8-NBS	RN	1			4				1
NBS_{RPW8}-LRR	$N_{RPW8}L$					6	5	4	8
NBS_{RPW8}	N_{RPW8}					3	1	1	0
Other					9	0	0	0	0
		121	278	303	87	571	289	337	465
Reference					Purwar et al. (2023)	Shao et al. (2014)	Shao et al. (2014)	Shao et al. (2014)	Shao et al. (2014)

Table 6 Distribution of NBS-encoding genes in legume genomes

Genes and Loci	Chickpea	*Vigna radiata*	*M. truncatula*	Pigeon Pea	Common Bean	Soybean
No. of chromosomes	8	11	8	11	11	20
No. of singleton loci (No. of NBS genes)	63(63)	23(21)	64 (64)	40 (40)	49 (49)	116 (116)
No. of clustered loci (No. of NBS genes)	23(58)	7(45)	72 (479)	29 (98)	22(129)	71 (343)
Clustered NBS genes/ singleton NBS genes	0.92	2.14	7.48	2.45	5.13	2.96
Scaffolds	28	21	–	–	3	5

genetic basis of disease resistance mechanisms, facilitating the development of improved legume varieties with enhanced resistance to pathogens and contributing to sustainable agriculture practices (Table 6).

Genes within the same cluster generally indicate similarity in sequences, increasing the likelihood of new variant development. This high selection pressure, combined with cluster formation and recombination, aids in the development of new resistance specificities against pathogens. Such clusters have been observed in various host plants.

For instance, in soybean, a significant number of NBS-LRR genes were found in clusters. Out of a total of 343 genes, 71 were present in clusters, while 116 genes existed as singletons. One cluster on Chromosome 16 contained 40 NBS-LRR genes, and another cluster on Chromosome 18 consisted of 32 NBS-LRR genes (Kang et al., 2012).

Similarly, in chickpea, out of 121 NBS-LRR genes, 58 genes were organized into 23 clusters, each containing two to four genes, while 68 genes were present as singletons. Chromosome 5 had the highest number of NBS-LRR genes (29), accounting for 27% of the mapped genes, while chromosome 8 had the lowest number (5) of NBS-LRR genes. At least one CNL gene was found on each chickpea chromosome, while the TNL class was limited to seven chromosomes, absent on chromosome 4 (Sagi et al., 2017). In *M. truncatula*, 72 gene clusters comprising 449 genes were reported, while 64 genes were singletons (Zhu et al., 2003). In common bean, 129 NBS-LRR genes belonging to 22 clusters were identified, with an average of 5.86 NBS-LRR genes per cluster. Chromosomes 10 and 11 had the highest number of clusters (4), and the number of genes per cluster ranged from 2 to 34. Cluster 21 had the highest number of members with 34 genes. Additionally, 49 NBS-LRR singletons were distributed across all chromosomes. The cluster sizes ranged from 9264 to 1,805,165 base pairs (Wu et al., 2017).

Vigna radiata had 45 genes located in 7 clusters, while the remaining 21 genes were present as singletons. These genes were distributed across all 11 chromosomes except chromosome 3 (Purwar et al., 2023). In pigeon pea, 98 genes were found in 29 clusters, while 49 genes were distributed as singletons on different chromosomes. In Arachis species, a total of 85 and 93 clusters were detected in *A.*

duranensis and *A. ipaënsis*, respectively. Chromosomes 2 contained the most clusters, while chromosomes 1 of A. duranensis A1 and chromosome 6 of *A. ipaënsis* had the fewest clusters (Mota et al., 2018). The number of clusters in *A. ipaënsis* was greater than that in *A. duranensis*, possibly due to more tandem duplication events occurring in *A. ipaënsis*. Approximately 57.14% and 84.62% of paralogous genes in *A. duranensis* and *A. ipaënsis*, respectively, were located within the clusters (Mota et al., 2018).

Comparing cluster genes to singleton genes, the ratio was greater than 1 in all cases except in chickpea. This suggests that the majority of genes are present in clusters, indicating that more tandem duplication events occur. The formation of NBS-LRR gene clusters appears to be a strategy for plants to rapidly adapt to a changing spectrum of pathogens (Mota et al., 2018).

5.11 Comparing the Phylogenetic Analysis

Comparative genomics enables the use of phylogenetic trees to analyze the evolutionary relationships among NBS-LRR genes in legumes. This approach helps categorize these genes into different clades or subfamilies based on their sequence similarities and evolutionary history, providing insights into the diversification and expansion of NBS-LRR gene families in legumes.

Phylogenetic analysis reveals that NBS-LRR genes can be classified into two clades in phylogenetic trees: TNL and CNL groups. In various legumes such as chickpea, *Vigna radiata*, AdNBS, AiNBS, *M. truncatula*, common bean, and soybean, phylogenetic trees consistently exhibit these two groups using ML and NJ methods. However, it is possible to find CNL sequences nested within the TNL group and vice versa. Similar findings were observed in *M. truncatula* and *Vigna radiata* NBS-LRR sequences, where eight TNL genes in *M. truncatula* were nested within the CNL group, and two CNL sequences were found to cluster with TNL proteins (Song & Nan, 2014). Based on phylogenetic tree analysis, *VrNBS-LRR* genes can be classified into three major clades: CNL, TNL, and RNL. The CNL clade contains 40 genes, including 36 from the CNL clade and 4 from the RNL clade, which possess RPW8 domains at the N-terminal ends. The TNL clade is clearly distinct from the CNL and RNL clades, but a few exceptions exist, such as three CNL genes (*VrNBS_Ncc-31*, *VrNBA_Ncc32*, and *VrNBS_P-52*), which are located within the TNL clade. The RNL clade consists of 27 genes, including 11 strictly from the RNL group (RxN RNL, NW, DN, and NWC) and 16 belonging to the CNL group, which carry RxN domains overlapping with CC domains at the N-terminal site. In-depth studies of the RNL clade have revealed that it contains the majority of genes with an RxN domain at the N-terminal, whether in RNL or CNL (Purwar et al., 2023).

However, in AdNBS, one CNL sequence is nested within the TNL group, and three TNL sequences cluster together with CNL proteins (Song et al., 2017). In pigeon pea and common bean, clear-cut distinction between CNL and TNL members of genes is observed (Wu et al., 2017). It is hypothesized that recombination events occur in the NBS domain. Recombination between some NBS domains from CNL and TNL proteins has been reported (Innes et al., 2008). In chickpea, an NJ phylogenetic tree of 117 complete NBS-LRR proteins reveals two distinct clades, separating the TNL class from the non-TNL class. The TNL clade consists of three TNL sub-clades, while the non-TNL clades are separated into CNL and NL sister clades. Moreover, the CNL clade is clearly divided into two sub-clades, consisting of the CNL clade and RPW8 clade. The phylogenetic clustering of genes with similar sequences from different chromosomes and the same chromosomes is also observed (Lozano et al., 2015. Identification and distribution of the NBS-LRR gene family in the Cassava genome. *BMC genomics*, *16*(1), 1–14.). The alignment of phylogenetic tree with gene structure revealed that exon–intron structure tends to remain the same within the genes present in the same clade reflecting strong conservation of the gene structure during evolution.

By phylogenetic analysis by ML and NJ, also identified orthologous and paralogous gene, for examples in Arachis species, identified 51 orthologous gene pairs, four paralogous AdNBS gene pairs, and six paralogous gene paired in AiNBS (Mota et al., 2018). Most of these 51 orthologous gene pairs were distributed in a similar locus on the corresponding chromosomes, except AdNBS2 (chromosome 2)—AiNBS274 (chromosome 3). Additionally, one gene pair contained both CNL (AdNBS104) and TNL (AiNBS144) genes, indicating that recombination was present between the CC and TIR domains. All of the paralogous gene pairs were located on one chromosome, indicating the tandem duplication is the main mechanism in forming NBS–LRR paralogs (Ratnaparkhe et al., 2011). Generally, tandem duplication produces novel resistant functions of NBS–LRR genes. In soybean and *Medicago*, tandem duplication played a primary role in NBS–LRR gene expansion.

5.12 *Pathogens Recognized by NBS-LRR Gene Products with Pathogen-Related Quantitative Trait Loci (QTLs) in Leguminous Plant*

The NBS-LRR (nucleotide-binding site–leucine-rich repeat) gene products in leguminous plants play a crucial role in recognizing and responding to various pathogens. These proteins specifically detect and interact with pathogen-derived molecules, known as effectors, to trigger defense responses (Table 7). The co-localization of NBS-LRR genes with pathogen-related quantitative trait loci (QTLs) has been observed in various legume species. This co-localization suggests a potential involvement of NBS-LRR genes in conferring resistance to specific pathogens in these legumes. The presence of NBS-LRR genes within or in close proximity to pathogen-related QTLs indicates their potential role in recognizing and responding to pathogen attacks, leading to enhanced resistance traits in legume crops. NBS-LRR (nucleotide-binding site-leucine-rich repeat) proteins play a crucial role in

Table 7 Selected examples of pathogens recognized by NBS-LRR gene products in leguminous plants

Gene or Protein	Plant species	Pathogen	Predicted domain	References
Rpg1-b	*Glycine max*	*Pseudomonas syringae* pv. glycinea	CC-NBS-LRR	Ashfield et al. (2014)
EIN3 and CaETR1	Chickpea	*Ascochyta blight*	NBS-LRR	Sagi et al. (2017)
AdNBS	*A. duranensis*	*Aspergillus flavus*	NBS–LRR	Song et al. (2017)
AiNBS	*A. ipaensis*	*Aspergillus flavus*	NBS–LRR	Song et al. (2017)
Arahy	*A. hypogaea* cv. Tifrunner	*Botic stress* (late leaf spot, bacterial wilt, tomato spotted wilt virus)	NBS–WRKY	Song et al. (2017)
Rj2 and *Rfg1*	Soybean	*B. japonicum* and *S. fredii*	NBS-LRR	Yang et al. (2010a, b)
ANT	Common bean	*Anthracnose*	NBS-LRR	Wu et al. (2017)
CBB	Common bean	*Common bacterial Blight*	NBS-LRR	Wu et al. (2017)

legume breeding. These proteins are involved in recognizing and responding to pathogen attacks by activating defense mechanisms. In legumes, NBS-LRR genes have been extensively studied and utilized to enhance disease resistance traits through breeding programs. Further studies and investigations are necessary to determine the specific mechanisms and interactions between NBS-LRR genes and the associated QTLs in legumes for a better understanding of plant–pathogen interactions and the development of resistant varieties. Based on the physical position of the SSR markers on the chickpea chromosomes, 16 QTLs for ascochyta blight resistance reported were physically mapped on chromosomes 2, 3, 4, 5, 6, and 8. Nine QTLs were co-localized with the NBS-LRR genes. Sagi et al., (2018) identified 30 NBS-LRR genes co-located within the flanking markers corresponding to these nine AB-QTLs. Among the co-localized NBS-LRR genes, 24 were complete genes, i.e., they carry all essential domains for their independent functions. Among these 24 genes, 13 belong to the TNL class, eight belong to the CNL class, and three belong to the NL class. Six co-localized genes were incomplete that belong to the RN class (1), CN class (2), and NBS class (3). The majority of the genes (17) co-localized with the AB-QTLs were present in clusters of 2 to 3 genes. In common bean (Wu et al., 2017) developed NBS-SSR markers near NBS genes, offering co-segregation markers for resistant genotypes with minimal recombination. Using these markers, their group identified ANT and CBB resistance genes, with nine markers associated with ANT resistance and seven markers with CBB resistance. In their study on cultivated peanut (*A. duranensis* and *A. ipaensis*), Song et al. (2019) discovered three QTLs associated with late leaf spot, involving 27 NBS-LRRs, four QTLs linked to tomato spotted wilt virus, involving six NBS-LRRs, and one QTL associated with bacterial wilt, involving eight NBS-LRRs. By identifying and incorporating specific NBS-LRR genes or QTLs (quantitative trait loci) into legume breeding, breeders can develop cultivars with improved resistance against various pathogens, such as fungi, bacteria, and viruses. NBS-LRR genes provide a genetic basis for durable resistance, as they often exhibit high levels of specificity to particular pathogens. Through marker-assisted selection (MAS) or genomic selection techniques, breeders can efficiently screen and select legume plants carrying desirable NBS-LRR genes or alleles associated with disease resistance. This enables the development of resistant cultivars that can help minimize yield losses caused by diseases, reduce the reliance on chemical pesticides, and promote sustainable agriculture practices. Overall, NBS-LRR genes and QTLs play a pivotal role in legume breeding by providing a valuable genetic resource for enhancing disease resistance and developing improved cultivars with enhanced pathogen defense mechanisms.

In conclusion, comparative analysis of the NBS-LRR gene family in legumes offers valuable insights into their evolutionary dynamics, functional diversity, and genomic organization. Such studies contribute to our understanding of plant immune systems, aid in the development of disease-resistant legume varieties, and provide a foundation for crop improvement efforts aimed at enhancing disease resistance in legumes.

6 Future Prospects of Utilizing R Genes for Developing Disease-Resistant Legume Varieties

The utilization of R genes for developing disease-resistant legume varieties holds great prospects for improving crop productivity and sustainability in legume-based agriculture. By incorporating R genes into breeding programs, breeders can develop legume varieties that are resistant to specific pathogens or have broad-spectrum resistance against multiple pathogens. This can reduce yield losses and the dependence on chemical pesticides, promoting sustainable agriculture. One of the key advantages of utilizing R genes is the potential for developing legume varieties with durable resistance. Durable resistance is long-lasting and remains effective against evolving pathogen populations. By harnessing the genetic diversity of R genes and deploying multiple R genes with different modes of action, breeders can develop legume varieties that exhibit durable resistance, reducing the risk of pathogen adaptation and breakdown of resistance over time.

The deployment of disease-resistant legume varieties developed using R genes can significantly reduce the need for chemical pesticides. This has important environmental benefits, including decreased chemical pollution, preservation of beneficial insects, and reduced health risks for farmers and consumers. Additionally, reduced pesticide use can contribute to cost savings for farmers, making legume production more economically viable.

By protecting legume crops from diseases, R genes can improve yield and overall crop productivity. Disease outbreaks can lead to significant yield losses, and the deployment of disease-resistant varieties ensures stable and higher crop yields. This not only benefits farmers by increasing their profitability but also contributes to global food security by ensuring an adequate supply of legume-based food products. Incorporating R genes into breeding programs provides an opportunity to tap into the genetic diversity present in wild relatives and landraces of legumes. This genetic diversity can be harnessed to enhance the adaptation of cultivated legume varieties to different environmental conditions and increase their resilience to biotic stresses. By utilizing R genes from diverse sources, breeders can improve the adaptability and sustainability of legume crops.

The utilization of disease-resistant legume varieties can lead to reduced input costs for farmers. By minimizing the need for costly disease management practices, such as frequent pesticide applications or fungicide treatments, farmers can save on input costs, making legume production economically viable. This can have a positive impact on the livelihoods of farmers, particularly in resource-constrained agricultural systems. Furthermore, disease-resistant legume varieties align with the growing consumer demand for sustainably produced and pesticide-free food. Consumers are increasingly conscious of the environmental and health impacts of conventional farming practices. Disease-resistant legume varieties not only offer improved quality and safety but also address consumer preferences for sustainable and environmentally friendly food options.

In conclusion, the utilization of R genes for developing disease-resistant legume varieties offers multiple benefits, including enhanced disease resistance, durable resistance, reduced pesticide use, increased yield and productivity, genetic diversity, reduced input costs, and improved marketability. Incorporating R genes into breeding programs can contribute to sustainable agriculture, addressing the challenges of plant diseases and promoting the production of healthy and environmentally friendly legume crops.

7 Terminology

Autotrophs: Autotrophs are organisms that can produce their own food, using materials from inorganic sources.

Chromosome: A chromosome is a long DNA molecule with part or all of the genetic material of an organism.

Gene families: A gene family is a set of several similar genes, formed by duplication of a single original gene, and generally with similar biochemical functions.

Gene: A gene is the basic physical and functional unit of heredity. Genes are made up of DNA. Some genes act as instructions to make molecules called proteins.

Hypersensitive response (HR): A mechanism used by plants to prevent the spread of infection by microbial pathogens. HR is characterized by the rapid death of cells in the local region surrounding an infection and it serves to restrict the growth and spread of pathogens to other parts of the plant.

NCBI: The **National Center for Biotechnology Information** advances science and health by providing access to biomedical and genomic information.

Nucleotide: A nucleotide is an organic molecule that is the building block of DNA and RNA. They also have functions related to cell signaling, metabolism, and enzyme reactions. Nucleotide is made up of three parts: a phosphate group, a 5-carbon sugar, and a nitrogenous base.

Nucleic acid sequence: A **nucleic acid sequence** is a succession of bases signified by a series of a set of five different letters that indicate the order of **nucleotides** forming alleles within a **DNA** (using GACT) or RNA (GACU) molecule. By convention, **sequences** are usually presented from the 5′ end to the 3′ end.

Pathogen: A pathogen is an organism that causes disease. Your body is naturally full of microbes.

Programmed cell death (PCD): Sometimes referred to as cellular suicide, is the death of a cell as a result of events inside of a cell, such as apoptosis or autophagy.

References

Ali, S., Ganai, B. A., Kamili, A. N., Bhat, A. A., Mir, Z. A., et al. (2018). Pathogenesis-related proteins and peptides as promising tools for engineering plants with multiple stress tolerance. *Microbiological Research, 213*, 29–37. https://doi.org/10.1016/j.micres.2018.04.008

Araújo, A. C. D., Fonseca, F. C. D. A., Cotta, M. G., Alves, G. S. C., & Miller, R. N. G. (2019). Plant NLR receptor proteins and their potential in the development of durable genetic resistance to biotic stresses. *Biotechnology Research and Innovation, 3*, 80–94. https://doi.org/10.1016/j.biori.2020.01.002

Arumuganathan, K., Earle, E. D. (1991). Nuclear DNA content of some important plant species. *Plant Molecular Biology Reporter, 9*(3), 208–218.

Ashfield, T., Redditt, T., Russell, A., Kessens, R., Rodibaugh, N., Galloway, L., Kang, Q., Podicheti, R., & Innes, R. W. (2014). Evolutionary relationship of disease resistance genes in soybean and Arabidopsis specific for the Pseudomonas syringae effectors AvrB and AvrRpm1. *Plant Physiology, 166*(1), 235–251. https://doi.org/10.1104/pp.114.244715. Epub 2014 Jul 17. PMID: 25034017; PMCID: PMC4149710.

Balint-Kurti, P. (2019). The plant hypersensitive response: Concepts, control and consequences. *Molecular Plant Pathology, 20*(8), 1163–1178. https://doi.org/10.1111/mpp

Bertioli, D. J., Cannon, S. B., Froenicke, L., Huang, G., Farmer, A. D., Cannon, E. K., Liu, X., Gao, D., Clevenger, J., Dash, S., Ren, L., Moretzsohn, M. C., Shirasawa, K., Huang, W., Vidigal, B., Abernathy, B., Chu, Y., Niederhuth, C. E., Umale, P., . . . Michelmore, R. (2016). The genome sequences of Arachis duranensis and Arachis ipaensis, the diploid ancestors of cultivated peanut. *Nature Genetics, 48*(4), 438–446. https://doi.org/10.1038/ng.3517

Bertioli, D. J., Jenkins, J., Clevenger, J., Dudchenko, O., Gao, D., Seijo, G., Leal-Bertioli, S. C. M., Ren, L., Farmer, A. D., Pandey, M. K., Samoluk, S. S., Abernathy, B., Agarwal, G., Ballén-Taborda, C., Cameron, C., Campbell, J., Chavarro, C., Chitikineni, A., Chu, Y., Dash, S., . . . Schmutz, J. (2019). The genome sequence of segmental allotetraploid peanut Arachis hypogaea. *Nature Genetics, 51*(5), 877–884. https://doi.org/10.1038/s41588-019-0405-z

Bezerra-Neto, J. P., Araújo, F. C., Ferreira-Neto, J. R., Silva, R. L., Borges, A. N., Matos, M. K., Silva, J. B., Silva, M. D., Kido, E. A., & Benko-Iseppon, A. M. (2020). NBS-LRR genes—Plant health sentinels: Structure, roles, evolution and biotechnological applications. In *Applied plant biotechnology for improving resistance to biotic stress* (pp. 63–120). https://doi.org/10.1016/B978-0-12-816030-5.00004-5

Caplan, J., Padmanabhan, M., & Dinesh-Kumar, S. P. (2008). Plant NB-LRR immune receptors: From recognition to transcriptional reprogramming. *Cell Host & Microbe, 3*(3), 126–135. https://doi.org/10.1016/j.chom.2008.02.010

Cesari, S. (2018). Multiple strategies for pathogen perception by plant immune receptors. *New Phytologist, 219*(1), 17–24. https://doi.org/10.1111/nph.14877

Chen, Z., Le Signor, C., She, Y., Kang, Y., & Verdier, J. (2021). Genome-wide association study identified candidate genes for seed size and seed composition improvement in M. truncatula. *Scientific Reports, 11*(1), 1–13. https://doi.org/10.1038/s41598-021-83581-7

Collier, S. M., & Moffett, P. (2009). NB-LRRs work a "bait and switch" on pathogens. *Trends in Plant Science, 14*(10), 521–529. https://doi.org/10.1016/j.tplants.2009.08.001. Epub 2009 Aug 31. PMID: 19720556.

Dangl, J. L., & Jones, J. D. (2001). Plant pathogens and integrated defence responses to infection. *Nature, 411*(6839), 826–833. https://doi.org/10.1038/35081161

Dubey, N., & Singh, K. (2018). Role of NBS-LRR proteins in plant defense. In A. Singh & I. Singh (Eds.), *Molecular aspects of plant-pathogen interaction*. Springer. https://doi.org/10.1007/978-981-10-7371-7_5

Fernandez-Gutierrez, A., & Gutierrez-Gonzalez, J. J. (2021). Bioinformatic-based approaches for disease-resistance gene discovery in plants. *Agronomy, 11*(11), 2259. https://doi.org/10.3390/agronomy11112259

Flor, H. H. (1942). Inheritance of pathogenicity in a cross between physiologic races 22 and 24 of Melampsora lini. *Phytopathology, 32*, 5.

Flor, H. H. (1971). Current status of the gene-for-gene concept. *Annual Review of Phytopathology, 9*, 275–296. https://doi.org/10.1146/annurev.py.09.090171.001423

Garg, R., Patel, R. K., Jhanwar, S., Priya, P., Bhattacharjee, A., Yadav, G., Bhatia, S., Chattopadhyay, D., Tyagi, A. K., & Jain, M. (2011). Gene discovery and tissue-specific transcriptome analysis in chickpea with massively parallel pyrosequencing and web resource development. *Plant Physiology, 156*(4), 1661–1678. https://doi.org/10.1104/pp.111.178616

Guo, L., You, C., Zhang, H., Wang, Y., & Zhang, R. (2022). Genome-wide analysis of NBS-LRR genes in Rosaceae species reveals distinct evolutionary patterns. *Frontiers in Genetics, 13*, 1052191. https://doi.org/10.3389/fgene.2022.1052191

Innes, R. W., Ameline-Torregrosa, C., Ashfield, T., Cannon, E., Cannon, S. B., Chacko, B., et al. (2008). Differential accu-mulation of retroelements and diversification of NB-LRR disease resistance genes in duplicated regionsfollowing polyploidy in the ancestor of soybean. *Plant Physiology, 148*, 1740–1759. https://doi.org/10.1104/pp.108.127902. PMID: 18842825.

Jupe, F., Witek, K., Verweij, W., Sliwka, J., Pritchard, L., Etherington, G. J., Maclean, D., Cock, P. J., Leggett, R. M., Bryan, G. J., Cardle, L., Hein, I., & Jones, J. D. (2013). Resistance gene enrichment sequencing (RenSeq) enables reannotation of the NB-LRR gene family from sequenced plant genomes and rapid mapping of resistance loci in segregating populations. *The Plant Journal, 76*(3), 530–544. https://doi.org/10.1111/tpj.12307

Kang, Y. J., Kim, K. H., Shim, S., Yoon, M. Y., Sun, S., Kim, M. Y., Van, K., & Lee, S. H. (2012). Genome-wide mapping of NBS-LRR genes and their association with disease resistance in soybean. *BMC Plant Biology, 9*(12), 139. https://doi.org/10.1186/1471-2229-12-139. PMID: 22877146; PMCID: PMC3493331.

Kushalappa, A. C., Yogendra, K. N., & Karre, S. (2016). Plant innate immune response: qualitative and quantitative resistance. *Critical Reviews in Plant Sciences, 35*(1), 38–55. https://doi.org/10.1080/07352689.2016.1148980

Kwon, J. M., & Goate, A. M. (2000). The candidate gene approach. *Alcohol Research & Health, 24*(3), 164–168. PMID: 11199286; PMCID: PMC6709736.

Li, M., Jiang, B., Han, T., Zhang, G., & Lam, H. (2021). Genomic research on soybean and its impact on molecular breeding. *Advances in Botanical Research, 102*, 1–42. https://doi.org/10.1016/bs.abr.2022.02.021

Liu, C., Wang, Y., Peng, J., Fan, B., Xu, D., Wu, J., Cao, Z., et al (2022). High-quality genome assembly and pan-genome studies facilitate genetic discovery in mung bean and its improvement. *Plant Communications, 3*(6).

Lozano, R., Hamblin, M. T., Prochnik, S., & Jannink, J. L. (2015). Identification and distribution of the NBS-LRR gene family in the Cassava genome. *BMC Genomics, 16*(1), 1–14.

Lozano-Torres, J. L., et al. (2012). Dual disease resistance mediated by the immune receptor Cf-2 in tomato requires a common virulence target of a fungus and a nematode. *Proceedings of the National Academy of Sciences of the United States of America, 109*, 10119–10124.

Macho, A. P., & Zipfel, C. (2014). Plant PRRs and the activation of innate immune signaling. *Molecular Cell, 54*(2), 263–272. https://doi.org/10.1016/j.molcel.2014.03.028

McHale, L., Tan, X., Koehl, P., et al. (2006). Plant NBS-LRR proteins: Adaptable guards. *Genome Biology, 7*, 212. https://doi.org/10.1186/gb-2006-7-4-212

Meng, S., Cao, J., He, Q., Xiong, L., Chang, E., Radovick, S., Wondisford, F. E., & He, L. (2014). Metformin activates AMP-activated protein kinase by promoting formation of the αβγ heterotrimeric complex. *Journal of Biological Chemistry, 290*(6), 3793–3802. https://doi.org/10.1074/jbc.M114.604421. Epub 2014 Dec 23. PMID: 25538235; PMCID: PMC4319043.

Meng, J., Yang, J., Peng, M., Liu, X., & He, H. (2020). Genome-wide characterization, evolution, and expression analysis of the leucine-rich repeat receptor-like protein kinase (LRR-RLK) gene family in *Medicago truncatula. Life (Basel), 10*(9), 176. https://doi.org/10.3390/life10090176. PMID: 32899802; PMCID: PMC7555646.

Meyers, B. C., Dickerman, A. W., Michelmore, R. W., Sivaramakrishnan, S., Sobral, B. W., & Young, N. D. (1999). Plant disease resistance genes encode members of an ancient and diverse protein family within the nucleotide-binding superfamily. *The Plant Journal, 3*, 317–332. https://doi.org/10.1046/j.1365-313x.1999.t01-1-00606.x

Meyers, B., Morgante, M., & Michelmore, R. (2002). TIR-X and TIR-NBS proteins: Two new families related to disease resistance TIR-NBS-LRR proteins encoded in Arabidopsis and other plant genomes. *The Plant Journal, 32*, 77–92. https://doi.org/10.1046/j.1365-313X.2002.01404.x

Meyer, S. E., Stewart, T. E., & Clement, S. (2010). The quick and the deadly: Growth vs virulence in a seed bank pathogen. *New Phytologist, 187*(1), 209–216. https://doi.org/10.1111/j.1469-8137.2010.03255.x

Mota, A. P. Z., Bruna, V., Danchin, E. G. J., Togawa, R. C., Leal-Bertioli, S. C. M., Bertioli, D. J., Araujo, A. C. G., Brasileiro, A. C. M., & Guimaraes, P. M. (2018). Comparative root transcriptome of wild Arachis reveals NBS-LRR genes related to nematode resistance. *BMC Plant Biology 18*, 1–16

Mukhtar, M. S., Carvunis, A. R., Dreze, M., Epple, P., Steinbrenner, J., Moore, J., Tasan, M., Galli, M., Hao, T., Nishimura, M. T., Pevzner, S. J., Donovan, S. E., Ghamsari, L., Santhanam, B., Romero, V., Poulin, M. M., Gebreab, F., Gutierrez, B. J., Tam, S., Monachello, D., Boxem, M., Harbort, C. J., McDonald, N., Gai, L., Chen, H., He, Y., European Union Effectoromics Consortium, Vandenhaute, J., Roth, F. P., Hill, D. E., Ecker, J. R., Vidal, M., Beynon, J., Braun, P., & Dangl, J. L. (2011). Independently evolved virulence effectors converge onto hubs in a plant immune system network. *Science, 333*(6042), 596–601. https://doi.org/10.1126/science.1203659. PMID: 21798943; PMCID: PMC3170753.

Muthamilarasan, M., Parida, S. K., & Prasad, M. (2014). Advances in wheat genomics and its potential in ensuring food security in the scenario of climate change. *Proceedings of the Indian National Science Academy, 80*(2), 325–331.

Nandety, R. S., Caplan, J. L., Cavanaugh, K., Perroud, B., Wroblewski, T., Michelmore, R. W., & Meyers, B. C. (2013). The role of TIR-NBS and TIR-X proteins in plant basal defense responses. *Plant Physiology, 162*(3), 1459–1472. https://doi.org/10.1104/pp.113.219162. Epub 2013 Jun 4. PMID: 23735504; PMCID: PMC3707564.

Purwar, S., Singh, C. M., Kumar, M. et al. (2023). Genome-wide identification and analysis of NBS-LRR-encoding genes in mungbean (*Vigna radiata* L. Wilczek) and their expression in two wild non-progenitors reveal their role in MYMIV resistance. *Journal of Plant Growth Regulation 42*, 6667–6680. https://doi.org/10.1007/s00344-023-10948-7

Ratnaparkhe, M. B., Wang, X., Li, J., Compton, R. O., Rainville, L. K., Lemke, C., Kim, C., Tang, H., & Paterson, A. H. (2011). Comparative analysis of peanut NBS-LRR gene clusters suggests evolutionary innovation among duplicated domains and erosion of gene microsynteny. *New Phytologist, 192*, 164–178. https://doi.org/10.1111/j.1469-8137.2011.03800.x

Sagi, M. S. (2018). Genetic analysis of NBS-LRR genes and their association with ascochyta blight resistance in chickpea (*Cicer arietinum* L.). PhD diss., University of Saskatchewan.

Sagi, M. S., Deokar, A. A., & Tar'an, B. (2017). Genetic analysis of NBS-LRR gene family in chickpea and their expression Profifiles in response to Ascochyta blight infection. *Frontiers in Plant Science, 8*, 838. https://doi.org/10.3389/fpls.2017.00838

Schmutz, J., McClean, P. E., Mamidi, S., et al. (2014). A reference genome for common bean and genome-wide analysis of dual domestications. *Nature Genetics, 46*(7):707–713. https://doi.org/10.1038/ng.3008

Sekhwal, M. K., Li, P., Lam, I., Wang, X., Cloutier, S., & You, F. M. (2015). Disease resistance gene analogs (RGAs) in plants. *International Journal of Molecular Sciences, 16*(8), 19248–19290. https://doi.org/10.3390/ijms160819248. PMID: 26287177; PMCID: PMC4581296.

Shao, Z. Q., Zhang, Y. M., Hang, Y. Y., Xue, J. Y., Zhou, G. C., Wu, P., Wu, X. Y., Wu, X. Z., Wang, Q., Wang, B., & Chen, J. Q. (2014). Long-term evolution of nucleotide-binding site-leucine-rich repeat genes: understanding gained from and beyond the legume family. *Plant Physiology, 166*(1), 217–234. https://doi.org/10.1104/pp.114.243626. Epub 2014 Jul 22. PMID: 25052854; PMCID: PMC4149708.

Shao, Z. Q., Xue, J. Y., Wu, P., Zhang, Y. M., Wu, Y., Hang, Y. Y., Wang, B., & Chen, J. Q. (2016). Large-scale analyses of angiosperm nucleotide-binding site-leucine-rich repeat genes reveal three anciently diverged classes with distinct evolutionary patterns. *Plant Physiology, 170*(4), 2095–2109. https://doi.org/10.1104/pp.15.01487. Epub Feb 2. PMID: 26839128; PMCID: PMC4825152.

Shiu, S. H., & Bleecker, A. B. (2001). Receptor-like kinases from Arabidopsis form a monophyletic gene family related to animal receptor kinases. *Proceedings of the National Academy of Sciences of the United States of America, 98*(19), 10763–10768. https://doi.org/10.1073/pnas.181141598. Epub 2001 Aug 28. PMID: 11526204; PMCID: PMC58549.

Singh, R., Kumar, K., Puryannur, S., & Verma, P. K. (2023). Genomics-assisted genetics of complex regions from chickpea chromosome 4 reveals two candidate genes for Ascochyta blight resistance. *Plant Science, 334*, 111781. ISSN 0168-9452,. https://doi.org/10.1016/j.plantsci.2023.111781

Song, H., & Nan, Z. (2014). Genome-wide analysis of nucleotide-binding site disease resistance genes in Medicago truncatula. *Chinese Science Bulletin, 59*, 1129–1138.

Song, H., Wang, P., Li, C., Han, S., Zhao, C., Xia, H., et al. (2017). Comparative analysis of NBS-LRR genes and their response to Aspergillus flavus in Arachis. *PLoS One, 12*(2), e0171181. https://doi.org/10.1371/journal.pone.0171181

Song, H., Guo, Z., Hu, X., et al. (2019). Evolutionary balance between LRR domain loss and young *NBS–LRR* genes production governs disease resistance in *Arachis hypogaea* cv. Tifrunner. *BMC Genomics, 20*, 844. https://doi.org/10.1186/s12864-019-6212-1

Van Es, M. A., van Vught, P. W., Blauw, H. M., Franke, L., Saris, C. G., et al. (2008). Genetic variation in DPP6 is associated with susceptibility to amyotrophic lateral sclerosis. *Nature Genetics, 40*(1), 29–31. https://doi.org/10.1038/ng.2007.52. Epub 2007 Dec 16. PMID: 18084291.

Van Loon, L. C., Rep, M., & Pieterse, C. M. (2006). Significance of inducible defense-related proteins in infected plants. *Annual Review of Phytopathology, 44*, 135–162. https://doi.org/10.1146/annurev.phyto.44.070505.143425. PMID: 16602946.

Van Ooijen, G., Mayr, G., Kasiem, M. M. A., Albrecht, M., Cornelissen, B. J. C., & Takken, F. L. W. (2008). Structure–function analysis of the NB-ARC domain of plant disease resistance proteins. *Journal of Experimental Botany, 59*(6), 1383–1397. https://doi.org/10.1093/jxb/ern045

Verma, S., & Sowdhamini, R. (2022). A genome-wide search of Toll/Interleukin-1 receptor (TIR) domain-containing adapter molecule (TICAM) and their evolutionary divergence from other TIR domain containing proteins. *Biology Direct, 17*, 24. https://doi.org/10.1186/s13062-022-00335-9

Wei, H., Liu, J., Guo, Q., Pan, L., Chai, S., Cheng, Y., Ruan, M., Ye, Q., Wang, R., Yao, Z., Zhou, G., & Wan, H. (2020). Genomic organization and comparative phylogenic analysis of NBS-LRR resistance gene family in Solanum pimpinellifolium and Arabidopsis thaliana. *Evolutionary Bioinformatics Online, 16*. https://doi.org/10.1177/1176934320911055

Weßling, R., Epple, P., Altmann, S., He, Y., Yang, L., Henz, S. R., McDonald, N., Wiley, K., Bader, K. C., Gläßer, C., et al. (2014). Convergent targeting of a common host protein-network by pathogen effectors from three kingdoms of life. *Cell Host & Microbe, 16*, 364–375.

Williams, S. J., Sohn, K. H., Wan, L., Bernoux, M., Sarris, P. F., Segonzac, C., et al. (2014). Structural basis for assembly and function of a heterodimeric plant immune receptor. *Science, 344*(6181), 299–303.

Wu, J., Zhu, J., Wang, L., & Wang, S. (2017). Genome-wide association study identifies NBS-LRR-encoding genes related with anthracnose and common bacterial blight in the common bean. *Frontiers in Plant Science, 8*, 1398.

Xun, H., Yang, X., He, H., et al. (2019). Over-expression of *GmKR3*, a TIR–NBS–LRR type *R* gene, confers resistance to multiple viruses in soybean. *Plant Molecular Biology, 99*, 95–111. https://doi.org/10.1007/s11103-018-0804-z

Yang, S., Tang, F., Gao, M., Krishnan, H. B., & Zhu, H. (2010a). R gene-controlled host specificity in the legume-rhizobia symbiosis. *Proceedings of the National Academy of Sciences of the United States of America, 107*(43), 18735–18740. https://doi.org/10.1073/pnas.1011957107

Yang, S. S., Valdés-López, O., Xu, W. W., et al. (2010b). Transcript profiling of common bean (*Phaseolus vulgaris* L.) using the GeneChip®Soybean Genome Array: Optimizing analysis by masking biased probes. *BMC Plant Biology, 10*, 85. https://doi.org/10.1186/1471-2229-10-85

Zhang, Y., Xia, R., Kuang, H., & Meyers, B. C. (2016a). The diversification of plant NBS-LRR defense genes directs the evolution of MicroRNAs that target them. *Molecular Biology and Evolution, 33*(10), 2692–2705. https://doi.org/10.1093/molbev/msw154.61

Zhang, Y. M., Shao, Z. Q., Wang, Q., Hang, Y. Y., Xue, J. Y., Wang, B., & Chen, J. Q. (2016b). Uncovering the dynamic evolution of nucleotide-binding site-leucine-rich repeat (NBS-LRR) genes in Brassicaceae. *Journal of Integrative Plant Biology, 58*, 165–177.

Zhu, H., Kim, D. J., Baek, J. M., Choi, H. K., Ellis, L. C., Kuester, H., McCombie, W. R., Peng, H-M., & Cook, D. R. (2003). Syntenic relationships between Medicago truncatula and Arabidopsis reveal extensive divergence of genome organization. *Plant Physiology, 131*(3), 1018–1026.

Genomics-Assisted Molecular Breeding for Pigeon Pea Improvement

Dharamsheela Thakur, Seema Kumari, Vikash Kumar Jha, and Ravi Shankar Singh

Abstract Pigeon pea is a multipurpose food legume cultivated predominantly in semiarid tropics and subtropics of Asia, Africa, and Latin America. It is the second most important pulse crop of India. Pigeon pea breeding is primarily centered around enhancing its biotic stress tolerance/resistance, and improvement of plant ideotype for traits like determinacy, earliness, and photosensitivity and toward development of high yielding hybrids. In the last few decades, a number of improved varieties have been developed through conventional breeding approaches but practical yield realized from these cultivars has still remained below the potential yield. Additionally, the time required to develop a new variety through conventional breeding is major constraint. Hence, integration of advanced genomics technologies is essential to produce high-yielding, stress-tolerant cultivars in the shortest possible time. Over the past few decades, pigeon pea genomic resources have been enriched with whole genome sequences along with germplasm resequencing and transcriptomics data. Genomic resources have been utilized for genome-wide molecular markers discovery, deciphering marker/gene–trait associations and development of high-throughput genotyping assays. Molecular characterization of germplasm and diverse mapping populations has led to the creation of a dense genetic maps, identification of quantitative trait loci (QTLs) and biotic and abiotic stress tolerance genes. The progress in modern breeding tools, including genomic selection, sequence-based breeding, and speed breeding, coupled with the availability of cost-effective high-throughput genotyping and automated phenotyping facilities, is poised to revolutionize pigeon pea crop improvement programs in the future.

Keywords Pigeon pea · Abiotic and biotic stress · Genomics · Hybrid · Photosensitivity · Transcriptomics · Molecular marker · Marker-assisted breeding · Speed breeding · Germplasm · Genomic selection · Sequence-based breeding · Genotyping-by-sequencing (GBS)

D. Thakur (✉) · S. Kumari · V. K. Jha · R. S. Singh
Department of Molecular Biology and Genetic Engineering, Bihar Agricultural University, Sabour, Bhagalpur, Bihar, India

A. Pratap et al. (eds.), *Integrated Improvement of Food Legumes*,
https://doi.org/10.1007/978-3-031-82629-0_9

Abbreviations

AFLP	Amplified fragment length polymorphism
BESs	BAC end sequences
CNV	Copy number variation
CAPS	Cleaved amplified polymorphic sequence
CMS	Cytoplasmic male sterility
DArT	Diversity arrays technology
DT	Determinate
ESTs	Expressed sequenced tags
FAOSTAT	Food and agriculture organization of United Nations
FW	Fusarium wilt
GS	Genomic selection
GoT	Grow-out test
IDT	Indeterminate
LTR	Long terminal repeat
PAV	Presence/absence variation
MABC	Marker-assisted backcrossing
MAS	Marker-assisted selection
MAGIC	Multiparent advanced generation intercross
NAM	Nested association mapping
NGS	Next-generation sequencing
nsSNP	Non-synonymous SNP
PPSMV	Pigeon pea sterility mosaic virus
QTL	Quantitative trait loci
REs	Repeat elements
RFLP	Restriction fragment length polymorphism
RAPD	Random amplified polymorphic DNA
SBB	Sequence-based breeding
Seq-BSA	NGS-based bulked segregant analysis
SB	Speed breeding
SNP	Single-nucleotide polymorphism
SSR	Simple sequence repeats
SMD	Sterility mosaic disease
SPC	Seed protein content
ORFs	Open reading frames
GBS	Genotyping-by-sequencing
WGRS	Whole-genome resequencing
SVs	Structural variations

1 Introduction

Pigeon pea [*Cajanus cajan* (L) Millspaugh], also known as "*arhar*" or "*tur*" in several regions, is a multipurpose nutritious legume. Pigeon pea seeds are consumed as *dal* (split peas) or ground into flour for making various dishes and its green pods are consumed as vegetable, thus contributing to the culinary diversity and uplifting nutritional status of many populations. Seeds and pods are rich in protein, dietary fiber, and essential nutrients like vitamins and minerals. Native to India, pigeon pea is cultivated in tropical and subtropical regions across the world. Globally pigeon pea is cultivated in 6.30 mha with production of 5.33 mt. India ranks first with production of 4.22 mt followed by Malawi(0.45 mt), Myanmar (0.33mt), Tanzania (0.16 mt), andKenya (0.084 mt) (FOASTAT, 2022 2018). 'Hardy and drought-resistant nature of pigeon pea makes it a crop of choice in drought-prone regions having erratic rainfall. Pigeon pea plays a crucial role in sustainable agriculture due to its ability to fix atmospheric nitrogen, restoring soil fertility. Beside this other plant parts are also useful; its seed husk and leaves are used as fodder, stem is used for making baskets, fences, and huts and also used as domestic firewood. In spite of its potential the average yield of pigeon pea has remained 882 kg/ha, which is much below its potential yield of over 2000 kg/ha. Poor yield of pigeon pea could be attributed to non-availability of quality seed of superior varieties and various abiotic (drought, salinity, and waterlogging) and biotic (diseases and insect pests) stresses.

Over the last five decades breeding efforts have led to development of 186 improved varieties for cultivation in India (Pratap, 2024). However, the actual yield potential could not be harnessed; so current breeding activities need to be integrated with genomics-assisted breeding research in order to develop new varieties with still higher potential yield. Availability of draft genome sequence along with large-scale high-throughput molecular marker and transcriptomics data has reoriented the research toward trait-specific mapping. Next-generation sequencing and genome-wide selection further increase selection efficiency and lead toward next-generation breeding for pigeon pea genetic enhancement. Availability of defined core and mini-core collection has enhanced the accessibility of breeders for existing genetic diversity and exploring them for developing biotic and abiotic stress resistance. In addition, cytoplasmic male sterility (CMS)-based hybrid system has also emerged as a promising tool to boost pigeon pea yield. Pigeon pea genomics resources enriched with whole-genome assemblies, high-throughput molecular markers, high-density maps, and molecular markers associated with the desired traits will accelerate pigeon pea improvement program. Adoption of recently developed breeding tools including genomic selection (GS), sequence-based breeding (SBB), and speed breeding (SB) will further enhance genetic gains in pigeon pea. This chapter emphasizes the ongoing genetic improvement program in pigeon pea and suggested need of utilization of genomic tools in combination with efficient use of genetic diversity, sequence-based breeding, and speed breeding for accelerating genetic gain in pigeon pea.

2 Limitations in Pigeon Pea Genetic Enhancement

Genetic improvement of pigeon pea faces many challenges, among them lack of genetic diversity, linkage drag, and photoperiod sensitivity are the major contributing factors. Genetic diversity within the primary gene-pool of pigeon pea is limited, so exploration of wild relatives for discovering useful traits is needed whenever utilized. Poor agronomic traits in wild relatives, their incomplete characterization, cross incompatibility, and limited collections of wild relatives add on to the constraints (Saxena et al., 2014). Pre-breeding is needed for introduction of useful biotic and abiotic stress resistance/tolerance and desirable agronomic traits like earliness and dwarfness in cultivated types from secondary and tertiary gene pools (Pratap et al., 2021). Linkage drag is a major problem associated with transferring the target genes from wild relatives into cultivated types due to tight linkage of desired traits with undesirable plant/seed characters. Another limitation in genetic enhancement of pigeon pea is photo and low temperature sensitivity. Interaction of day and night temperature along with optimum photoperiod is pivotal for flower induction in pigeon pea. Photosensitivity and low temperature sensitivity hinder expansion of pigeon pea to higher altitudes and latitudes (Vales et al., 2012).

3 Current Scenarios in Pigeon Pea Breeding

Presently pigeon pea breeding is mainly focused toward introduction of abiotic and biotic stress tolerance/resistance, improvement of agronomic traits like earliness, photo- and thermoperiod insensitivity, cleistogamy, determinancy, and development of CMS-based hybrids. Molecular markers and QTLs governing certain traits have been deciphered and research for mapping QTLs for other traits is still under progress.

3.1 Biotic Stress Resistance

In the Indian context, Fusarium wilt, Phytophthora blight, and Sterility mosaic stand as the primary diseases affecting pigeon pea, while *Helicoverpa armigera* and *Maruca vitrata* have emerged as the most destructive insect-pests. It is imperative to prioritize the development of resistant sources against these diseases and pests to address the current losses poised by these pests.

Fusarium wilt affects pigeon pea crop during early and mid vegetative stage causing yield loss ranging from 30% to 100%. Wilt symptoms include loss of turgidity, chlorosis in inter-veinal region, browning of xylem vessels, and a purple band on stem extending upwards from the base (Fig. 1). First wilt resistant variety released for cultivation in India was ICP 8863 (Maruti) followed by ICPL 87119

Fig. 1 Fusarium Wilt infested pigeon pea plant

(Asha). Recently, a number of wilt-resistant genotypes, including ICPL 20109, ICPL 20096, ICPL 20115, ICPL 20116, ICPL 20102, and ICPL 20094, have been identified after a thorough screening of a large number of pigeon pea accessions (Sharma et al., 2016). Saxena et al. (2017b) employed a genotyping-by-sequencing (GBS) approach to identify three quantitative trait loci (QTLs)-qFW11.1, qFW11.2, and qFW11.3 associated with Fusarium wilt resistance. These QTLs hold potential for developing Fusarium wilt-resistant pigeon pea genotypes. Furthermore, Saxena et al. (2021) identified two specific genes (*Ccajan_03691* and *Ccajan_18888*) that confer resistance against Fusarium wilt.

Phytophthora blight is another soil borne disease of pigeon pea account up to 98% yield loss under severe infestation. Phytophthora resides in soil as dormant mycelium and on infected plant debris. The incidence of Phytophthora blight tends to escalate during periods of heavy rainfall and in areas with inadequate surface water drainage (Bisht et al., 1988). Wild pigeon pea relatives *Cajanus platycarpus* and *Cajanus sericeus* have shown resistance to the P2 isolate but susceptibility to the P3 isolate (Reddy et al., 1991). Certain pigeon pea accessions, including BDN 627, ICPL 1871, Sehore 1971, ICPL 84052, ICPL 84023, and ICPL 88009, have been identified as resistant source of Phytophthora (Gupta, Singh, Reddy, & Bajpai, 1997). Pande et al. (2012) have also reported ICP 11376–5, ICP 12730, ICP 12751, ICP 12755, ICPL 20093, ICPL 20100, ICPL 20101, ICPL 20104, ICPL 20105, and ICPL 20109 as resistant line for Phytophthora. Resistance genes and markers associated with Phytophthora blight resistance have not been reported till date.

Sterility mosaic disease (SMD) is another devastating disease, potentially causing complete yield loss of up to 100% under severe infestations (Fig. 2). Pigeon pea sterility mosaic virus (PPSMV) is transmitted by the mite *Aceria cajani*. Controlling mite populations through insecticide spray can help mitigate the spread of the disease only up to certain extent so there is dire need to explore the resistance source in gene pool. Some resistant SMD genotypes have been identified and study on inheritance pattern has highlighted the involvement of four independent loci,

Fig. 2 Sterility mosaic disease of pigeon pea

consisting of two duplicate dominant genes (Sv1 and Sv2) and two duplicate recessive genes (*sv3* and *sv4*), in the inheritance of resistance to sterility mosaic disease (Saxena, 2008). Six QTLs associated with SMD resistance were identified on Linkage Groups 7 and 9 (Gnanesh et al., 2011. Saxena et al. (2017a) subsequently identified a significant QTL for SMD resistance on Linkage Group 11, which holds promise for molecular breeding efforts aimed at developing superior lines with enhanced resistance to SMD. Recently four candidate genes (*Ccajan_07858*, *Ccajan_20995*, *Ccajan_21801*, and *Ccajan_17341*) were identified to be associated with SMD resistance (Saxena et al., 2021).

Helicoverpa armigera is the most notorious insect-pest of pigeon pea (Fig. 3), so there is dire need to develop resistance against this pest. Consequently, there is an urgent need to develop resistance mechanisms against this pest. As of now, sources of resistance against *Helicoverpa armigera* are not available among cultivated pigeon pea species. However, breeding populations created through interspecific crosses, utilizing the wild relative of pigeon pea, "*Cajanus acutifolius*," as the donor parent, have demonstrated considerable resistance to pod borer damage (Jadhav et al., 2012; Mallikarjuna et al., 1997). Additionally, transgenic pigeon pea carrying the chimeric *cry1AcF* gene, which encodes *cry1Ac* and *cry1F* domains, have exhibited resistance to *H. armigera* (Ramu et al., 2012). The presence of pod trichomes plays a significant role in the plant's defense system against Helicoverpa. The orientation, density, type, and length of trichomes found in wild pigeon pea species differ from those in released cultivars, thereby favoring the plant's defense mechanism (Aruna et al., 2005).

Brown spotted pod borer *Maruca vitrata* has also emerged as a serious pest in India (Sharma & Franzmann, 2000), accounting to yield loss up to the tune of 84% (Mahalle & Taggar, 2018). Unfortunately, the desired level of resistance against *Maruca vitrata* is unavailable within the cultivated gene pool. Consequently, farmers are left with no choice but to rely solely on chemical insecticides for managing

Fig. 3 *Helicoverpa armigera* infested pod of pigeon pea

pod borers in pigeon pea. It becomes imperative to explore resistance genes/QTLs from wild relatives of pigeon pea reported to possess considerable levels of resistance to this pest (Khoury et al., 2015).

The bruchid (*Callosobruchus maculatus* F.) is a major storage pest of pigeon pea. Resistance to this pest has not been observed in cultivated pigeon pea. Delayed bruchid lifecycle as a result of the antibiosis mechanism of resistance to bruchids has been observed in hybrids developed from the cross involving wild pigeon pea *Cajanus lanceolatus* (Mallikarjuna et al., 2017).

3.2 *Abiotic Stress Tolerance*

Drought, waterlogging, and salinity are the major abiotic stresses significantly hindering pigeon pea production and productivity. To address this challenge, it is crucial to identify tolerant sources for these stresses among pigeon pea germplasm and incorporate them into pigeon pea improvement programs focused on enhancing abiotic stress tolerance.

3.2.1 Drought Stress Tolerance

Pigeon pea is primarily cultivated as rainfed crop during the kharif season. While it can withstand drought during the vegetative stage, the reproductive phase is susceptible to terminal water deficits. Drought stress not only limits pigeon pea productivity but also impairs its ability for symbiotic nitrogen fixation (Kumar et al., 2014). Physiological traits such as leaf area, relative water content, dehydration tolerance, and osmotic adjustments play a crucial role in conferring drought stress tolerance. Additionally, it is important to consider agronomic traits like the number of pods per plant, seeds per pod, seed size, seed yield per plant, as well as the presence of a deep root system while breeding for drought tolerance (Choudary et al., 2011). Molecular mechanism of drought tolerance and identification of candidate genes have been accomplished through utilization of genomic resources (Sinha et al., 2016). A large number of proteins homologous to drought-responsive universal stress proteins have been identified by Varshney et al. (2012). Candidate gene *CcCYP* encoding cyclophilin has been identified for drought, salinity, and extreme temperature tolerance (Sekhar et al., 2010). Transgenic Arabidopsis lines expressing *CcCYP* showed tolerance against multiple stress viz., drought, salinity, and extreme temperature as evidenced by increase in plant survival, biomass, chlorophyll content, and profuse root growth under stress.

3.2.2 Water Logging Stress Tolerance

Pigeon pea is grown in areas with medium to heavy rainfall. Pigeon pea sowing is dependent on monsoon and germination to early vegetative stage coincides with heavy to very heavy rainfall. Prolonged rain leads to collection of water on soil surface without infiltrating into the soil resulting in waterlogged condition. Waterlogging leads to induction of hypoxic (low oxygen) and later creates anoxic (complete absence of oxygen) condition in root zone due to poor solubility of gas in water. Prolonged waterlogging condition causes inhibition of aerobic respiration and shifts to anaerobic respiration thus creating nutrient imbalance. Waterlogged plants also face oxidative stress, mineral nutrient deficiencies, and microelement toxicities. Carbohydrate starvation and oxidative stress cause detrimental effects on plants. Waterlogged condition can be avoided by agronomic management such as the use of raised bed, planting on ridges but it is not economically viable. Developing water logging tolerant genotypes would be the most economical, easiest, and effective way to minimize losses. A number of water logging tolerant genotypes have been identified till date. Water logging tolerance is reported to be governed by a single dominant gene (Perera et al., 2001; Sarode et al., 2007), thus it can be incorporated in lines for development of water-logging tolerant hybrids or genotypes. Pigeon pea lines developed from the cross employing wild relative *C. acutifolius* showed formation of lenticels under waterlogged condition leading to better survival of these lines (Hingane et al., 2015; Mallikarjuna et al., 2017).

3.2.3 Salinity Stress Tolerance

Growth and development of pigeon pea are adversely affected under saline soil especially in irrigated and dryland condition (Choudary et al., 2011). Soil salinity leads to the generation of osmotic stress, accumulation of toxic levels of salts inside cells, and interference with the uptake of essential mineral nutrients. Additionally, salinity results in delayed days to 50% flowering, extended peak flowering period, and reduction in both the number and weight of seeds (Promila & Kumar, 2000). There is lack of source of salinity tolerance among cultivated species while wild relatives of pigeon pea, *C. scarabaeoides*, *C. albicans*, and *C. platycarpus*, have demonstrated tolerance to salinity. The potential transfer of salinity tolerance from *C. albicans* to *C. cajan* is considered feasible due to cross compatibility and the expression of salinity tolerance as a dominant genetic trait (Choudary et al., 2011).

3.3 CMS-Based Hybrids

The first cytoplasmic male sterility (CMS) line in pigeon pea was developed using cytoplasm of a wild relative *C. scarabaeoides* by Reddy and Faris in 1981; further intensive selection and subsequent backcrossing by Saxena et al. (1996) resulted in the identification of a promising CMS line. Wild cytoplasm interacts with cultivated nuclear genome to produce male sterility. Nine CMS systems have been reported in pigeon pea by Singh et al. (2017a), namely *C. sericeus* (A1), *C. scarabaeoides* (A2), *C. volubilis* (A3), *C. cajanifolius* (A4), *C. cajan* (A5), *C. lineatus* (A6), *C. platycarpus* (A7), *C. reticulatus* (A8), and *C. lanceolatus* (A9). Among these A4 system is the most preferred system since cytoplasm of this system shows stability under various agro-climate zones and availability of good maintainers and restores. Hybrids developed using A4 system produce good pollen load and pod set.

The world's first successful commercial pigeon pea hybrid, ICPH 2671 (a cross between ICPA 2043 and ICPL 87119), was released in 2010. It demonstrated a remarkable 47% yield advantage over the national check, Maruti (Kumar et al., 2016a). Notably, ICPH 2671 exhibited resistance to Fusarium wilt (FW) and sterility mosaic disease (SMD), as well as tolerance to drought and waterlogging stresses. Subsequently, two more hybrids, ICPH 3762(A4) and ICPH 2740, were released by Odisha University of Agriculture and Technology in 2014, and Professor Jayashankar Telangana State Agriculture University, Hyderabad in 2015, respectively. ICPH 3762(A4) demonstrated a yield advantage of 20%–67% over local checks and also displayed resistance to FW and SMD. On the other hand, ICPH 2740 exhibited a superiority of 42% over the national check variety, Asha (Kumar et al., 2016a). Recently, three high-yielding early maturing pigeon pea hybrids, namely IPH10-02, IPH 15-03, and IPH 09-5, have been developed by the Indian Institute of Pulses Research for the North Western Plain Zone. Pigeon pea hybrids with high vigor and disease resistance have brought about a significant revolution in pigeon pea cultivation.

Identification of stable restorers and maintainers is crucial for hybrid development in pigeon pea. Fertility restoration in pigeon pea is regulated by two dominant *Rf* genes that render a hybrid fully fertile and stable across different environmental conditions. Four major quantitative trait loci (QTLs) associated with fertility restoration have been identified, namely QTL-RF-1, QTL-RF-2, QTL-RF-3, and QTL-RF-4, explaining 24% of the phenotypic variance. Additionally, ten markers linked to these fertility restoration QTLs have been identified in pigeon pea, which can expedite hybrid breeding through marker-assisted selection (Bohra et al., 2012).

To ensure hybrid purity, a hybrid purity testing kit has been developed by Saxena et al. (2010). This kit facilitates a quick assessment of hybrid purity, empowering both public and private seed companies to identify any off-types within commercial hybrid seed lots (Kumar et al., 2016b). The availability of such kits has greatly streamlined the quality assurance process in hybrid seed production.

3.3.1 Temperature-Sensitive Male-Sterility System

Environmental factors including photoperiod, temperature, and stress govern the expression of genes controlling male sterility or fertility in various plant species. Conversion of male sterility to fertility and the vice versa due to variation in climate has been reported in both GMS and CMS system (Kaul, 1988). It is desirable to breed temperature-sensitive male-sterility system in pigeon pea. Temperature-sensitive male-sterile lines were developed by Saxena et al. (2014). When these lines were grown under low temperature (<24 °C), they turn into male fertile and produce self-pollinated seeds; hence, such male-sterile lines will not require any maintainer line. In contrast, when these lines were grown under high temperature (>25 °C), they remain male sterile, and hence can be used for hybrid seed production with assistance from insect pollinators. This temperature-sensitive system offers a flexible approach to pigeon pea breeding programs.

3.4 Photo-Insensitivity and Earliness

Development of early-maturing and photo-insensitive pigeon pea cultivars is highly desirable to integrate them effectively into cropping systems. Prevailing cultivars of pigeon pea are mostly of long duration and photo-sensitive. Introducing photo-insensitive super-early pigeon pea varieties with high productivity per day and stress tolerance can seamlessly fit into wheat-pulse cropping patterns as well as rice fallows (Shruthi et al., 2017). This can be particularly beneficial for optimizing agricultural rotations. Furthermore, the accelerated generation turnover in such cultivars enables a swift introgression of desired traits. Study of genetics of biotic and abiotic stress tolerance will also be accelerated due to rapid development of mapping population (Vales et al., 2012). Therefore, there is a clear need to prioritize the development of photo-insensitive early-maturing pigeon pea cultivars for fitting into existing cropping system.

3.5 Plant Type

Determinate (DT) and indeterminate (IDT) are the two major plant types of pigeon pea (Mir et al., 2012). Indeterminate (IDT) is dominant trait but breeding of DT type is preferred over IDT type due to high initial vigor, tolerance to drought; water logging, and synchronous maturity. Synchronous maturity in DT type makes them suitable for mechanical harvesting. Some Indel-derived molecular markers were identified to differentiate DT/IDT lines by Saxena et al. (2017c) and *CcTFL1* was identified as candidate gene for determinant growth habit by Mir et al. (2014). Mechanism of transformation from indeterminate growth habit to determinate growth habit was unraveled by whole-genome scanning through SNP and DArT assays (Mir et al., 2012). Restructuring of the growth habit and flowering time is desirable to breed climate-smart pigeon pea.

3.6 Cleistogamous Flower

In varietal breeding programs of pigeon pea natural out-crossing is considered a major limitation. Pigeon pea is an often cross-pollinated crop with extent of out crossing up to 25%–30% (Saxena & Sharma, 1990) so lot of resources are needed to maintain the true to types in terms of isolation distance, installation of insect-proof cages, rouging and removal of mixture from seed. At ICRISAT a novel flower trait called cleistogamy was identified which could help in maintaining the genetic purity of a variety. This trait could be incorporated into desirable cultivars to maintain genetic purity. Cleistogamy trait is governed by a single recessive gene. Partial cleistogamous line ICPL 87154 was developed with low natural out crossing (<1%) (Kumar et al., 2016b). This trait was found to be linked with shriveled seed and reduced seed size. QTL explaining up to 50.6% of phenotypic variance for cleistogamous flower, shriveled seed, and seed size was identified using RIL population (Yadav et al., 2019).

4 Genetic Enhancement Through Genomic Intervention

Pigeon pea was considered orphan crop in terms of genomic resources until 2011, when the first draft genome sequence was published by Singh et al. in 2012. Later on, bulk of genomic resources were generated by resequencing number of pigeon pea accessions making pigeon pea a genomic resource-rich legume crop. These genomic resources are being utilized for trait mapping and developing trait linked molecular markers. High-density genetic maps have been generated using molecular markers. Genetic maps along with association mapping, marker-based QTL mapping, transcriptomics, and whole-genome sequencing have been utilized to identify genes/QTLs responsible for important targeted traits. These information are being exploited for genetic improvement of pigeon pea (Mir et al., 2017).

Modern genomic tools like next-generation sequencing (NGS) technology, transcriptome/genome assemblies along with genome-wide molecular-markers discovery have led to enrichment of genomic resources. These information are being utilized to enhance the pace for genomics-assisted pigeon pea breeding. Cytoplasmic male sterility system was unraveled with the help of whole mitochondrial genome sequence which could be utilized for hybrid breeding in pigeon pea.

4.1 Draft Genome Sequence

Pigeon pea is the first grain legume crop to get whole genome sequence in 2012. Singh et al. (2012) published draft genome sequence of the genotype Asha (ICPL 87119) using 454 GS-FLX platform and assembled (de novo) 510 Mb of genome. Another group at ICRISAT sequenced and assembled 60,578 Mb of the genome for the same genotype Asha (ICPL 87119) using Illumina platform (Varshney et al., 2012). The genome assembly of the draft genome sequence leads to identification of large number of genes providing insights into important traits such as biotic and abiotic stress resistance/tolerance as well as the information about the pigeon pea genes affected by domestication and breeding. Genetic improvement of pigeon pea could be done by utilizing the genome sequence information. Genome sequence is also utilized for marker discovery and assembly of sequenced genome and other NGS-based applications. Deeper investigations of genetic diversity and identification of allelic variation controlling important traits have been made possible by sequencing/resequencing of multiple genomes (Tao et al., 2019; Varshney et al., 2017). These information will be of great use in terms of varietal identification and development.

4.2 Molecular Markers

Molecular markers are indispensable for genetic enhancement of any crop through marker-assisted selection (MAS). In pigeon pea, initially first-generation molecular markers, restriction fragment length polymorphism (RFLP), random amplified polymorphic DNA (RAPD), and amplified fragment length polymorphism (AFLP) were developed. These markers were used mainly for genetic diversity studies. Second generation SSR markers were developed by mining the genomic sequences (g-SSRs, Burns et al., 2001), expressed sequence tags (ESTs-SSRs), and bacterial artificial chromosome (BAC)-end sequences (BES-SSRs) (Table 1). SSRs were preferred markers for genetic mapping and diversity study due to their abundance across the genome, multi-allelic, co-dominant nature and ease of genotyping. Later with the development of next-generation sequence (NGS) technology third generation high-throughput automated marker system like diversity arrays technology (DArT; Yang et al., 2006) and single-nucleotide polymorphism (SNP) came into existence. In recent past, the availability of genome sequence and reduction in sequencing cost has made genotyping-by-sequencing (GBS) and whole-genome resequencing (WGRS)/skim sequencing the most preferred ways to discover and genotype thousands of SNPs (Saxena et al., 2017a). GBS approach leads to identification of huge number of SNPs in segregating populations and diverse pigeon pea accessions. SNP analysis across the genome of diverse pigeon pea germplasm has unraveled domestication and demographic history of the crop (Saxena et al., 2014). SNP marker has now become the most preferred marker system due to their higher abundance, ubiquitous distribution throughout the genome, and accessibility to automation and cost effectiveness. Development of advanced genotyping platforms such as KASP technology (Saxena et al., 2012), GoldenGate assay (Kassa et al., 2012), and BeadXpress (Roorkiwal et al., 2013) facilitates high-throughput SNP genotyping. Axiom Cajanus SNP array with 56,512 unique and informative sequence variations (Saxena et al., 2018a) with the inclusion of 1554 SNPs and 385 insertion/deletion (InDel) polymorphisms potentially associated with some key agronomic traits was developed by WGRS of diverse germplasm and advanced

Table 1 Molecular markers developed in pigeon pea

Molecular marker	Type	References
DNA markers	Simple sequence repeats (SSRs)	Burns et al. (2001), Odeny et al. (2007, 2009), Saxena et al. (2010), Bohra et al. (2011, 2017), Dutta et al. (2011)
	Diversity arrays technology (DArT)	Yang et al. (2006, 2011)
	Single-nucleotide polymorphisms (SNPs)	Kumar et al. (2016c), Saxena et al. (2012, 2017a, b, c, 2018a, b)
	Intron spanning region	Kudapa et al. (2012)
	Single feature polymorphisms	Saxena et al. (2011)
	Large structural variations (CNV, PAV, InDels)	Singh et al. (2017b), Varshney et al. (2017)
DNA markers associated with traits	Sterility mosaic disease	Gnanesh et al. (2011), Singh et al., 2016, Saxena et al. (2017a)
	Fusarium wilt	Singh et al., 2016, Patil et al., 2017, Saxena et al. (2017b),
	CMS restoration	Bohra et al. (2012), Saxena et al. (2018c)
	Seed traits (protein content/size)	Obala et al. (2019), Yadav et al. (2019)
	Plant type/growth habit	Kumawat et al. (2012), Mir et al. (2014), Saxena et al. (2017c)

breeding lines. Array-based technologies could be utilized for identification of new haplotypes for associated traits as well as allele mining and characterization of germplasm for genetic improvement in pigeon pea. WGRS of 20 *Cajanus* accessions lead to generation of 55 million genome-wide variants along with SNPs (4.6 millions), InDels (0.7 million), and large structural variations (SVs) like copy number variation (CNV) and presence/absence variation (PAV) (Kumar et al., 2016c). WGRS of 292 accessions including landraces, elite breeding lines, and wild accessions leads to divulgence of large SVs (≥1000 bp) (Varshney et al., 2017).

4.3 Transcriptomics Resources

Plethora of transcriptomics resources have been generated through introduction of high-throughput sequencing platforms. Kudapa et al. (2012) generated 21,434 transcript assembly contigs (TACs), and Dubey et al. (2011) (48,726 TACs) and Dutta et al. (2011) generated 48,726 TACs. Transcriptomics data has been utilized for unraveling the spatial and temporal expression of major genes, their regulation, and participation in biological processes. Beside this transcriptomics resource has led to generation of genic and functional molecular markers. Transcript profiling in combination with genome variants also leads to identification of expression quantitative loci (eQTLs). Candidate genes for Fusarium wilt (FW: 19 genes) and sterility mosaic disease (SMD: 20 genes) were identified using transcriptomics resource (9468 high-quality ESTs). Based on homology search and gene prediction program output, gene functions are assigned to the transcriptome assembly. Recent gene expression studies in pigeon pea in response to stress response leads to generation of a set of stable reference genes which will facilitate study on expression under abiotic stress (Sinha et al., 2015a, b). *Cajanus cajan* gene expression atlas (CcGEA) having set of 28,793 genes expressed during different pigeon pea developmental stages from embryo to senescence has been developed (Pazhamala et al., 2017). Gap between whole-genome sequence and the plant phenotypes may be filled by global gene expression analysis. Three genes linked with flower-related traits were identified based on analysis of co-expressing network and focusing on gene clusters involved in common biological function. Upgrading the current CcGEA will help in accelerating basic and applied research in pigeon pea.

4.4 High-Density Genetic Maps

High-density genetic maps are important genomic resources for marker trait association studies and fine mapping of gene(s)/QTLs. High-density genetic maps also facilitate comparative genomics studies and identification of sequence scaffolds (Hu et al., 2018). In past construction of genetic map was difficult due to unavailability of sufficient polymorphic markers and low genetic variability.

Initially three SSR-based genetic maps were developed using same interspecific F_2 population developed by crossing *Cajanus cajan* and *Cajanus scarabaeoides* (Bohra et al., 2011; Yang et al., 2011). A series of genetic maps were developed for cultivated pigeon pea; however, the marker density of these maps remained extremely poor. Two genetic maps were generated by Gnanesh et al. (2011) from intra-specific cross using 120 and 78 SSR markers spanning a distance of 53,489 cM and 46,697 cM distances, respectively. A consensus genetic map was further developed by merging six intra-specific genetic maps and it comprised of 339 SSR loci with an average inter-marker distance of 31 cM spanning 1059 cM of the genome (Bohra et al., 2012). The first high-density genetic map of pigeon pea with average inter marker distance of 111 cM was developed using SNP markers for an interspecific F_2 population (Saxena et al., 2012) followed by other high-density genetic maps of pigeon pea including 1101 loci (Saxena et al., 2017a), 964 loci (Saxena et al., 2017b), and 787 loci (Saxena et al., 2017c). The most comprehensive genetic map generated till date harbors 6818 SNP loci that span 974 cM of the genome (Yadav et al., 2019). Availability of such high-density genetic maps is a key to understand the genomic architecture of essential agronomic traits for fine mapping of gene(s)/QTL, facilitating comparative genomics and anchoring of sequence scaffolds (Hu et al., 2018).

4.5 Next-Generation Genetic Map

Genetic maps based on biparental mapping population give poor resolution due to narrow genetic base and limited recombination events. Advanced mapping techniques and mating designs are applied to overcome the shortcomings of conventional procedures of genetic mapping. Complex gold standard experimental designs based on inter-mating multiple parents resulted in profuse genome reshuffling and high resolution (Wallace et al., 2018). Multi-parent advanced generation intercross (MAGIC) and nested association mapping (NAM) are two such types of design and are suitable to map adaptive loci due to their "balanced" nature and separation of population structure from the adaptation. MAGIC panel created using eight parents with seven funnels (Huang et al., 2015) is being used for high-density mapping.

4.6 Advanced Trait Mapping

Next-generation trait mapping includes NGS-based bulked segregant analysis (BSA) in combination with non-synonymous (ns) SNP substitution. This approach led to discovery of eight nsSNPs (four associated with FW and four with SMD in seven genes; Singh et al., 2016). Two genes *Ccajan_01839* and *Ccajan_03203* have been reported to be involved in SMD and FW resistance, respectively. Three InDels linked with FW resistance and two controlling SMD resistances were identified in pigeon pea through InDel Seq approach and WGRS data (Singh et al., 2017b). Candidate genes associated with high seed protein content were identified using common variant analysis of WGRS data of high seed protein content (SPC) lines (HPL 24, ICP 5529), a low SPC line (ICP 11605), and the draft genome of ICPL 87119 (low SPC) (Obala et al., 2019). Sequence cleaved amplified polymorphic sequence (CAPS)/derived CAPS (dCAPS) markers associated with SPC trait were developed using sequence variation in these genes and validated in F_2 population.

Genome-wide association study (GWAS) is another approach of precise QTL mapping. It relies on historical recombination and enables dissection of complex phenotypes in a set of distantly related individuals with higher number of alleles and greater mapping resolution (Ingvarsson & Street, 2011; Liu & Yan, 2019). Genome-wide association study of 94 pigeon pea genotypes with 6144 DArT and 768 SNP markers leads to detection of significant marker trait association of 19 SNPs with determinacy trait (Mir et al., 2012). Candidate gene-based analysis of 142 pigeon pea germplasm coupled with the biparental (ICPA2039 × ICPR2447) QTL analysis and gene expression studies suggested strong association of *CcTFL1* with the determinacy trait (Mir et al., 2014). GWAS of 89 FW-responsive pigeon pea genotypes leads to identification of six SSR markers associated with FW resistance (Patil et al., 2017). GWAS of 286 resequenced pigeon pea accessions for various agronomic traits leads to identification of 241 MTAs (Varshney et al., 2017). With access to high-throughput genotyping and wide-scale phenotyping, GWAS will be the most preferred approach for high resolution trait analysis.

4.7 Genomics-Assisted Hybrid Breeding

Identification of a natural source of genetic male sterility (GMS) in a germplasm line (Reddy et al., 1978) followed by breeding for cytoplasmic–nuclear male sterility (Saxena et al., 2005) paved the way to hybrid development program. Hybrid is a promising technology but it faces challenges of purity maintenance of hybrid seed and parental lines (CMS, maintainer and restorer lines) as well as ascertaining fertility restorers. Maintenance of parental lines and hybrid purity are inevitable to harness the benefit of hybrid technology. Grow-out test (GoT) is generally performed on representative samples of the seed lot to assess the purity of hybrid seeds which is time consuming and in pigeon pea it takes almost a full year to assess the purity of hybrid seeds. Limitations of conventional GoT can be overcome by applying molecular markers. Having molecular markers/genes linked CMS, fertility restoration, heterosis, and purity can boost this promising technology.

4.7.1 Molecular Markers for CMS

Mitochondrial genome sequencing of a CMS line (A line, ICPA 2039), its maintainer line (B line, ICPB 2039), and wild species (*C cajanifolius* ICPW 29) have led to the identification of candidate genes in A-line and its maintainer line (Tuteja et al., 2013). PCR-based InDel marker was developed targeting 10 bp deletion in *nad7a* gene in the CMS line (*nad7a*_del) and is being used for marker-based genetic purity testing of A4-derived CMS lines (Sinha et al., 2015c). Two markers for fertility restorer (CcLG08_RFQI1" and "CcLG08_RFQI4) were developed from the QTL regions in a segregating F_2 population and are being used for differentiating restorer lines from non-restorer lines in pigeon pea (Saxena et al., 2018b). These markers will eliminate the need of phenotyping activities done for the identification of CMS and restorer lines.

4.7.2 Markers for Hybrid Seeds Testing

Grow out test (GoT) of representative seed lot is p for hybrid purity testing, but this method needs lot of time and resources as well as phenotype is affected by environment. In order to overcome these limitations, SSR-based molecular markers kit has been developed (Bohra et al., 2015; Patel et al., 2012; Saxena et al., 2010) for faster and efficient purity testing of hybrid seeds. SSR-based hybrid testing is costly so the research has now been oriented toward development of low cost SNP-based marker. A large number of SNPs for hybrid breeding have been identified and is being used in routine breeding programs.

4.8 Genomics-Assisted Diversity Divulgence

Genetic improvement program of pigeon pea can be accelerated by the use of diverse genetic resources in the breeding program along with understanding of the genomic diversity in the gene pools. Genetic diversity could be studied based on morphological observations but due to influence of environment on phenotype even multi-location data have some degree of errors in selection. DNA markers have overcome the limitations associated with phenotypic-based genetic diversity study but information about key adaptive traits could not be generated by them (Jackson et al., 2011). In pigeon pea gene-based functional markers and WGRS of diverse accessions have been employed for estimation of genetic diversity and generation of information about key adaptive traits.

Genome-wide SNP markers help in understanding germplasm diversity as well as establishing relationship among the breeding materials. WGRS data of 20 accessions showed frequent occurrence of gene loss events in domesticated pigeon pea as compared to the wild species (Kumar et al. 2016a, b, c). WGRS data of 292 accessions including breeding lines, landraces, and wild species (*C. cajanifolius*,

C. scarabaeoides, and *C. platycarpus*) showed selection sweeps on LG 9. Analysis of 63 released cultivars and 40 germplasm/landraces of pigeon pea on 50 K SNP array revealed loss of diversity among pigeon pea genotypes (Saxena et al., 2018a). Pigeon pea yield can be enhanced through reintroduction of the diversity lost during domestication and selection process. High-density genotyping/sequencing data of diverse germplasm panels in combination with phenotypic data could be utilized to discover novel alleles associated with breeding traits. Besides this, breeding for the adaptive traits requires a targeted germplasm collection strategy and comprehensive analysis of the gene bank.

4.9 Genomics-Assisted Breeding

Mapping of important traits including disease resistance (FW and SMD), plant type (determinacy), fertility restoration, and other agronomically important traits has revealed a number of QTLs/candidate gene(s). These QTLs could be introgressed through marker-assisted breeding into elite pigeon pea genotypes for development of the genotypes with desired trait. Complex traits controlled by many gene(s)/QTL could not be addressed by MABC; so genomic selection (GS) is the method of choice for studying traits having polygenic inheritance. Genome-wide marker data is used for the selection of superior individuals to avoid overestimation of marker effects circumventing need of the trait linked markers like MAS (Duhnen et al., 2017). In chickpea accurate introgression of Fusarium wilt resistance loci has been obtained through MAS (Bharadwaj et al., 2022; Pratap et al., 2017). In soybean moderate to high accuracies were obtained through GS for the traits including yield, protein content, and aschochyta blight disease resistance (Carpenter et al., 2018; Duhnen et al., 2017; Jarquín et al., 2014). Accuracy of genomic prediction could be improved by incorporating G × E interactions and epistasis data with GS models (Duhnen et al., 2017; Roorkiwal et al., 2018).

4.10 Sequence-Based Breeding

Pigeon pea improvement program can gain momentum by replacing the first generation molecular breeding technique like MABC by sequence-based breeding (SBB) (Varshney et al., 2019). iS. SBB is based on the availability of phenotypic data and in-depth sequencing of founder parents/germplasm followed by identification of individuals with large number of favorable alleles through GWAS. In SBB superior genotypes are crossed to generate segregating populations followed by selection of the genotypes based on the GEBVs obtained through GS models trained from data on founder parents/germplasm. Individuals having higher number of favorable alleles and showing better performance could be identified to be released as variety directly or can serve as starting material to start next breeding cycle.

Integrating genomic resources with practical breeding programs will thus leverage functional genomics and will help to bridge the genotype–phenotype gap.

5 Speed Breeding

Genetic improvement program of pigeon pea is time consuming due to lengthy breeding cycle. Breeding cycle length can be shortened using advanced technology like rapid generation turnover (RGT) or speed breeding (SB). This technique can improve genetic gains in pigeon pea by accelerating generation turnover (Hickey et al., 2019). In this technology expansion of daily exposure to light is done under controlled temperature coupled with harvesting of premature seeds to accelerate generation turnover. Four generations of early maturing pigeon pea genotypes (ICPL 4, ICPL 151, ICPL 85024, and ICPL 87093) were obtained in a year using RGT (Saxena et al., 2017d). Shortening of breeding cycle was accomplished by harvesting immature seeds from 35-day-old crop grown in controlled condition followed by application of single pod descent method for generation advancement. Utility of SB for shortening breeding cycles in medium and long duration pigeon pea is under investigation. In pigeon pea 12-h day length is established as the optimum photoperiod. Delay in flower onset was observed if day length is exceeded beyond 12 h and the delay was found to be greater in the case of long-duration types (Carberry et al., 2001). Temperature variation also influences flowering in pigeon pea. Rapid flowering was observed in the extra-short duration at 24.7 °C, short duration at 23.1 °C, medium duration at 22.2 °C, and long duration at 18.3 °C (Silim et al., 2007). Light quality, also impacting flowering induction in other cool-season grain legumes, needs exploration in pigeon pea (Croser et al., 2016). Flower onset can be hastened by combining in vitro seed germination with in vivo conditions and needs to be optimized for SB. The SB protocols with single seed/pod descent are very much useful for preserving genetic variability. SB could be combined with the MAS or GS (Voss-Fels et al., 2018) for enhancing genetic gain in pigeon pea.

6 Summary and Conclusions

Breeding efficiency can be improved through application of genomic technologies. Genomics-assisted divulgence of genetic diversity and development of advanced mapping population along with the application of next-generation breeding technologies like sequence-based breeding and speed breeding have accelerated genetic gain in pigeon pea. Exploitation of natural diversity emerges as a key strategy for developing superior genotypes. The resequencing of germplasm accessions provides a comprehensive understanding of existing genetic diversity, enabling the identification of genes linked to desirable traits. It is imperative to develop next-generation mapping populations like MAGIC and NAM for high-density genetic mapping and allelic richness. Availability of high-throughput NGS platforms, draft genome sequence, and WGRS data in pigeon pea provide a great opportunity to explore nucleotide-level diversity in cultivated, landraces, and wild species accessions. Pigeon pea hybrid breeding program can be accelerated through application of MAS or GS. High-throughput genotyping proves instrumental in early generation screening, while marker-assisted backcrossing (MABC) and MAS are essential for refining hybrid development. The incorporation of cutting-edge technologies, such as sequence-based breeding and speed breeding, deserves further attention and utilization in the ongoing efforts to enhance genetic improvement in pigeon pea.

Glossary of Technical Terms

Photosensitivity Plant sensitivity to light signal that activates special photoreceptors which controls plant growth and development by activating or inhibiting plant hormones.

Linkage drag It refers to the association of undesirable genes linked to the desired genes or QTL. Linkage drag of unwanted genes needs to be separated from the good QTLs through recombination.

Cytoplasmic male sterility Total or partial male sterility in plants as a result of specific nuclear and mitochondrial gene interactions.

Determinate (DT) plant type In DT plant type inflorescence is short and the apical bud develops into the flower, the sequence of inflorescence production is basipetal, and the flowers occur more or less in the same plane simultaneously.

Indeterminate (IDT) plant type In IDT plant type inflorescence is large, with a vegetative apical bud resulting in continuous growth.

Cleistogamy It is a type of automated self-pollination of plants that can propagate by using non-opening, self-pollinating flowers.

Genomics It is the study of whole genomes of organisms and uses a combination of DNA sequencing methods, and bioinformatics to sequence, assemble, and analyze the structure and function of genomes.

Transcriptomics It is the study of the transcriptome that is the complete set of RNA transcripts that are produced by the genome, under specific circumstances, or in a specific cell by high-throughput technologies.

Molecular markers It is a fragment of DNA that is associated with a certain location within the genome and is used in molecular biology to identify a particular sequence of DNA in a pool of unknown DNA.

Expressed sequence tags Small pieces of DNA sequence generated by sequencing either one or both ends of an expressed gene. The idea is to sequence bits of DNA that represent genes expressed in certain cells, tissues, or organs from different organisms and use these "tags" to fish a gene out of a portion of chromosomal DNA by matching base pairs.

QTL mapping A quantitative trait locus (QTL) is a locus (section of DNA) that correlates with variation of a quantitative trait in the phenotype of a population of organisms. QTL mapping is done by identifying molecular markers that correlate with an observed trait.

Genome-wide association study It is an observational study of a genome-wide set of genetic variants in different individuals to see if any variant is associated with a trait.

Genotyping-by-sequencing It is a method to discover single-nucleotide polymorphisms (SNP) in order to perform genotyping studies, such as genome-wide association studies (GWAS). It uses restriction enzymes to reduce genome complexity and genotype multiple DNA samples.

Sequence-based breeding The sequence-based genotyping technology is a powerful tool in which SNP discovery and genotyping are performed simultaneously.

Speed breeding A technique to shorten the breeding cycle by extending the duration of plants, daily exposure to light, combined with early seed harvest, to cycle quickly from seed to seed, thereby reducing the generation times.

References

Aruna, R., Manohar Rao, D., Reddy, L. J., et al. (2005). Inheritance of trichomes and resistance to pod borer (*Helicoverpa armigera*) and their association in interspecific crosses between cultivated pigeonpea (*Cajanus cajan*) and its wild relative *C. scarabaeoides*. *Euphytica, 145*, 247–257.

Bharadwaj, C., Jorben, J., Rao, A., Roorkiwal, M., Patil, B. S., Jayalakshmi, Ahammed, S. K., Saxena, D. R., Yasin, M., Jahagirdar, J. E., Sontakke, P. L., Pithia, M. S., Chudasama, M. K., Swarup, I., Singh, R. K., Nitesh, S. D., Chitikineni, A., Singh, S., Singh, I., Pratap, A., Dixit, G. P., Srivastava, A. K., & Varshney, R. K. (2022). Development of high yielding Fusarium wilt resistant cultivar by pyramiding of "genes" through marker-assisted backcrossing in Chickpea (*Cicer arietinum* L.). *Frontiers in Genetics*. https://doi.org/10.3389/fgene.2022.924287

Bisht, V. S., Kannaiyan, J., & Nene, Y. L. (1988). Methods of metalaxyl application to control of Phytophthora blight in pigeonpea. *International Pigeonpea Newsletter, 8*, 9–10.

Bohra, A., Dubey, A., Saxena, R. K., et al. (2011). Analysis of BAC-end sequences (BESs) and development of BES-SSR markers for genetic mapping and hybrid purity assessment in pigeonpea. *BMC Plant Biology, 11*, 56.

Bohra, A., Saxena, R. K., Gnanesh, B. N., et al. (2012). An intra-specific consensus genetic map of pigeonpea [Cajanus cajan(L) Millspaugh] derived from six mapping populations. *Theoretical and Applied Genetics, 125*, 1325–1338.

Bohra, A., Singh, I. P., Yadav, A. K., et al. (2015). Utility of informative SSR markers in the molecular characterization of cytoplasmic genetic male sterility based hybrid and its parents in pigeonpea. *National Academy Science Letters, 38*, 13–19.

Bohra, A., Jha, R. P., Patil, G., et al. (2017). New hypervariable SSR markers for diversity analysis, hybrid purity testing and trait mapping in pigeonpea [*Cajanus cajan* (L) Millspaugh]. *Frontiers in Plant Science, 8*, 1–15.

Burns, M. J., Edwards, K. J., Newbury, H. J., et al. (2001). Development of simple sequence repeat (SSR) markers for the assessment of gene flow and genetic diversity in pigeonpea (*Cajanus cajan*). *Molecular Ecology Notes, 1*, 283–285.

Carberry, P. S., Ranganathan, R., Reddy, L. J., et al. (2001). Predicting growth and development of pigeonpea: Flowering response to photoperiod. *Field Crop Research, 69*, 151–162.

Carpenter, M. A., Goulden, D. S., Woods, C. J., et al. (2018). Genomic selection for ascochyta blight resistance in pea. *Front Plant Science, 9*, 1878.

Choudary, A. K., Sultana, R., Pratap, A., et al. (2011). Breeding for abiotic stress in pigeonpea. *Journal of Food Legumes, 24*(3), 165–174. ISSN: 0976-2434.

Croser, J. S., Pazos-Navarro, M., & Bennett, R. G. (2016). Time to flowering of temperate pulses in vivo and generation turnover in vivo–in vitro of narrow-leaf lupin accelerated by low red to far-red ratio and high intensity in the far-red region. *Plant Cell, Tissue and Organ Culture, 127*, 591–599.

Dubey, A., Farmer, A., Schlueter, J., et al. (2011). Defining the transcriptome assembly and its use for genome dynamics and transcriptome profiling studies in pigeonpea (Cajanus cajan L.). *DNA Research, 18*, 153–164.

Duhnen, A., Gras, A., & Teyssèdre, S. (2017). Genomic selection for yield and seed protein content in soybean: A study of breeding program data and assessment of prediction accuracy. *Crop Science, 57*, 1–13.

Dutta, S., Kumawat, G., Singh, B. P., et al. (2011). Development of genic-SSR markers by deep transcriptome sequencing in pigeonpea [*Cajanus cajan*(L) Millspaugh]. *BMC Plant Biology, 11*, 17.

FAOSTAT. (2022). Food and agriculture organization of United Nations. http://faost.atfao.org

Gnanesh, B. N., Bohra, A., Sharma, M., et al. (2011). Genetic mapping and quantitative trait locus analysis of resistance to sterility mosaic disease in pigeonpea [*Cajanus cajan* (L) Millsp]. *Field Crops Research, 123*, 53–56.

Hickey, L. T., Hafeez, A. N., Robinson, H., et al. (2019). Breeding crops to feed 10 billion. *Nature Biotechnology, 37*, 744–754.

Hingane, A. J., Saxena, K. B., Patil, S. B., et al. (2015). Mechanism of water-logging tolerance in pigeonpea. *Indian Journal of Genet and Plant Breed, 75*(2), 208.

Hu, X. H., Zhang, S. Z., Miao, H. R., Cui, F. G., et al. (2018). High-density genetic map construction and identification of QTLs controlling oleic and linoleic acid in peanut using SLAF-seq and SSRs. *Scientific Reports, 8*, 5479.

Huang, B. E., Verbyla, K. L., Verbyla, A. P., et al. (2015). MAGIC populations in crops: Current status and future prospects. *Theoretical and Applied Genetics, 128*, 999–1017.

Ingvarsson, P. K., & Street, N. R. (2011). Association genetics of complex traits in plants. *The New Phytologist, 189*, 909–922.

Jackson, S. A., Iwata, A., Lee, S. H., et al. (2011). Sequencing crop genomes: Approaches and applications. *The New Phytologist, 191*, 915–925.

Jadhav, D. R., Mallikarjuna, N., Sharma, H. C., & Saxena, K. B. (2012). Introgression of Helicoverpa armigera resistance from *Cajanus acutifolius*-a wild relative from secondary gene pool of pigeon pea (*Cajanus cajan*). *Asian Journal of Agriculture and Science, 4*, 242–248.

Jarquín, D., Crossa, J., Lacaze, X., Du Cheyron, et al. (2014). A reaction norm model for genomic selection using high-dimensional genomic and environmental data. *Theoretical and Applied Genetics, 127*, 595–607.

Kassa, M. T., Penmetsa, R. V., Carrasquilla-Garcia, N., et al. (2012). Genetic patterns of domestication in pigeonpea (*Cajanus cajan* (L,) Millsp,) and wild Cajanus relatives. *PLoS One, 7*, e39563.

Kaul, M. L. H. (1988). *Male sterility in higher plants*. Springer Verlag Berlin, Heidelberg.

Khoury, C. K., Castañeda-Alvarez, N. P., Achicanoy, H. A., et al. (2015). Crop wild relatives of pigeonpea [*Cajanus cajan* (L.) Millsp.]: Distributions, ex situ conservation status, and potential genetic resources for abiotic stress tolerance. *Biological Conservation, 184*, 259–270.

Kudapa, H., Bharti, A. K., Cannon, S. B., et al. (2012). A comprehensive transcriptome assembly of pigeonpea (*Cajanaus cajan* L.) using Sanger and second generation sequencing platforms. *Molecular Plant, 5*, 1020–1028.

Kumar, C. V. S., Mula, M. G., Singh, I. P., et al. (2014). Pigeonpea perspective in India. In *Proceedings of Phillippine congress* (pp. 1–5).

Kumar, C. V. S., Wani, S. P., Nagesh Kumar, M. V., et al. (2016a). Hybrid technology–a new vista in pigeonpea breeding. *The Journal of Research PJTSAU, 44*(4), 1–13. ISSN 2395-5945.

Kumar, C. V. S., Singh, I. P., Patil, S. B., et al. (2016b). Recent advances in pigeonpea [*Cajanus cajan* (L) Millspaugh] research. In *Open access repository* (pp. 1–7). ICRISAT.

Kumar, V., Khan, A. W., Saxena, R. K., et al. (2016c). First generation hapmap in *Cajanus spp* reveals untapped variations in parental lines of mapping populations. *Plant Biotechnology Journal, 14*, 1673–1681.

Kumawat, G., Raje, R. S., Bhutani, S., et al. (2012). Molecular mapping of QTLs for plant type and earliness traits in pigeonpea (*Cajanus cajan* L Millsp). *BMC Genetics, 13*, 84.

Liu, H. J., & Yan, J. (2019). Crop genome-wide association study: A harvest of biological relevance. *The Plant Journal, 97*, 8–18.

Mahalle, R. M., & Taggar, G. K. (2018). Yield loss assessment and establishment of economic threshold level of Maruca vitrata in pigeonpea. *Journal of Food Legumes, 31*, 36–44.

Mallikarjuna, N., Sharma, H. C., & Upadhyaya, H. D. (1997). Exploitation of wild relatives of pigeonpea and chickpea for resistance to *Helicoverpa armigera. Jounal of SAT Agricultural Research, 3*, 4.

Mallikarjuna, N., Saxena, R. K., Byre Gowda, M., & Varshney, R. K. (2017). Wide crossing Technology for Pigeonpea Improvement. In R. K. Varshney, R. K. Saxena, & S. A. Jackson (Eds.), *The Pigeonpea genome* (pp. 4–54). Springer International Publishing AG.

Mir, R. R., Saxena, R. K., Saxena, K. B., et al. (2012). Whole-genome scanning for mapping determinacy in pigeonpea (*Cajanus spp*). *Plant Breeding, 132*, 472–478.

Mir, R. R., Kudapa, H., Srikanth, S., et al. (2014). Candidate gene analysis for determinacy in pigeonpea (Cajanus spp). *Theoretical and Applied Genetics, 127*, 2663–2678.

Mir, R. R., Rather, I. A., Bhat, M. A., et al. (2017). Molecular mapping of genes and QTLs in pigeonpea. In R. K. Varshney, R. K. Saxena, & S. A. Jackson (Eds.), *The Pigeonpea genome* (pp. 41–54). Springer International Publishing AG.

Obala, J., Saxena, R. K., & Singh, V. (2019). Development of sequence-based markers for seed protein content in pigeonpea. *Molecular Genetics and Genomics, 294*, 57–68.

Odeny, D. A., Jayashree, B., & Ferguson, M. (2007). Development, characterization and utilization of microsatellite markers in pigeonpea. *Plant Breeding, 126*, 130–136.

Odeny, D. A., Jayashree, B., Gebhardt, C., & Crouch, J. (2009). New microsatellite markers for pigeonpea (*Cajanus cajan* (L) Millsp). *BMC Research Notes, 2*, 35.

Pande, S., Sharma, M., Mangla, U. N., et al. (2012). Phytophthora blight of Pigeonpea [*Cajanus cajan* (L) Millsp]: An updating review of biology, pathogenicity and disease management. *Crop Protection, 30*, 951–957.

Patel, K. A., Acharya, S., Prajapati, N., & Patel, J. B. (2012). Molecular identification of cytoplasmic male sterility based hybrid GTH 1 and its parents in pigeonpea. *Indian Journal of Genetics and Plant Breeding, 72*, 94–96.

Patil, P. G., Dubey, J., Bohra, A., et al. (2017). Association mapping to discover signifcant marker-trait associations for resistance against Fusarium wilt variant 2 in pigeonpea [*Cajanus cajan*(L) Millspaugh] using SSR markers. *Journal of Applied Genetics, 58*, 307–319.

Pazhamala, L. T., Shilp, S., Saxena, R. K., et al. (2017). Gene expression atlas of pigeonpea and its application to gain insights into genes associated with pollen fertility implicated in seed formation. *Journal of Experimental Botany, 68*, 2037–2054.

Perera, A. M., Pooni, H. S., & Saxena, K. B. (2001). Components of genetic variation in short duration pigeonpea crosses under waterlogged conditions. *Journal of Genetics & Breeding, 55*, 21–38.

Pratap, A. (2024). *Project coordinator's report 2023-24, All India Coordinated Research Project on Kharif Pulses* (p. 135). ICAR-Indian Institute of Pulses Research. ISBN 978-93-340-6544-2.

Pratap, A., Chaturvedi, S. K., Tomar, R., Rajan, N., Malviya, N., Thudi, M., Saabale, P. R., Prajapati, U., Varshney, R. K., & Singh, N. P. (2017). Marker-assisted introgression of resistance to fusarium wilt race 2 in Pusa 256, an elite cultivar of *desi* chickpea. *Molecular Genetics and Genomics, 292*, 1237. https://doi.org/10.1007/s00438-017-1343-z.292

Pratap, A., Das, A., Kumar, S., & Gupta, S. (2021). Current perspectives on introgression breeding in food legumes. *Frontiers in Plant Science, 11*. https://doi.org/10.3389/fpls.2020.589189

Promila, K., & Kumar, S. (2000). *Vigna radiata* seed germination under salinity. *Biologia Plantarum, 43*, 423–426.

Ramu, S. V., Rohini, S., Keshavareddy, G., et al. (2012). Expression of a synthetic cry1AcF gene in transgenic Pigeonpea confers resistance to *Helicoverpa armigera*. *Journal of Applied Entomology, 136*, 675–687.

Reddy, L. J., & Faris, D. G. (1981). A cytoplasmic-nuclear male sterile line in pigeonpea. *International Pigeonpea Newsletter, 1*, 16–17.

Reddy, B. V. S., Green, J. M., & Bisen, S. S. (1978). Genetic male-sterility in pigeonpea. *Crop Science, 18*, 362–364.

Reddy, M. V., Sarkar, N., Nene, Y. L., & Raju, T. N. (1991). Predisposing factors for Phytophthora blight of Pigeonpea. *Indian Phytopathology, 45*(2), 268–270.

Roorkiwal, M., Sawargaonkar, S. L., Chitikineni, A., et al. (2013). Single nucleotide polymorphism genotyping for breeding and genetics applications in chickpea and pigeonpea using the BeadXpress platform. *Plant Genome, 6*, 1–10.

Roorkiwal, M., Jarquin, D., Singh, M. K., et al. (2018). Genomic-enabled prediction models using multi-environment trials to estimate the efect of genotype×environment interaction on prediction accuracy in chickpea. *Scientific Reports, 8*, 11701.

Sarode, S. B., Singh, M. N., & Singh, U. P. (2007). Genetics of waterlogging tolerance in pigeonpea [*Cajanus cajan*(L) Millsp]. *Indian Journal of Genet and Plant Breed, 67*, 264–265.

Saxena, K. B. (2008). Genetic improvement of pigeonpea–A review. *Tropical Plant Biology, 1*, 159–178.

Saxena, K. B. (2014). Temperature-sensitive male sterility system in pigeonpea. *Current Science, 107*, 277–281.

Saxena, K. B., & Sharma, D. (1990). Pigeonpea: Genetics. In Y. L. Nene, S. D. Hall, & V. K. Sheila (Eds.), *The Pigeonpea* (pp. 137–157). CAB International.

Saxena, K. B., Rao, A. N., Singh, U., & Remanandan, P. (1996). Intraspecies variation in Cajanus platycarpus for some agronomic traits and crossability. *International Chickpea and Pigeonpea Newsletter, 3*, 49–51.

Saxena, K. B., Kumar, R. V., Srivastava, N., & Shiying, B. A. (2005). Cytoplasmic-nuclear male-sterility system derived from a cross between *Cajanus cajanifolius* and *Cajanus cajan*. *Euphytica, 145*, 289–294.

Saxena, R. K., Saxena, K. B., & Varshney, R. K. (2010). Application of SSR markers for molecular characterization of hybrid parents and purity assessment of ICPH 2438 hybrid of pigeonpea [*Cajanus cajan* (L) Millspaugh]. *Molecular Breeding, 26*, 371–380.

Saxena, R. K., Cui, X., Thakur, V., et al. (2011). Single feature polymorphisms (SFPs) for drought tolerance in pigeonpea (*Cajanus spp*). *Functional & Integrative Genomics, 11*, 651–657.

Saxena, R. K., Penmetsa, R. V., Upadhyaya, H. D., et al. (2012). Large-scale development of cost-effective single-nucleotide polymorphism marker assays for genetic mapping in pigeonpea and comparative mapping in legumes. *DNA Research, 19*, 449–461.

Saxena, R. K., Wettberg, E. V., Upadhyaya, H. D., et al. (2014). Genetic diversity and demographic history of Cajanus spp, Illustrated From genome-wide SNPs. *PLoS One, 9*, 1–9.

Saxena, R. K., Kale, S. M., Kumar, V., et al. (2017a). Genotyping-by-sequencing of three mapping populations for identification of candidate genomic regions for resistance to sterility mosaic disease in pigeonpea. *Scientific Reports, 7*, 1813.

Saxena, R. K., Singh, V. K., Kale, S. M., et al. (2017b). Construction of genotyping-by-sequencing based high-density genetic maps and QTL mapping for Fusarium wilt resistance in pigeonpea. *Scientific Reports, 7*, 1911.

Saxena, R. K., Obala, J., Sinjushin, A., et al. (2017c). Characterization and mapping of Dt1 locus which co-segregates with CcTFL1 for growth habit in pigeonpea. *Theoretical and Applied Genetics, 130*, 1773–1784.

Saxena, K. B., Saxena, R. K., & Varshney, R. K. (2017d). Use of immature seed germination and single seed descent for rapid genetic gains in pigeonpea. *Plant Breeding, 136*, 954–957.

Saxena, R. K., Rathore, A., Bohra, A., et al. (2018a). Development and application of high density Axiom® Cajanus SNP Array with 56 K SNPs to understand the genome architecture of released cultivars and founder genotypes for redefining future pigeonpea breeding programs. *Plant Genome, 11*, 3.

Saxena, R. K., Patel, K., & Kumar, S. C. V. (2018b). Molecular mapping and inheritance of restoration of fertility (Rf) in A4 hybrid system in pigeonpea (*Cajanus cajan* (L) Millsp). *Theoretical and Applied Genetics, 131*, 1605–1614.

Saxena, R.K., Patel, K., Kumar S.C.V., et al. (2018c) Molecular mapping and inheritance of restoration of fertility (Rf) in A4 hybrid system in pigeonpea (Cajanus cajan (L.) Millsp.). *Theorical and Applied Genetics, 131*, 1605–1614.

Saxena, R. K., Hake, A., Bohra, A., et al. (2021). A diagnostic marker kit for Fusarium wilt and sterility mosaic diseases resistance in pigeonpea. *Theoretical and Applied Genetics, 134*, 367–379.

Sekhar, K., Priyanka, B., Reddy, V. D., & Rao, K. V. (2010). Isolation and characterization of a pigeonpea cyclophilin (CcCYP) gene and its over expression in Arabidopsis confers multiple abiotic stress tolerance. *Plant, Cell and Environment, 33*, 1324–1338.

Sharma, H. C., & Franzmann, B. A. (2000). Biology of the legume pod borer, Maruca vitrata and its damage to pigeonpea and adzuki bean. *Insect Science and its Application, 20*, 99–108.

Sharma, M., Ghosh, R., Telangare, R., et al. (2016). Environmental influences on pigeonpea-Fusarium udum interactions and stability of genotypes. *Frontiers in Plant Science, 7*, 1–9.

Shruthi, H. B., Hingane, A. J., & Reddi, S. M. (2017). Super early pigeonpea for sustainable cropping system intensification and diversification. In *National symposium on pulses for nutritional security and agricultural sustainability from December 2-4 at IIPR Kanpur.*

Silim, S., Gwata, E. T., Coeb, R., & Omanga, P. A. (2007). Response of pigeonpea genotypes of different maturity duration to temperature and photoperiod in Kenya. *African Crop Science Journal, 15*, 73–81.

Singh, N. K., Gupta, D. K., Jayaswal, P. K., et al. (2012). The first draft of the pigeonpea genome sequence. *Journal of Plant Biochemistry and Biotechnology, 21*, 98–112.

Singh, V. K., Khan, A. W., Saxena, R. K., Kumar, V., et al. (2016). Next-generation sequencing for identification of candidate genes for Fusarium wilt and sterility mosaic disease in pigeonpea (*Cajanus cajan*). *Plant Biotechnology Journal, 14*, 1183–1194.

Singh, I. P., Sameer Kumar, C. V., Byregowda, M., et al. (2017a). Exploitation of heterosis and new plant types in pigeonpea. In *National symposium on pulses for nutritional security and agricultural sustainability from December 2–4 at IIPR Kanpur* (pp. 37–44).

Singh, V. K., Khan, A. W., Saxena, R. K., et al. (2017b). Indel-seq: A fast-forward genetics approach for identifcation of trait-associated putative candidate genomic regions and its application in pigeonpea (*Cajanus cajan*). *Plant Biotechnology Journal, 15*, 906–914.

Sinha, P., Singh, V. K., Suryanarayana, V., et al. (2015a). Evaluation and validation of housekeeping genes as reference for gene expression studies in pigeonpea (*Cajanus cajan*) under drought stress conditions. *PLoS One, 10*, e0122847.

Sinha, P., Saxena, R. K., Singh, V. K., & Varshney, R. K. (2015b). Selection and validation of housekeeping genes as reference for gene expression studies in pigeonpea (*Cajanus cajan*) under heat and salt stress conditions. *Frontiers in Plant Science, 6*, 1071.

Sinha, P., Saxena, K. B., & Saxena, R. K. (2015c). Association of *nad7a* gene with cytoplasmic male sterility in pigeonpea. *Plant Genome, 8*, 1–12.

Sinha, P., Pazamala, L. T., & Singh, V. K. (2016). Identification and validation of selected universal stress protein domain containing drought-responsive genes in pigeonpea (*Cajanus cajan*). *Frontiers in Plant Science, 6*, 1–9.

Tao, Y., Zhao, X., & Mace, E. (2019). Exploring and exploiting pan-genomics for crop improvement. *Molecular Plant, 12*, 156–169.

Tuteja, R., Saxena, R.K., Davila, J., et al. (2013). Cytoplasmic male sterility-associated chimeric open reading frames identified by mitochondrial genome sequencing of four Cajanus genotypes. *DNA Research, 20*, 485–495.

Vales, M. I., Srivastava, R. K., Sultana, R., et al. (2012). Breeding for earliness in pigeonpea: Development of new determinate and nondeterminate lines. *Crop Science, 52*(6), 2507–2516.

Varshney, R. K., Chen, W., Yupeng, L., et al. (2012). A Draft genome sequencing of pigeonpea (Cajanus cajan) an orphan legume crop of resource-poor farmers. *Nature Biotechnology, 30*(1), 83–89.

Varshney, R. K., Saxena, R. K., & Upadhyaya, H. D. (2017). Whole-genome resequencing of 292 pigeonpea accessions identifes genomic regions associated with domestication and agronomic traits. *Nature Genetics, 49*, 1082–1088.

Varshney, R. K., Pandey, M. K., Bohra, A., et al. (2019). Towards sequence-based breeding in legumes in postgenome sequencing era. *Theoretical and Applied Genetics, 132*, 797–816.

Voss-Fels, K. P., Herzog, E., & Dreisigacker, S. (2018). Speed GS to accelerate genetic gain in spring wheat. In T. Miedaner & V. Korzun (Eds.), *Applications of genetic and genomic research in cereals* (1st ed.). Woodhead Publishing.

Wallace, J. G., Rodgers-Melnick, E., & Buckler, E. S. (2018). On the road to breeding 40: Unraveling the good, the bad, and the boring of crop quantitative genomics. *Annual Review of Genetics, 52*, 421–444.

Yadav, P., Saxena, K. B., Hingane, A., et al. (2019). An "Axiom Cajanus SNP Array" based high density genetic map and QTL mapping for high-selfing flower and seed quality traits in pigeonpea. *BMC Genomics, 20*, 235.

Yang, S. Y., Pang, W., Harper, J., et al. (2006). A low level of genetic diversity in cultivated pigeonpea compared to its wild relatives is revealed by diversity arrays technology (DArT). *Theoretical and Applied Genetics, 113*, 585–595.

Yang, S. Y., Saxena, R. K., Kulwal, P. L., et al. (2011). The first genetic map of pigeonpea based on diversity arrays technology (DArT) markers. *Journal of Genetics, 90*, 103–109.

Genomics and Transcriptomics Approaches for Developing Biotic and Abiotic Stress Tolerant Lentil Cultivars

Anil Kumar, Pooja Yadav, Archana Kushwaha, Ravi Ranjan Kumar, Chandan Kishore, Anand Kumar, Seema Prajapati, Kumari Rajani, Sunita Pandey, Prem Chand Gyani, and Jitesh Kumar

Abstract Lentil (*Lens culinaris* Medik. ssp. *culinaris*) is one of the oldest cool season food legumes. It is a highly desirable pulse crop providing protein, iron, selenium, and zinc when included in the diet. Lentil originated from *L. culinaris* Medik. ssp. *orientalis* in the Near East, although subsequent domestication and selection over several years have reduced the genetic variation in lentil cultivars. Thus, characterization and utilization of crop wild relatives and prebreeding require attention in the crop. Other than a narrow genetic base, low and unstable yield, longer maturity, poor seedling vigor, and flower drop accompanied with abiotic and biotic stresses are the bottlenecks in lentil breeding and cultivation. Conventional breeding techniques have been deployed well for improving traits which are mostly

A. Kumar (✉) · P. Yadav · C. Kishore · A. Kumar
Department of Plant Breeding and Genetics, Bihar Agriculture University, Sabour, Bihar, India

A. Kushwaha
College of Community Science, G.B.P.U.A. &T., Pant Nagar, Uttarakhand, India

R. R. Kumar · J. Kumar
Department of Molecular Biology and Genetic Engineering, Bihar Agriculture University, Sabour, Bihar, India

S. Prajapati
Department of Molecular Biology and Genetic Engineering, Bihar Agriculture University, Sabour, Bihar, India

Department of Agronomy, Bihar Agricultural University, Sabour, Bhagalpur, Bihar, India

K. Rajani
Department of Seed Science and Technology, Bihar Agriculture University, Sabour, Bihar, India

S. Pandey
Department of Genetics and Plant Breeding, Agriculture University, Jodhpur, Rajasthan, India

P. C. Gyani
Division of Genetics, ICAR-Indian Agriculture Research Institute, New Delhi, India

A. Pratap et al. (eds.), *Integrated Improvement of Food Legumes*,
https://doi.org/10.1007/978-3-031-82629-0_10

monogenic. However, the majority of economic traits, such as yield, are governed by polygenic inheritance and G × E interaction. Recent advances in plant biotechnological tools, high-throughput genotyping, and phenotyping procedures and their applications are a way forward for improvement of such complex traits. Mutation breeding is another tool that has complemented genetic improvement in this self-pollinated crop. After the success of genome sequencing in lentils, new genomic resources are being developed, which will aid lentil researchers. The parallel utilization of genetic and genomic resources in the crop will lead to the accelerated development of improved varieties. This chapter provides an overview of the importance, current status, and future opportunities of lentil genetic improvement.

Keywords Biodiversity · Genetic resources · Mutation breeding · Biotechnological advances · Omics · Genomic tools · Stress resistance · Crop Improvement

Abbreviations

AFLP	Amplified fragment length polymorphism
CAS	CRISPR-associated proteins
CRISPR	Clustered regularly interspaced palindromic repeats
DArT	Diversity Arrays Technology marker
GAB	Genomics-assisted breeding
ISSR	Inter simple sequence repeats
LD	Linkage disequilibrium
MABC	Marker-assisted backcrossing
MAS	Marker-assisted selection
NGS	Next-generation sequencing
PCR	Polymerase chain reaction
QTL	Quantitative trait locus
RAPD	Random amplified polymorphic DNA
RFLP	Restriction fragment length polymorphism
RIL	Recombinant inbred line
SSR	Simple sequence repeat
TILLING	Target-induced local lesions in genomes

1 Introduction

Lentil (*Lens culinaris* Medikus) is a highly nutritious protein- and iron-rich pulse crop that fits into the class of target food crops or the smart crops grown throughout the subtropics. It is one of the oldest grain legumes that originated in the Near East from *L. culinaris* Medik. ssp. *orientalis*. Production of lentil has increased over the last few decades; however, the recent change in climate due to human activities and practices such as domestication and selection posed a threat to the genetic diversity of indigenous lentil germplasm across the world, including the Indian subcontinent. Of late, the unprecedented growth in human population has resulted in the reduction of per capita food availability. By the year 2050, the global population is expected to cross nine billion, which will require 70% more demand to feed the population. This calls for urgent demands to attain enhanced crop productivity (FAO, 2025). In addition, enhanced nutrition and better keeping quality are also matters of concern. The national and international gene banks have approximately 37,000 accessions maintained. A stable high seed yield in addition to resistance against abiotic and biotic stresses is the major concern of lentil-producing countries. The improvement of many monogenic traits has been carried out by conventional breeding techniques such as selection and recombination. However, these improvements still lack in addressing economic traits such as yield being governed by polygenes and G × E interaction. Diverse species of the genus *Lens* can aid in the enhancement of genetic variation for target traits in lentil in order to produce new varieties or modify existing ones. The present state of decline in agricultural land as a result of a decrease in water resources and addition to urbanization and industrialization demands has led to accelerated targets of food production with limited agricultural inputs. Therefore, the need of the time is to generate high-yielding improved crop varieties that are genetically diverse, input use-efficient, climate-change resilient, and are also biofortified and versatile to various farming practices and agroecosystems. The implication of plant breeding strategies for the development of diverse suite of such crop varieties for crucial food crops will be a solution to fight the global food crisis. The limited genetic variability of grain legumes is a major bottleneck to crop improvement. Therefore, it is necessary to induce innovative breeding tools for generating new genetic variability for desirable traits in these crops. The national and international gene banks have approximately 37,000 accessions maintained. These genetic resources can be evaluated and characterized for the identification of trait-specific doors and put to use for introgression breeding of desirable genes. Induced mutagenesis is another vital breeding method, which can ease the availability of lentil genomic resources. Various success stories in the application of molecular markers and biotechnological techniques will further speed up the development of improved varieties. This chapter reviews the status of genetic improvement in lentil and its genetic and genomic resources and provides a compendium of important aspects of the crop's origin, distribution, diversity and taxonomy, conservation and maintenance of germplasm, and its uses and limitations in plant breeding with the way forward.

2 Importance

Lentil is a highly nutritious and easily digestible pulse crop used as food since ancient times, being among the first foods cultivated (Sarker & Erskine, 2006). This pulse crop has great significance in cereal-based cropping systems due to its nitrogen-fixing ability and high-protein seeds and nutritious plant parts used in food, feed, and fodder. Lentil is known by many names in different countries where it is an essential legume, indicating its widespread cultivation and importance. It is a source of essential macro- and micronutrients of dietary importance for human welfare. The presence of insignificant quantities of cholesterol, fat, and antinutritional factors found in lentil makes it the most desired protein source for human consumption. Table 1 presents the nutritional value of lentil. Furthermore, short cooking time of lentils leads to a decreased nutrient loss. Different lentil recipes are used throughout South Asia. Lentil *dal* or curry is a frequently consumed recipe of everyday diet in the Indian subcontinent generally eaten together with rice. Other dishes prepared from rice and lentil include *Mejadra* or *mujaddra* mostly made in

Table 1 Nutritional value of lentil, raw (dry weight) per 100 g

Nutrient	Value
Water	10.40 g
Energy	343 kcal
Protein	25.80 g
Total lipid (fat)	1.06 g
Carbohydrate	60.08 g
Fiber, total dietary	30.50 g
Sugars, total	2.03 g
Minerals	
Calcium, Ca	56 mg
Iron, Fe	7.54 mg
Magnesium, Mg	122 mg
Phosphorus, P	451 mg
Potassium, K	955 mg
Sodium, Na	6.00 mg
Zinc, Zn	4.78 mg
Vitamins	
Vitamin C, total ascorbic acid	4.40 mg
Thiamin	0.873 mg
Riboflavin	0.211 mg
Niacin	2.605 mg
Vitamin B6	0.540 mg
Folate, DFE	479 μg
Vitamin A, RAE	2.00 μg
Vitamin A, IU	39 IU
Vitamin E (alpha-tocopherol)	0.49 mg
Vitamin K (phylloquinone)	5.0 μg
Lipids	
Fatty acids, total saturated	0.156 g
Fatty acids, total monounsaturated	0.189 g
Fatty acids, total polyunsaturated	0.516 g

Source: USDA, National Nutrient Database for Standard Reference 2018

Arab countries, *Khichdi*, etc. Lentil flour is also commonly used to make stews, soups, and purees. Breads and cakes are also prepared from a mixture of lentil flour with cereals.

Lentil consumption offers numerous benefits to human beings, viz., eases oxidative stress, improves serum antioxidant capacity, and reduces concentrations of lipoprotein cholesterol, thus reducing the risks of cardiovascular diseases. Moreover, the potential application of lentil protein hydrolysates as hypotensive ingredients containing angiotensin I-converting enzyme inhibitory peptides has been identified. Therefore, lentils could be considered a valuable source of cardioprotective compounds. A significant decrease in blood glucose was reported on adding 50 g of cooked lentil to a diabetic patient's diet. Lentil seeds are also used for treating type II diabetes, skin burns, and infections, lowering blood pressure.

3 History and Origin

The genus Lens Miller is part of the family Fabaceae (Leguminosae) from the subfamily Faboideae, tribe Fabeae (Soltis et al., 2011). Another report says its subfamily is Papilionaceae, tribe Vicieae. Lentil is a diploid ($2n = 2x = 14$) annual plant with self-pollinating behavior and with an estimated genome size of 4063 Mbp/C (Arumuganathan & Earle, 1991). It seems to be domesticated from its wild progenitor *Lens culinaris* ssp. *orientalis* in Syria and Turkey approximately 8500 BC (Cubero, 1981). Some report that its history dates back to the pre-Pottery Neolithic period 11,000–10,000 years ago. The archeological studies still do not provide direct diagnostic traits (such as indehiscent pod) to ensure the origin of lentil domestication, though lentil seed size suggests selection. It has also been reported that lentil derivation was not certain. Until 1973, Barulina's (1930) dogma was accepted, which proposes the origin of lentil cultivation between the Hindu Kush and the Himalaya. Later, on the basis of the presence of pulses among the charred plant remains retrieved from archeological sites, Kislev and Bar-Yosef (1988) believed lentils were the earliest domesticated plants of the Near East although they too cautioned that there was not sufficient evidence to support this. Morphological change cannot be the first indication of domestication for a longer time; in spite of the fact that a prolonged period of increasingly intensive human management typically precedes the manifestation of archeologically detectable morphological change in managed crops. Agriculture in the Near East arose in the context of broad-based systematic human efforts of cultivating plant resources. There is a current controversy over the slow or fast rate (duration) of the process of domestication. A fast rate of plant domestication is supported by biological evidence of Near Eastern wild and domesticated lentil. Initial domestication of lentils occurred in southeastern Turkey or northern Syria based on genetic and archeological evidence. Lentils spread quickly as a crop from there to the southern Levant. A monophyletic origin was hypothesized by Zohary (1999), who applied this theory to lentils being "very likely taken into cultivation only once or – at most – a very few times, without considering allozyme data published". Recently, studies based on data of eight founder species suggested domestication in a small region of the southern Levant. Also, botanical, genetic, and archeological evidences point to a core area of domestication in southeastern Turkey and present-day northern Syria, near the Tigris and Euphrates rivers. A study of wild and cultivated lentil further supports the hypothesis of a polycentric origin of domestication (Alo et al., 2011). Abbo et al. (2012) expressed that "only detailed phylogenetic studies of representative collections of wild and domesticated forms can determine the place of origin and their phylogeny."

Wild *Lens* taxa are widely distributed in the Mediterranean basin, Aegean, and southwestern Turkey. An in situ and ex situ gap analysis using taxonomic, ecological, geographic, and conservation information for 672 wild *Lens* collated from ICARDA (International Center for Agricultural Research in the Dry Areas) and GBIF (Global Biodiversity Information Facility) was performed by Maxted et al. (2010) on data sets collected by the authors over 25 years. Gap analysis is a process by which the distribution of taxon and vegetation types is compared to identify biodiversity for its in situ or ex situ protection. Maxted's gap analysis refined the regions of highest *Lens* species richness (three to four species) to the Crimea Peninsula and along southeastern Turkey through the eastern Mediterranean countries such as Syria, Jordan, Israel, and Palestinian Authority.

Lens culinaris ssp. *orientalis* (Boiss.) Ponert has an eastern distribution from Turkey, Cyprus, and Palestine across to Uzbekistan. *Lens culinaris* subsp. *odemensis* (Ladiz.) M.E. has a more restricted distribution in the east, extending from Turkey southwards to Syria and Palestine. A single population of *Lens culinaris* subsp. *tomentosus* (ladiz.) M.E. has been found in Libya. *Lens ervoides* (Brign.) Grande has a broad distribution from Spain to Ukraine and south to Jordan. *L. nigricans* has a western distribution from Spain to Turkey and south to Morocco (Ferguson et al., 1996) and east to Crimea and the eastern shore of the Mediterranean Sea. *Lens nigricans* (M. Bieb.) Godr. grows in small diffused colonies on stony hillsides and shallow rocky soils in pine forest clearings. *Lens lamottei* (Czefr.) grows in Morocco. The distributions of all wild taxa overlap only in Aegean and southwestern portions of Turkey. Unfortunately, like other Mediterranean countries, Turkey is also suffering from rapid loss of valuable genetic resources, which have the potential to provide useful genetic material for plant breeding efforts. But they are being eroded mostly by habitat destruction. Ferguson et al. (1996) noted "the poor competitive ability and palatability of *Lens* species, together with the fact that they occur in small disjunct populations, intensifies this threat."

Molecular diversity evaluations of ex situ germplasm collections include studies completed with DNA-based markers, such as random amplified polymorphic DNA (RAPDs; Sultana & Ghafoor, 2008; Hoque & Hasan, 2012), intersimple sequence repeats (ISSRs; Fikiru et al., 2010, 2011), amplified fragment length polymorphisms (AFLPs), and simple sequence repeats (SSRs). Further, international germplasm collection molecular diversity is presented in three published studies. Central Asian and Caucasian cultivated lentil germplasm were genotyped with five SSR markers and clustered into six groups. Fourteen newly reported SSR markers were used to examine the molecular diversity of 109 accessions including both cultivated lines and wild *Lens* species (Hamwieh et al., 2009). They found that the wild accessions were rich in alleles (151 alleles) compared to cultigens (114 alleles). These lines also clustered into two groups, one cultivated and the other wild germplasm. A large study published to date consisted of 133 domesticated lentil and 175 wild lentil accessions conducted by ICARDA using 22 cross-genera SSR markers (Alo et al., 2011). Structure analysis revealed eight haplotype groups ($K = 8$). All groups consisted of one taxon except one, which had all taxa except domesticated *Lens* (Alo et al., 2011). Linkage disequilibrium (LD) was calculated and varied across the individual groups, with the higher LD in the cultivated lines as found in other domesticated crop species. However, other biochemical genetic diversity research has been conducted on lentil genetic resources. Examples include a study on the diversity of lentil seed starch and complex carbohydrates, where the diversity discovered invites researchers, especially breeders, to exploit the variability uncovered. Two studies have looked at the seed protein profiles of 144 accessions, mainly landraces of Pakistan (Sultana et al., 2006) and 14 cultivars of Turkey (Yüzbaşioğlu et al., 2008). Both studies have identified useful diversity. The lentil seed proteome was determined for two Italian landraces; multivariate analysis of 145 differentially expressed protein spots demonstrated that 52 proteins are required to discriminate. Taxonomically, lentil holds an intermediate position between *Vicia* and *Lathyrus*. Zohary (1972) reported five species of cultivated *L. culinaris* Medikus. [*L. esculenta* Moench] and four wild species: *Lens monbretti* (Fisch. & May) Davis and Plitm. [*L. kotschyanum* (Boiss.) Nab.; *L. kotschyaya* (Boiss.) Alef.]; *Lens nigricans* (Bieb.) Godr. [*Ervum nigricans* Bieb.]; *Lens ervoides* (Brign.) Grande [*L. lenticula* (Schreb,) Alef.] and *Lens orientalis* (Boiss.) Hand.-Mazz. During that time, all the morphological evidence indicated five lentil species. *L. monbretti* (Fisch. & Mey.) is morphologically and cytologically different from the other *Lens* species and was moved back to the genus *Vicia*. Pinkas et al. (1985) proposed five *Lens* species, namely *L. culinaris*, *L. orientalis*, *L. odemensis*, *L. ervoides*, and *L. nigricans*, based on allozyme divergence. Hoffman and Smith (1986) proposed two species and five taxa, namely *L. culinaris* with three subspecies *Lens culinaris* subsp. *culinaris*, *L. culinaris* subsp. *orientalis* and *L. culinaris* subsp. *odemensis*; *L. nigricans* with two subspecies *L. nigricans* subsp. *nigricans* and *L. nigricans* subsp. *ervoides*. Ladizinsky updated the *Lens* taxa (1997) and defined four species by reducing *L. orientalis* to a subspecies and creating two new *Lens* species, namely *L. lamottei* Czefr. and *L. tomentosus* Ladiz. Chloroplast DNA marker variation briefly concluded there are six species in the *Lens* taxa.

Further molecular phylogeny analysis clarifies and confuses *Lens* taxonomy regarding species and subspecies. From the period between 1979 and 2005, numerous studies focused on the phylogeny of *Lens* using the molecular tools of various marker classes, including isozymes (Ferguson et al., 1998), restriction fragment length polymorphism (RFLPs) (Havey & Muehlbauer, 1989), RAPDs, and AFLPs. Fortunately, all the studies indicate that *Lens culinaris* spp. *orientalis* is the closest progenitor of cultivated lentil. Ferguson et al. (2000) proposed the taxonomy of four species, reducing *L. odemensis* and *L. orientalis* to subspecies of *L. culinaris* based on morphological, isozyme, and RAPD marker data combined (Table 7.1). The contemporary literature is fraught with differing interpretations of the exact number of taxa and splits (e.g., Tahir et al., 2012). The taxonomy is understandably difficult given the close relationships between the *Lens* taxa (Ferguson et al., 2000). This taxonomic description for *Lens* is accepted by the USDA for use in GRIN. This study is given due weightage by the taxonomic community as it combines the molecular characterization with botanical descriptors of the species and subspecies for the classification of the herbarium samples. Many taxonomists worked in order to classify lentils, but the most accepted latest classification recommends the following seven taxa in four *Lens* species (Ferguson and Erskine 2001; Ferguson et al. 2000). In 1787, Medikus, a German botanist, has given the credit to the scientific name of lentil as *Lens culinaris* (Fratini et al. 2014).

1. *Lens culinaris* Medik.

    ssp. *culinaris*
    ssp. *orientalis* (Boiss.) Ponert
    ssp. *tomentosus* (Ladiz.) M.E. Ferguson& al.
    ssp. *odemensis* (Ladiz.) M.E. Ferguson& al.

2. *Lens ervoides* (Brign.) Grande
3. *Lens nigricans* (M. Bieb.) Godron
4. *Lens lamottei* Czefr.

4 Lentil Distribution and Production Statistics

Lentil is cultivated in an area of 5.67 million ha globally, giving a production of 7.07 million metric tons. Countries such as Canada, India, Australia, and Turkey are the top producers of this pulse crop (Fig. 1; FAO, 2025).

The average productivity in India comes to about 952 kg/ha, which is still lesser compared to the global average of 1245 kg/ha (FAO, 2025). Figure 2 gives a graphical representation of trends in the area and production of lentil in the world based on 10 years of data (from 2014 to 2023). It can be seen that both area and production trends are the crop increased throughout the decade. However, toward the increase, it slowed down and became stagnant after 2016. This calls for an alert to the researchers of the crop to investigate the reasons behind this lag, which is discussed in more detail in the following subsection.

Figure 3 represents major continents involved in lentil production, which gives an overview of are as to be targeted for further studies in this grain legume. In India, lentil is grown on over 1.64 million ha land giving a production of 1.56 million tons, with Madhya Pradesh, Uttar Pradesh, Bihar, West Bengal and Chhatisgarh are the major growing states.

The lentils were classified into two groups based on seed size by Barulina (1930) viz. macrosperma/small-seeded type (<6 mm diameter) and macrosperma/large-seeded types (>6 mm diameter). Lentil grains are an important source of nutritious

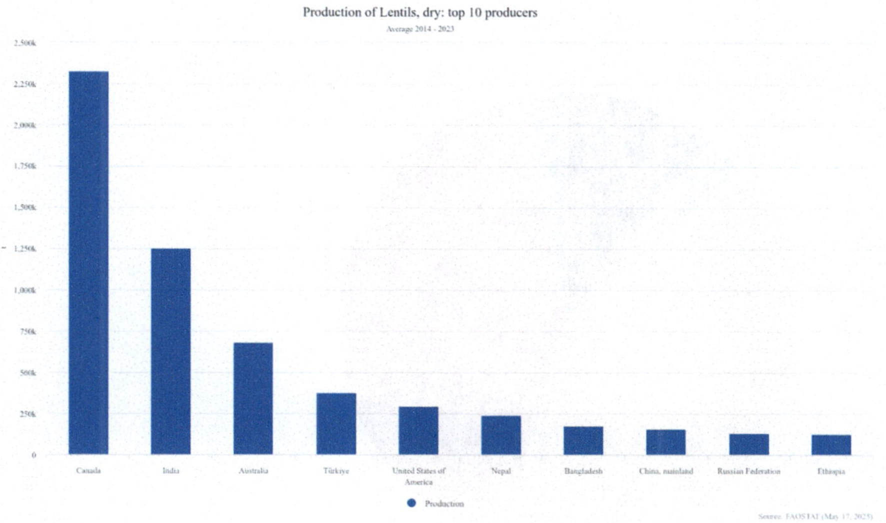

Fig. 1 Production of Lentil: top 10 producers

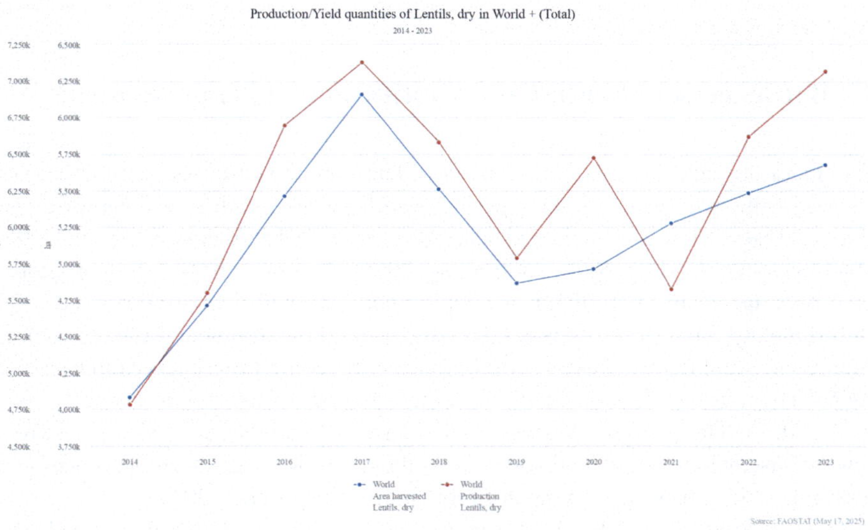

Fig. 2 Production/Yield quantities of Lentils in World + (Total)

dietary protein (22–35%), carbohydrates, minerals, and vitamins in human nutrition. They are mainly consumed as dal, fried snacks, in the form of soup, or sprouts. Lentil grains and straw are also utilized as animal feed. It is considered a high-value crop option for farmers as it provides the benefits of crop rotation favorable to weed and pest management.

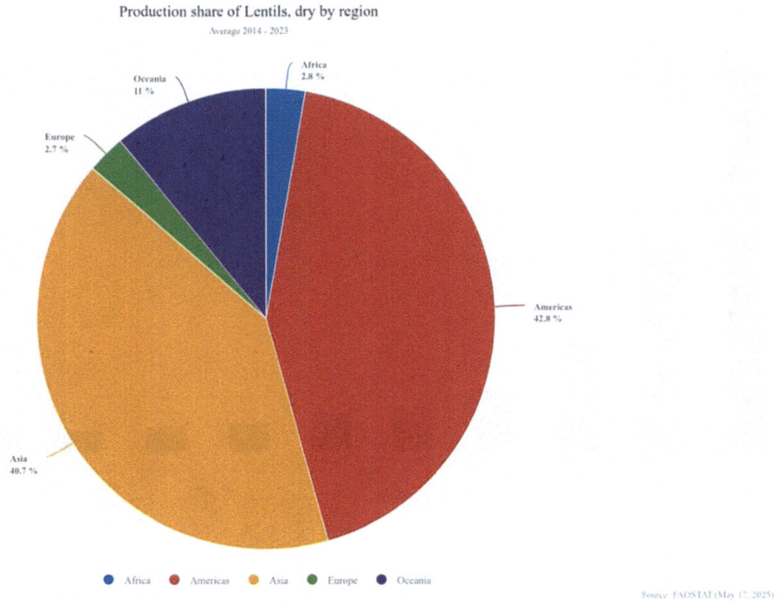

Fig. 3 Production share of Lentils by region

5 Bottlenecks in Lentil Crop Cultivation and Improvement

The major obstacles to good productivity of lentil are the narrowed genetic base of released cultivars, shorter growing periods, biotic stresses such as wilt and rust, and abiotic stresses such as drought and heat (Amin et al., 2015; Asnake & Bejiga, 2003; Muehlbauer et al., 2006; Sarker & Erskine, 2006). A number of breeding programs have been taken up in lentil for developing cultivars with increased productivity. Unfortunately, little progress was made due to high environmental effects, G × E interactions, and repeated use of similar lines in breeding programs (Kumar & Ali, 2006). The domestication in lentil led up to 40% loss of its genetic diversity (Alo et al., 2011). The pedigree analysis of Indian lentil varieties proved their narrow genetic base, which could be broadened by the use of Mediterranean landraces and crossable wild subspecies having desirable traits (Kumar et al. 2014).

Heat stress is another barrier for lentil-growing farmers. It can be defined as a period of prolonged hot temperature causing irreversible damage to the growth and development of crop plants (Delahunty et al. 2018). In the key lentil-growing areas, the crop is exposed to >35 °C temperature at flowering and pod-filling stages during maturity (Delahunty et al., 2015). This problem becomes even more serious in rice–lentil crop rotations, since the late harvesting of rice delays lentil sowing. As a result, the reproductive phase of lentil is placed under high-temperature stress (Subbarao et al., 2001). High temperature results in drastic losses of yields attributable to poor grain filling in lentil. In addition to the above, high temperature deteriorates physiological processes leading to forced grain maturity (Redden et al., 2014). A yield reduction of 87% was observed in lentil plants grown under field conditions in the pots when exposed to a temperature of >38/23 °C (average day/night temperature) during the reproductive phase (Bhandari et al., 2016). This could be worse, as higher temperatures by reason of global warming are ever challenging for lentil production and productivity (Kaur et al., 2015).

6 Germplasms Conservation

Germplasm accessions of the genus *Lens* comprise wild relatives, landraces, and advanced breeding materials. Globally, *Lens* has 41 leading collectors together holding 43,214 accessions. Official world germplasm collections of lentils for breeding purposes are maintained by ICARDA, and the ICAR-National Bureau of Plant Genetic Resources (NBPGR), the Vavilov Research Institute of Plant Industry, and the USDA collection at the Regional Plant Introduction Station, Pullman, Washington are the other major lentil germplasm repositories. The Global Crop Diversity Trust (https://www.croptrust.org) is coming up with new conservation strategies. ICARDA leads in conserving the global *Lens* germplasm accessions with 24% of the total accessions including 583 wild accessions. The South Asian subregion conserved 3022 lentil accessions, with the largest collection at NBPGR, India and the working collection at ICAR-Indian Institute of Pulses Research, Kanpur, India. The Global Seed Vault in Svalbard, Norway, famous as the Doomsday Vault, is also preserving a wide variety of plant seeds including lentil seeds collected from operational gene banks worldwide since 2008. A number of other gene banks with large lentil collections are working to conserve diverse accessions of lentil genetic resources to be assessed for future breeding and research.

7 Breeding Objectives

The key goals in lentil breeding are higher and stable seed yield, disease resistance, and better seed quality (Muehlbauer et al., 1995). Genetically diverse parents are used in lentil breeding to generate new gene combinations with traits of interest such as increased yield and stress resistance (Sarker et al., 2009). Pureline selection in heterogeneous landraces was the most common cultivar development method during initial phases of lentil improvement (Muehlbauer, 1992). The application of pureline selection developed genetically uniform and locally adaptable lentil cultivars from local landraces or introduced germplasm. Species from these selection efforts had decreased pod dehiscence and dormant seeds, with a considerable increase in erect plant habits and seed size (Zohary, 1996). Hybridization and selection then took over as methods preferred for the quick development of new cultivars/genotypes in lentil over the past few decades. Lentil being predominantly self-pollinating species has very low outcrossing rate. Therefore, the breeding methods in lentil alike other self-pollinated crops viz., pureline selection or hybridization followed by the bulk method, the pedigree method, the single seed descent, or a modification of these procedures along with mutation breeding (Muehlbauer et al., 2009; Toker et al., 2007). Genetic improvement of any trait through plant breeding is only successful if a crop gene pool is available with sufficient variability for a trait.

7.1 Plant Introduction

Plant introduction is a method of obtaining high yield and wide adaptability of lentil cultivars from within or outside the country. The soil and climatic conditions of the new location where the cultivar is introduced alter the success of plant introduction. Homozygous purelines are preferred as they are adaptable compared to a heterozygous segregating population. This is because the latter requires recognizing productive line with target traits. Plant introduction is a quick and economic way of developing cultivars. For developing countries, it is even more feasible due to the availability of lesser area under crop cultivation, economic curtailments, or the lack of skilled personnel. Lentil is reported to be introduced into South Asia around 2000 BC from West Asian regions via Afghanistan (Cubero, 1981; Materne & Siddique, 2009). Selection, introduction, and acclimatization of lentil are being done at different institutes globally, ICARDA being the leading one. In order to widen the narrow genetic base of small-seeded Pilosae type, bold-seeded Precoz type, of Argentine origin, was introduced from ICARDA (Erskine et al., 1998). Lentil cultivars Vipasha and VL 507 in India; Simal, Sikhar, Khajura Masuro 1, and Khajura Masuro 2 in Nepal; and Mansehra 89 and Shiraz 96 in Pakistan were also introduced from ICARDA (Rahman et al., 2009).

7.2 Hybridization

Hybridization is a method of integrating desirable traits from two or more parents into progeny cultivar. The key objective of hybridization is to elevate the degree of genetic variation.

7.2.1 Selection of Parents

The selection of appropriate parents plays a crucial role in the success of hybridization. Depending on the objective, the choice of parents is made. One or two parents are selected if higher yield and wider adaptability are the improvement criteria. If broadening of genetic base is desired, diverse parents are selected. Biometrical approaches can be deployed to analyze the diversity and combining ability of genotypes to be used in hybridization. DNA-based markers facilitate precise estimation of genetic diversity through germplasm characterization. Lentil diversity assessment studies in recent times using SNP markers suggested the selection of contrasting parental genotypes for lentil breeding programs (Lombardi et al., 2014).

7.2.2 CrossBreeding Techniques

Ladizinsky (1992) defined crossability as the potential of creating fertile F_1 hybrid from intercrossing between individuals belonging to different taxa. Crossbreeding in lentil is a cumbersome task with a success range of a mere 20–50% owing to tiny and fragile flowers being prone to injury during emasculation and pollination. Favorable results in artificial hybridization also depend upon the genotypes involved and climatic conditions prevailing such as temperature and humidity. For successful hybridization,

(i) Selection of appropriate size flower buds should be done
(ii) Lateral buds should be preferred over the terminal ones (Sindhu et al., 1981)
(iii) Complete care should be ensured during emasculation and pollination to keep away any mechanical injury to the floral parts
(iv) Timings of pollination and fertilization are key determining factors for the success rate. Low temperatures, afternoon emasculation, and pollination during the following morning are recommended. In case of high temperatures, morning emasculation followed by prompt pollination at same time is also successful (Bejiga & Tessema, 1981; Pundir & Reddy, 1984)

Interspecific hybridization within the genus *Lens* shows crossability barriers both between and within the species (Ferguson et al., 2000; Ladizinsky, 1997; Pratap et al., 2021). The high crossability potential for accessions of *L. culinaris* ssp. *orientalis* and *L. culinaris* ssp. *culinaris* suggested ssp. *orientalis* as the wild progenitor of lentil (Ladizinsky et al., 1984). Davey et al. (2005) suggested that interspecific hybridization hurdles can be overcome via techniques such as somatic hybridization and protoplast fusion. However, there are limited successful reports on the application of these in lentil.

7.3 *Mutation Breeding*

Prolonged use of traditional breeding methods has narrowed the genetic variability. Other than recombination, mutation includes all types of heritable genetic changes in an organism. Mutations can aid plant breeding programs by creating novel genetic variability. Induced mutation breeding techniques are swift and cost-effective methods for augmenting the genetic base and recreating genetic variability in lentil. Many successful studies have been carried out in this field (Toker et al., 2007;

Nakagawa et al., 2011; Shu et al., 2012; Mba, 2013; Tomlekova et al., 2014; Riaz & Gul, 2015; Oladosu et al., 2016; Jankowicz-Cieslak et al., 2017). The success of mutation breeding depends upon the productiveness of induced mutations, a desirable frequency and spectrum of mutation, and then the accuracy of breeders in screening them effectively (Khan & Siddiqui, 1992a, b; Manju & Gopimony, 2009). The advantage of this method is that shorter time is required to breed improved varieties, and the process does alter only a small portion and not the whole genome. The technique thus became a highly convenient tool for plant breeders to generate the much-needed genetic variation in different crop species, thereby providing them access to unexplored allelic combinations within the crop genome. In lentil mutation breeding, careful selection of mutants with altered morphological architecture is focused on to find elite breeding lines for yield and yield-attributing traits. Induced mutagenesis is believed to be the best approach done in lentils for broadening the genetic base and to produce useful new mutants overcoming the genetic bottlenecks. Among the different plant breeding methods for self-pollinated crops, mutation breeding along with advanced modern genomic techniques is highly recommended to broaden the genetic base of lentil (Erskine et al., 1998; Toker et al., 2007). Some studies have confirmed the effectiveness of chemical mutagenesis in lentil (Toker et al., 2007; Gaikwad & Kothekar, 2004; Sarker & Sharma, 1989; Solanki & Sharma, 1994, 1999). A combination of physical and chemical mutagens also have shown useful results for the improvement of lentil and other pulses.

7.3.1 Past Achievements

In the current era of climate change, sustainable agricultural intensification is a necessity to meet food demand and supply. Lentil is a great dietary staple owing to its high-protein content and nutrient density. Generating polygenic variability for quantitative traits in lentils, such as yield, is highly recommended. Mutation breeding has contributed to the development of 18 mutant varieties worldwide. Two cultivars, Ranjan and Rajendra Masoor 1, released from India, have different improved traits such as high yield and spreading type, tolerance to low temperatures, early maturity, and suitability for late sowing.

8 Advances Through In Vitro Culture

The application of in vitro culture techniques can be an efficient tool for the management of genetic variability and acceleration of the conventional breeding process in pulse crops (Gatti et al., 2016). One of the essential necessities for success in genetic transformation using in vitro culture is a reliable regeneration protocol. In comparison to the success attained in other grain legumes, this technique is difficult in lentil due to its recalcitrant nature (Sarker et al., 2003; Gatti et al., 2016). In the past few decades, techniques have improved progressively. The first partial success of using meristem tips as explants and obtaining lentil regenerants was achieved by Bajaj and Dhanju (1979). A lentil regeneration protocol from the hypocotyl and epicotyl-derived callus cells was put forth by Williams and McHughen (1986). Polanco et al. (1988) reported multiple shoot formation from shoot tips. As the techniques in such kinds of studies advanced, better work was seen. A quick, effective, and reproducible protocol for in vitro shoot regeneration by employing various explants and different concentrations of BAP (6-benzylaminopurine, benzyl adenine) was reported by Omran et al. (2008). The results revealed successful in vitro shoot regeneration with a slight modification in the Murashige and Skoog (MS) medium. To a great extent, higher levels of BAP facilitated shoot regeneration in lentil genotypes. Also, decapitated embryos were the ideal explants for the highest shoot regeneration. MS medium supplemented with 25 mg/l indole butyric acid (IBA) showed 30% rooting efficiency (Sarker et al., 2003). The use of cytokinin in the initial explants for multiple shoots regeneration has been linked to problems of root induction at a later stage (Mohamed et al., 1992). In vitro induction of flowering followed by pod formation and seed set directly from in vitro regenerated shoots was successfully demonstrated in lentil, which is remarkable for improving and shortening the breeding process (Sarker et al., 2012; Das et al., 2012). The in vitro embryo rescue technique has been applied for introgression of resistance to diseases such as anthracnose, ascochyta blight, and stemphylium blight (Fiala et al., 2009; Tullu et al., 2013; Saha et al., 2015). The *Agrobacterium tumefaciens* biolistic transformation including electroporation and particle bombardment is reported in lentil (Chowrira et al., 1996; Lurquin et al., 1998; Gulati et al., 2002; Mahmoudian et al., 2002). Transgenic lentils were produced with genes such as *nptII*, *gus*A, and *DREB1A* successfully (Akcay et al. 2009; Khatib et al. 2011). Transgenic lentil shoots were developed in two microsperma varieties using *A. tumefaciens* (Subroto et al., 2012). An efficient and reproducible protocol for in vitro shoot regeneration from cotyledonary node explants was reported by Bermejo et al. (2012), from which subsequent transgenic lentils were developed later (Bermejo et al., 2016).

9 Lensomics: Advances in Lentil Genomics and Molecular Breeding

Lentil breeders have succeeded in many aspects in the crop's improvement for a number of monogenic traits using the conventional plant breeding techniques via selection and recombination. However, major economic traits such as seed yield are quantitative with a complex polygenic mode of inheritance. Thus, execution of conventional plant breeding techniques for the improvement of these traits faces difficulties in precision, management, and time durations. With the advances in biotechnological techniques such as recombinant DNA technologies, molecular marker-based technologies, and bioinformatics, lentil breeders can now use novel genetic variability in the cultivated gene pool. They can identify the background genetic network and introgress desirable traits into cultivars using genomics-assisted breeding (GAB) and genetic engineering techniques more precisely in lesser time.

9.1 Marker-Assisted Selection (MAS)

Marker is a tag that can identify a particular location within the plant genome. MAS gives an exceptional and efficient path to breeders by which they can select plants with elite gene combinations. A number of molecular markers for genetic analysis, viz. RFLP, RAPD, AFLP, SSR, and ISSR, have been used by lentil geneticists and breeders (Sharma and Lavania 1996; Havey & Muehlbauer, 1989; Ferguson et al., 1998; Sonnante & Pignone, 2001; Toklu et al., 2009; El-Nahas et al., 2011; Alghamdi et al., 2014; Dikshit et al., 2015; Idrissi et al., 2015; Tsanakas et al., 2018). However, due to inadequate availability and slow development of lentil genomic resources, the application of MAS in lentil-breeding programs is hindered and remains limited compared to other major legumes (Kumar et al., 2014). Linkage analysis experiments were initiated in lentils by Zamir and Ladizinsky (1984), while the first DNA marker-based genetic map was developed by the efforts of Havey and Muehlbauer (1989). The advent of PCR-based markers expanded genetic map studies in lentil. Eujayl et al. (1998) first published an extensive linkage map in lentil utilizing RAPD, AFLP, RFLP, and morphological markers with an interspecific cross between *Lens* ssp. *culinaris* and ssp. *orientalis*. The first intraspecific lentil map comprised 114 RAPD, inter-simple sequence repeats (ISSR), and resistance gene analog (RGA) markers (Rubeena & Taylor, 2003). The first genomic library was constructed using lentil cultivar ILL5588 (Hamwieh et al., 2005). SSR markers are preferred over other DNA markers for studying genetic diversity, population structure, phylogenetic relationships, construction of framework linkage maps, QTL interval mapping, map-based gene cloning, MAS, etc., owing to their reproducibility, codominant multiallelic nature, high degree of polymorphism, locus-specific inheritance, relative abundance, and better genome coverage (Powell et al., 1996; Hendre et al., 2007). Several polymorphic and functional SSR markers were also evolved for genetic diversity studies and genomic library construction in lentil (Hamwieh et al., 2009; Verma et al., 2014; Andeden et al., 2015).

A molecular linkage map with 11 linkage groups of lentils was reported by Tanyolac et al. (2010), making use of AFLP, ISSR, and RAPD markers covering 1396.3 cM with an average map distance of 8.4 cM between markers. Several genetic linkage maps have been constructed from various lentil mapping populations, but the marker density is low and the span in cM is higher, making them impractical. Ates et al. (2018a) used diversity arrays technology (DArT) markers to construct a high-density consensus linkage map comprising of seven linkage groups, representing seven chromosomes on lentil genome covering a total of 977.47 cM with an average inter-marker distance of 0.10 cM. Altogether, genetic maps reported in lentil have helped in the understanding of the lentil genome (Eujayl et al., 1998; Kahraman et al., 2004; Tanyolac et al., 2010; Saha et al., 2010a, b; Ates et al., 2018a, b).

Qualitative markers were identified for morphological traits such as cotyledon (Yc), anthocyanin in the stem (Gs), pod indehiscence (Pi), seed coat pattern (Scp), flower color (W), radiation frost tolerance locus (Rf), early flowering (Sn), and ground color of the seed (Gc), as they exhibited monogenic dominant inheritance (Tullu et al., 2003; Duran et al., 2004; Hamwieh et al., 2005). Quantitatively inherited traits were mapped by Duran et al. (2004), which could identify QTLs for height of the first ramification, flowering time, plant height, pod dehiscence, shoot number, and seed diameter. A study using RIL population of 106 lines derived from WA8649090 × Precoz identified QTLs that established winter survival and winter injury (Kahraman et al., 2004). A QTL that conveyed resistance to *Stemphylium* blight and rust diseases was identified using RIL populations (Saha et al. 2010a,b). Such advanced breeding populations give higher efficacy in mapping studies, and there are still fewer such materials developed in lentil till date. With advancements in biotechnological tools and molecular markers, there are novel concepts coming up like consensus linkage maps using multiple mapping populations, although the possibility and success of these still need to be worked in lentil. In present times, DNA chip-based markers using single nucleotide polymorphisms (SNPs) are gaining popularity over PCR-based markers in next-generation sequencing (NGS) approaches with several success reports of identifying candidate genes for desirable traits (Sharpe et al., 2013; Temel et al., 2014; Bett et al., 2016). This headway will assist breeders to implement MAS in lentil breeding (Table 2).

Table 2 Web databases containing genetic and genomic data of lentil

Databases	Source institute/Group	Website	Tools
LIS	Legume Information System, National Center for Genome Resources, Santa Fe, NM, USA	http://lencu.comparative-legumes.org/	GBrowse sequenced legumes and other legumes
KnowPulse	KnowPulse, hosted by University of Saskatchewan Pulse Crop Research Group	http://knowpulse2.usask.ca/portal/	GBrowse with lentil track
CSFL genome	Cool Season Food Legume Genome Database, hosted by Washington State University	http://coolseasonfoodlegume.org/	GBrowse with lentil track
IBP	Integrated Breeding Platform	http://www.integratedbreeding.net/	Lentil crop and genomic information (under construction)

(Source: Coyne & McGee, 2013)

9.2 Next-Generation Technologies

In the past, breeders' efforts in bringing success to breeding programs were not maximized due to a narrow genetic base and unavailability of genomic resources in lentil. The progress in next-generation sequencing (NGS) technology has provided new opportunities for lentil genome sequencing projects and speedy sequence-based marker development. Owing to the next-generation DNA sequencing, some genome assemblies, genomic map, and QTLs have also been reported in different studies (Gupta et al., 2012; Bett et al., 2016). Transcriptome profiling of lentil has revealed key defense response genes against *Ascochyta* blight via RNA-seq analysis and quantitative real-time PCR (RT-qPCR approaches) (Khorramdelazad et al., 2018; Sari et al., 2018). Identification of QTLs for Fe and Mn concentrations in the lentil genome through NGS will enhance future biofortification attempts in the crop (Aldemir et al., 2017; Ates et al. 2018b). The high frequency of genomic variations induced by physical and chemical mutagens is accurately detectable by screening SNPs at the molecular level. High-throughput technique, viz. target-induced local lesions in genomes (TILLING), can identify mutations in single nucleotide, contributing significantly to an understanding of gene(s) function.

10 Conclusion and Future Prospects

Lentil is predominantly a self-pollinating legume crop having minimal out-crossing under special circumstances. Major breeding objectives in lentil are higher and stable yield in addition to earliness, wider adaptability, high-protein quality in grains, and biotic and abiotic stress resistance. Direct selection for seed yield is mostly ineffective as it is a complex polygenic trait governed by multiple minor genes. To attain this, breeders need to identify new traits with prime focus on yield. Studies decoding complex traits through advanced tools and techniques such as gene tagging of target traits are in demand. The advent of cutting-edge advanced techniques has added pace to plant breeding through utilizing molecular markers, DNA sequencing, genomics, and statistical and bioinformatic-based tools. Crop improvement work plans and strategies have become automated, cost effective, and reliable. DNA sequencing and various analysis methods are now permitting larger sample size in shorter with higher precision. The last decade witnessed the development of linkage maps and the identification of QTLs for the traits of interest in lentil. In recent times, clustered regularly interspaced palindromic repeats (CRISPR) and CRISPR-associated protein (Cas) systems have gained much interest from plant geneticists and crop scientists. The incorporation of genetic variability is of major concern in order to successfully run a plant breeding program to develop a crop genotype with target traits. The count of use of markers for MAS and marker-assisted backcrossing (MABC) in lentil breeding is still on fingertips. These advances will simultaneously encourage lentil breeders in developing countries to incorporate MAS and MABC into lentil breeding programs. The ultimate aim is to select lines with improved traits like high yield, multipodding, determinate growth habit, tolerance to stresses, climate adaptability, and biofortified lentil.

Glossary

Abiotic stress A nonbiological environmental factor adversely affecting crop performance.

Acclimatization The process (natural selection) leading to adaptation of a variety, line, or population to a new environment.

AFLP The genomic DNA is digested with two restriction enzymes; the cut ends of fragments are modified by ligating them with adapters.

Biofortification The approach to enhance nutritional value of crop.

Biotic stress An organism adversely affecting the performance of another organism/crop, e.g., pathogen, weed, insect pest, etc.

Bulk method F_2 and subsequent generation are grown in bulk without artificial selection. Finally, pureline varieties are developed through individual plant selection.

Domestication Bringing wild species under human management.

Explant A small piece of living tissue used to initiate an in vitro culture.

Gene cloning Production and introduction of recombinant DNA into a suitable host to produce multiple copies of recombinant DNA or protein encoded by cloned gene.

Genetic base Genetic variability present among cultivated varieties of a crop species.

Genome A complete set of chromosomes of a diploid species.

Genomics Science dealing with structure and function of complete genomes of organisms.

Genotyping Screening of mapping population using polymorphic molecular markers.

Germplasm The hereditary material transmitted from parents to offspring through germ cells.

Hybridization The process of obtaining progeny by union of gametes from two different strains.

In vitro Biological processes occurring outside the organism, e.g., in test tube.

Linkage map A linear map of genes showing linkage with each other depicting sequence of location on a chromosome.

MAS An indirect form of selection, when DNA markers are used for selection of traits which are otherwise difficult to select through phenotype-based visual observations.

Monogenic Governed by or involving a single gene.

Mutagen A chemical or physical agent that induces mutation.

Mutation A sudden heritable change in a character of an organism which is not a result of segregation or recombination.

Narrow genetic base Cultivated varieties of species having similarities in their genotypes as a consequence of common parents in their ancestry.

Phenotyping Recording the phenotypic expression of target traits in the individuals of a mapping population.

Phylogeny The evolutionary history of a species.

Polygenes Genes having small and cumulative effect on a character that is quantitative.

PCR Amplification of a DNA segment by using DNA polymerase and two primers specific to 3′ end of this segment.

Prebreeding Accessions with desirable traits are initially crossed with modern varieties or breeding lines to improve their agronomic traits. The lines thus derived are used in breeding programs instead of original accessions or genotype.

Pureline The progeny of a single homozygous self-pollinated plant.

QTL Chromosome segments or positions containing one or more polygene controlling quantitative trait(s).

QTL mapping Linkage mapping of QTLs using biparental mapping populations.

Recombination The production of new combination of genes or traits not present in parents of a cross or generation.

RFLP Differences in the length of fragments produced by cleavage with the relevant restriction enzyme of the same DNA stretch different individuals.

RILs Homozygous lines derived from individual F_2 plants of a suitable cross.

Single seed descent method A method of advancing generation by obtaining one seed from each F_2 plant of a cross and subsequent generations is bulked to raise the next generation.

Transgene A gene isolated from an organism and transferred into another through genetic engineering.

References

Abbo, S., Lev-Yadun, S. & Gopher, A. (2012). Plant domestication and crop evolution in the Near East: On events and processes. *Critical Reviews in Plant Sciences, 31*(3), 241–257.

Akcay, U. C., Mahmoudian, M., Kamci, H., Yucel, M., & Oktem, H. A. (2009). *Agrobacterium tumefaciens*-mediated genetic transformation of a recalcitrant grain legume, lentil (*Lens culinaris* Medik.). *Plant Cell Reports, 28*, 407–417.

Aldemir, S., Ateş, D., Temel, H. Y., Yağmur, B., Alsaleh, A., Kahriman, A., Özkan, H., Vandenberg, A., & Tanyolaç, M. B. (2017). QTLs for iron concentration in seeds of the cultivated lentil (*Lens culinaris* Medic.) via genotyping by sequencing. *Turkish Journal of Agriculture and Forestry, 41*, 243–255.

Alghamdi, S. S., Migdadi, H. M., Khan, M. A., Ammar, M. H. & El-Harty, E. H. (2014). Assessment of genetic diversity in faba bean (Vicia faba L.) using agro-morphological and molecular markers. *Saudi Journal of Biological Sciences, 21*(2), 136–145.

Alo, F., Furmann, B. J., Akhunov, E., Dvorak, J., & Gepts, P. (2011). Leveraging genomic resources of model species for the assessment of diversity and phylogeny in wild and domesticated lentil. *The Journal of Heredity, 102*, 315. https://doi.org/10.1093/jhered/esr015

Amin, R., Laskar, R. A., & Khan, S. (2015). Assessment of genetic response and character association for yield and yield components in lentil (*Lens culinaris* L.) population developed through chemical mutagenesis. *Cogent Food & Agriculture, 1*, 1000715. https://doi.org/10.1080/23311932.2014.1000715

Andeden, E. E., Baloch, F. S., Derya, M., & Kilian, B. (2015). Genetic diversity and population structure of Turkish lentil (Lens culinaris Medik.) landraces and wild relatives assessed by SSR markers. *Plant Genetic Resources, 13*(2), 111–119.

Arumuganathan, K., & Earle, E. D. (1991). Nuclear DNA content of some important plant species. *Plant Molecular Biology Reporter, 9*(3), 208–218.

Asnake, F., & Bejiga, G. (2003). Breeding lentil for wider adaptation in forage and food legumes of Ethiopia: Progress and prospects. In A. Kemal, K. Gemechu, & A. Seid (Eds.), *Proceedings of the workshop on food and forage legumes* (pp. 80–86). EIAR and ICARDA.

Ates, D., Aldemir, S., Alsaleh, A., Erdogmus, S., Nemli, S., Kahriman, A., Ozkan, H., Vandenberg, A., & Tanyolac, B. (2018a). A consensus linkage map of lentil based on DArT markers from three RIL mapping populations. *PLoS One, 13*(1), e0191375. https://doi.org/10.1371/journal.pone.0191375

Ates, D., Aldemir, S., Yagmur, B., Kahraman, A., Ozkan, H., Vandenberg, A., & Tanyolac, M. B. (2018b). QTL mapping of genome regions controlling manganese uptake in lentil seed. *G3: Genes, Genomes, Genetics, 8*(5), 1409–1416.

Bajaj, Y. P., & Dhanju, M. S. (1979). Regeneration of plants from apical meristem tips of some legumes. *Current Science, 84*(20), 906–907.

Barulina, H. (1930). Lentils of the USSR and other countries. *Bulletin of Applied Botany, Genetics & Plant Breeding, Leningrad, 40*, 265–304.

Bejiga, G., & Tessema, T. (1981). Identification on the optimum time of emasculation and pollination to increase percentage of hybrid seed set in chickpea. *The Journal of Agricultural Science, 3*, 129–134.

Bermejo, C., Espósito, M., Cravero, V., Anido, F. L., & Cointry, E. (2012). *In vitro* plant regeneration from cotyledonary nodes of recombinant inbred lines of lentil. *Scientia Horticulturae, 134*, 13–19.

Bermejo, C., Gatti, I., & Cointry, E. (2016). *In vitro* embryo culture to shorten the breeding cycle in lentil (*Lens culinaris* Medik). *Plant Cell, Tissue and Organ Culture, 127*, 585–590.

Bett, K., Ramsay, L., Chan, C., Sharpe, A. G., Cook, D. R., Penmetsa, R. V., Chang, P., Coyne, C., McGee, R., Main, D., Doleze, J., Edwards, D., Kaur, S., Kumar, S., Udupa, S., & Vandenberg, A. (2016). The lentil genome—From the sequencer to the field. In *International conference on pulses*, Marrakesh, Morocco, 18–20 April, Abstract.

Bhandari, K., Siddique, K. H., Turner, N. C., Kaur, J., Singh, S., Agrawal, S. K., & Nayyar, H. (2016). Heat stress at reproductive stage disrupts leaf carbohydrate metabolism, impairs reproductive function and severely reduces seed yield in lentil. *Journal of Crop Improvement, 30*, 118–151.

Chowrira, G. M., Akella, V., Fuerst, P. E., & Lurquin, P. F. (1996). Transgenic grain legumes obtained by in plant a electroporation-mediated gene transfer. *Molecular Biotechnology, 5*, 85–96.

Coyne Clarice and McGee Rebecca. (2013). *Lentil, In eds; Mohar Singh Hari Upadhyaya I.* Elsevier.

Cubero, J. I. (1981). Origin, domestication and evolution. In C. Webb & G. C. Hawtin (Eds.), *Lentils* (pp. 15–38). Commonwealth Agricultural Bureaux.

Das, S. K., Shethi, K. J., Hoque, M. I., & Sarker, R. H. (2012). Agrobacterium mediated genetic transformation in lentil (*Lens culinaris* Medik.) followed by in vitro flowering and seed formation. *Plant Tissue Culture Biotechnology, 22*, 13–26.

Davey, M. R., Anthony, P., Power, J. B., & Lowe, K. C. (2005). Plant protoplasts: Status and biotechnological perspectives. *Biotechnology Advances, 23*, 131–171.

Delahunty, A., Nuttall, J., Nicolas, M., & Brand, J. (2015). Genotypic heat tolerance in lentil. In *Proceedings of the 17th ASA Conference*, 2015 September, pp. 20–24.

Delahunty, A., Nuttall, J., Nicolas, M., & Brand, J. (2018). Response of lentil to high temperature under variable water supply and carbon dioxide enrichment. *Crop & Pasture Science, 69*(11), 1103–1112.

Dikshit, H. K., Singh, A., Singh, D., Aski, M., & Singh, D. (2015). Genetic diversity in lentil (Lens culinaris Medik.) germplasm assessed by SSR markers. *Indian Journal of Genetics and Plant Breeding, 75*(3), 300–306.

Durán, Y., Fratini, R., Garcia, P., & De la Vega, M. P. (2004). An intersubspecific genetic map of *Lens*. *Theoretical and Applied Genetics, 108*, 1265–1273.

El-Nahas, A. I., El-Banna, A. N., & El-Kady, M. A. (2011). Genetic diversity among lentil genotypes using RAPD markers. *Journal of Agricultural Research, 37*(2), 123–130.

Erskine, W., Chandra, S., Chaudhry, M., Malik, I. A., Sarker, A., Sharma, B., Tufail, M., & Tyagi, M. C. (1998). A bottleneck in lentil: Widening the genetic base in South Asia. *Euphytica, 101*, 207–211.

Eujayl, I., Baum, M., Powell, W., Erskine, W., & Pehu, E. (1998). A genetic linkage map of lentil (*Lens* sp) based on RAPD and AFLP markers using recombinant inbred lines. *Theoretical and Applied Genetics, 97*, 83–89.

FAO. (2025). http://faostat.fao.org/site/291/default.aspx

Ferguson, M. E., & Erskine, W. (2001). Lentils (*Lens* L.). In N. Maxted & S. J. Bennett (Eds.), *Plantgenetic resources of legumes in the Mediterranean* (pp. 132–157). KluwerAcademic Publishers.

Ferguson, M. E., Robertson, L. D., Ford-Lloyd, B. V., & Newbury, H. J. (1996). Genetic diversity in chickpea (Cicer arietinum L.) germplasm revealed by RAPD markers. *Euphytica, 97*(2), 177–183.

Ferguson, M., Ford-Lloyd, B. V., Robertson, L. D., Maxted, N., & Newbury, H. J. (1998). Mapping the geographical distributionof genetic variation in the genus *Lens* for the enhanced conservation of plant genetic diversity. *Molecular Ecology, 7*, 1743–1755.

Ferguson, M. E., Maxted, N., Van Slageren, M., & Robertson, L. D. (2000). A re-assessment of the taxonomyof *Lens* Mill. (Leguminosae, Papilionoideae, Vicieae). *Botanical Journal of the Linnean Society, 133*(1), 41–59.

Fiala, J. V., Tullu, A., Banniza, S., Séguin-Swartz, G., & Vandenberg, A. (2009). Interspecies transfer of resistance to anthracnose in lentil (*Lens culinaris* Medik.). *Crop Science, 49*, 825–830.

Fikiru, E., Tesfaye, K., & Bekele, E. (2010). Genetic diversity and population structure ofEthiopian lentil *(Lens culinaris* Medikus) landraces as revealed by ISSR marker. *African Journal of Biotechnology, 6*(12), 1460–1468.

Fikiru, E., Tesfaye, K., & Bekele, E. (2011). Morphological and molecular variation in Ethiopian lentil (*Lens culinaris* Medikus) varieties. *African Journal of Biotechnology, 3*(4), 60–67.

Fratini, R. M., de la Vega, M. P., & Sánchez, M. L. R. (2014). Lentil. In M. Singh, I. S. Bisht, & M. Dutta (Eds.), *Broadening the genetic base of grain legumes* (pp. 115–148). Springer.

Gaikwad, N. B., & Kothekar, V. S. (2004). Mutagenic effectiveness and efficiency of ethylmethane sulphonateand sodium azide in lentil (*Lens culinaris* Medik.). Indian. *Journal of Genetics, 64*(1), 73–74.

Gatti, I., Guindón, F., Bermejo, C., Espósito, A., & Cointry, E. (2016). *In vitro* tissue culture in breeding programs of leguminous pulses: Use and current status. *Plant Cell, Tissue and Organ Culture, 127*(3), 543–559.

Gulati, A., Schryer, P., & McHughen, A. (2002). Production of fertile transgenic lentil (*Lens culinaris*Medik.) plants using particle bombardment. *In Vitro Cellular & Developmental Biology. Plant, 38*, 316–324.

Gupta, D., Taylor, P. W. J., Inder, P., Phan, H. T. T., Ellwood, S. R., Mathur, P. N., Sarker, A., & Ford, R. (2012). Integration of EST-SSR markers of *Medicago truncatula* into intraspecific linkage map of lentil and identification of QTL conferring resistance to ascochyta blight at seedling and podstages. *Molecular Breeding, 30*, 429–439.

Hamwieh, A., Udupa, S. M., Choumane, W., Sarker, A., Dreyer, F., Jung, C., & Baum, M. (2005). A genetic linkage map of *Lens* sp based on microsatellite and AFLP markers and the localization of fusarium vascular wilt resistance. *Theoretical and Applied Genetics, 110*, 669–677.

Hamwieh, A., Udupa, S. M., Sarker, A., Jung, C., & Baum, M. (2009). Development of new microsatellite markers and their application in the analysis of genetic diversity in lentils. *Breeding Science, 59*, 77–86.

Havey, M. J., & Muehlbauer, F. J. (1989). Linkages between restrictions fragment length, isozyme andmorphological markers in lentil. *Theoretical and Applied Genetics, 77*, 395–401.

Hoffman, D. L., & Smith, D. H. (1986). Genetic diversity among lentil cultivars and landraces. *Crop Science, 26*(3), 526–530.

Hoque, M. M., & Hasan, M. J. (2012). Genetic diversity in lentil (Lens culinaris Medik.) germplasm. *Bangladesh Journal of Agricultural Research, 37*(3), 389–398.

Idrissi, O., Udupa, S. M., De Keyser, E., & Baum, M. (2015). Genetic diversity and population structure of lentil (Lens culinaris Medik.) landraces and wild relatives from the Mediterranean Basin. *Plant Genetic Resources, 13*(3), 283–291.

Jankowicz-Cieslak, J., Tai, T. H., Kumlehn, J., & Till, B. J. (2017). *Biotechnologies for plant mutation breeding protocol*. International Atomic Energy Agency.

Kahraman, A., Kusmenoglu, I., Aydin, N., Aydogan, A., Erskine, W., & Muehlbauer, F. J. (2004). QTL mapping of winter hardness genes in lentil. *Crop Science, 44*, 13–22.

Kaur, R., Bains, T. S., Bindumadhava, H., & Nayyar, H. (2015). Responses of mungbean (*Vigna radiata* L.) genotypes to heat stress: Effects on reproductive biology, leaf function and yield traits. *Scientia Horticulturae, 197*, 527–541.

Khan, S., & Siddiqui, B. A. (1992a). Effect of selection for improvement of quantitative characters inmutated population of mungbean (*Vigna radiata* (L.) Wilczek). *Journal of the Indian Botanical Society, 71*, 69–71.

Khan, S., & Siddiqui, B. A. (1992b). Mutagenic effectiveness and efficiency of chemical mutagens in *Vigna radiata* (L.) Wilczek. *Thai Journal of Agricultural Science, 25*, 291–297.

Khatib, F., Makris, A., Yamaguchi-Shinozaki, K., Kumar, S., Sarker, A., Erskine, W., & Baum, M. (2011). Expression of the DREB1A gene in lentil (*Lens culinaris* Medik. subsp *culinaris*) transformed with the *Agrobacterium* system. *Crop and Pasture Science, 62*, 488–495.

Khorramdelazad, M., Bar, I., Whatmore, P., Smetham, G., Bhaaskaria, V., Yang, Y., Bai, S. H., Mantri, N., Zhou, Y., & Ford, R. (2018). Transcriptome profiling of lentil (*Lens culinaris*) through the first 24 hours of *Ascochyta lentis* infection reveals key defence response genes. *BMC Genomics, 19*, 108.

Kislev, M. E. & Bar-Yosef, O. (1988). The legumes: The earliest domesticated plants in the Near East? *Current Anthropology, 29*(1), 175–179.

Kumar, S., & Ali, M. (2006). GE interaction and its breeding implications in pulses. *The Botanica, 56*, 31–36.

Kumar, H., Dikshit, H. K., Singh, A., Jain, N., Kumari, J., Singh, A. M., Singh, D., Sarker, A., & Prabhu, K. V. (2014). Characterization of grain iron and zinc in lentil (Lens culinaris Medikus culinaris) and analysis of their genetic diversity using SSR markers. *Australian Journal of Crop Science, 8*(7), 1005–1012.

Ladizinsky, G. (1992). Crossability relations. *Monographs on Theoretical and Applied Genetics, 16*, 15–31.

Ladizinsky, G. (1997). A new species of *Lens* from south-east Turkey. *Botanical Journal of the Linnean Society, 123*, 257–260.

Ladizinsky, G., Braun, D., Goshen, D., & Muehlbauer, F. J. (1984). The biological species of the genus *Lens*. *Botanical Gazette, 154*, 253–261.

Lombardi, M., Materne, M., Cogan, N. O. I., Rodda, M., Daetwyler, H. D., Slater, A. T., Forster, J. W., & Kaur, S. (2014). Assessment of genetic variation within a global collection of lentil (*Lens culinaris* Medik,) cultivars and landraces using SNP markers. *BMC Genetics, 15*, 150.

Lurquin, P. F., Cai, Z., Stiff, C. M., & Fuerst, E. P. (1998). Half-embryo cocultivation technique for estimatingthe susceptibility of pea (*Pisum sativum* L.) and lentil (*Lens culinaris* Medik.) cultivars to *Agrobacterium tumefaciens*. *Molecular Biotechnology, 9*, 175–179.

Mahmoudian, M., Yucel, M., & Oktem, H. A. (2002). Transformation of lentil (*Lens culinaris* M) cotyledonarynodes by vacuum infiltration of *Agrobacterium tumefaciens*. *Plant Molecular Biology Reporter, 20*, 251–257.

Manju, P., & Gopimony, R. (2009). Anjitha—A new okra variety through induced mutation in interspecifichybrids of *Abelmoschus* spp. In Q. Y. Shu (Ed.), *Induced plant mutation in the genomic era* (pp. 87–90). FAO.

Materne, M., & Siddique, K. H. M. (2009). Agroecology and crop adaptation. In W. Erskine, F. J. Muehlbauer, A. Sarker, & B. Sharma (Eds.), *The lentil—Botany, production and uses* (pp. 47–63). CABI.

Maxted, N., Hargreaves, S., Kell, S. P., Amri, A., Street, K., Shehadeh, A., Piggin, J., & Konopka, J. (2010). *Temperate forage and pulse legume genetic gap analysis* (pp. 22–26). XIII OPTIMA Meeting in Antalya.

Mba, C. (2013). Induced mutations unleash the potentials of plant genetic resources for food and agriculture. *Agronomy, 3*, 200–231.

Mohamed, M. F., Read, P. E., & Coyne, D. P. (1992). Plant regeneration from in vitro culture of embryonic axis explants in common and tepary beans. *Journal of the American Society for Horticultural Science, 117*, 332–336.

Muehlbauer, F. J. (1992). Use of introduced germplasm in cool-season food legume cultivar development. In H. L. Shands & L. E. Weisner (Eds.), *Use of plant introductions in cultivar development. Part 2. Crop Science Society of America (CSSA) Special Publication* (Vol. 20, pp. 49–73). CSSA.

Muehlbauer, F. J., Kaiser, W. J., Clement, S. L., & Summerfield, R. J. (1995). Production and breeding of lentil. *Advances in Agronomys, 54*, 283–332.

Muehlbauer, F. J., Cho, S., Sarker, A., McPhee, M. E., Coyne, C. J., Rajesh, P. N., & Ford, R. (2006). Application of biotechnology in breeding lentil forresistance to biotic and abiotic stress. *Euphytica, 147*, 149–165.

Muehlbauer, F. J., Mihov, M. A., Vandenberg, A., Tullu, A., & Materne, M. (2009). Improvement in developed countries. In W. Erskine, F. J. Muehlbauer, A. Sarker, & B. Sharma (Eds.), *Lentil-botany, production and uses* (pp. 137–154). CABI.

Nakagawa, H., Annai, T., Okabe, A., Takahashi, K., Hajika, M., & Takagi, Y. (2011). Mutation breeding of soybean in Japan. In S. Khan & M. I. Kozgar (Eds.), *Breeding of pulse crops* (pp. 55–84). Kalyani Publishers.

Oladosu, Y., Rafii, M. Y., Abdullah, N., Hussin, G., Ramli, A., & Rahim, H. A. (2016). Principle and application of plant mutagenesis in crop improvement: A review. *Biotechnology & Biotechnological Equipment, 30*(1), 1–16.

Omran, V. G., Bagheri, A., & Moshtaghi, N. (2008). Direct *in vitro* regeneration of lentil (*Lens culinaris* Medik). *Pakistan Journal of Biological Sciences, 11*, 2237–2242.

Pinkas, Y., Gera, A., & Tsror, L. (1985). Fusarium wilt of chickpea in Israel. *Phytoparasitica, 13*(3):229–230.

Polanco, M. C., Pelaez, M. I., & Ruiz, M. L. (1988). Factors affecting callus and shoot formation from *in vitro* cultures of *Lens culinaris* Medik. *Plant Cell, Tissue and Organ Culture, 15*(2), 175–182.

Pundir, R. P., & Reddy, K. N. (1984). Comparison of cross-pollination methods in chickpea. *International Chickpea and Pigeonpea Newsletter, 11*, 9–11.

Rahman, M. M., Sarker, A., Kumar, S., Asghar, A., Yadav, N. K., & Rahman, M. L. (2009). Breeding for short season environments. In M. F. J. ErskineW, A. Sarker, & B. Sharma (Eds.), *Lentil—Botany, production and uses* (pp. 121–136). CABI.

Redden, R. J., Hatfield, P. V., Prasad, V., Ebert, A. W., Yadav, S. S., & O'Leary, G. J. (2014). Temperature, climate change, and global food security. In K. A. Franklin & P. A. Wigge (Eds.), *Temperature and plant development* (pp. 181–202). John Wiley & Sons, Inc.

Riaz, A., & Gul, A. (2015). Plant mutagenesis and crop improvement. In K. R. Hakeem (Ed.), *Crop production and global environmental issues* (pp. 181–210). Springer International Publishing Switzerland.

Rubeena, F. R., & Taylor, P. W. J. (2003). Construction of an intraspecific linkage map of lentil (*Lens culinaris*ssp. *culinaris*). *Theoretical and Applied Genetics, 107*, 910–916.

Saha, G. C., Sarker, A., Chen, W., Vandemark, G. J., & Muehlbauer, F. J. (2010a). Inheritance and linkage map positions of genes conferringresistance to stemphylium blight in lentil. *Crop Science, 50*, 1831–1839.

Saha, G. C., Sarker, A., Chen, W., Vandemark, G. J., & Muehlbauer, F. J. (2010b). Identification of markers associated with genes for rust resistance in *Lens culinaris* Medik. *Euphytica, 175*, 261–265.

Saha, S., Tullu, A., Yuan, H. Y., Lulsdorf, M. M., & Vandenberg, A. (2015). Improvement of embryo rescue technique using4-chloroindole-3-acetic acid in combination with in vivo grafting to overcome barriers inlentil interspecific crosses. *Plant Cell, Tissue and Organ Culture, 120*, 109–116.

Sari, E., Bhadauria, V., Ramsay, L., Borhan, M. H., Lichtenzveig, J., Bett, K. E., Vandenberg, A., & Banniza, S. (2018). Defense responses of lentil (*Lens culinaris*) genotypes carrying non-allelic ascochyta blight resistance genes to *Ascochyta lentis* infection. *PLoS, One13 (9)*, e0204124.

Sarker, A., & Erskine, W. (2006). Recent progress in the ancient lentil. *Journal of Agriculture Science (Cambridge), 144*, 1–11.

Sarker, A., & Sharma, B. (1989). Effect of mutagenesis on M1 parameters in lentil. *LENS, 16*(2), 8–10.

Sarker, R. H., Mustafa, B. M., Biswas, A., Mahbub, S., Nahar, M., Hashem, R., & Hoque, M. I. (2003). *In vitro* regeneration in lentil (*Lens culinaris* Medik.). *Plant Tissue Culture, 13*, 155–163.

Sarker, A., Aydogan, A., Chandra, S., Kharrat, M., & Sabaghpour, S. (2009). Genetic enhancement for yield and yield stability. In W. Erskine, F. J. Muehlbauer, A. Sarker, & B. Sharma (Eds.), *Lentil-botany, production and uses* (pp. 102–120). CABI.

Sharma, S. K., & Lavania, U. C. (1996). Enhanced frequency of direct shoot regeneration from leaf explants of Nicotiana tabacum L. *Plant Cell Reports, 15*(9), 666–669.

Sharpe, A. G., Ramsay, L., Sanderson, L. A., Fedoruk, M. J., Clarke, W. E., Li, R., Kagale, S., Vijayan, P., Vandenberg, A., & Bett, K. E. (2013). Ancient orphan crop joins modern era: Gene based SNP discovery and mapping in lentil. *BMC Genomics, 14*, 192.

Shu, Q. Y., Forster, B. P., & Nakagawa, H. (2012). *Plant mutation breeding and biotechnology*. CABI.

Sindhu, J. S., Lal, S. B., & Singh, R. P. (1981). Studies on the factors determining crossing success in chickpea (*Cicer arietinum*). *Pulse Crops News, 0*(1), 21–22.

Solanki, I. S., & Sharma, B. (1994). Mutagenic effectiveness and efficiency of gamma rays, ethyl imineand N-nitroso-N-ethyl urea in macrosperma lentil (*Lens culinaris* Medik.). *Indian Journal of Genetics, 54*, 72–76.

Solanki, I. S., & Sharma, B. (1999). Induction and isolation of morphological mutations in different damage groups in lentil (*Lens culinaris* Medik.). Indian. *Journal of Genetics, 59*(4), 479–485.

Soltis, D. E., Smith, S. A, Cellinese, N., Wurdack, K. J., Tank, D. C., Brockington, S. F., Refulio-Rodriguez, N. F., Walker, J. B., Moore, M. J., & Carlsward, B. S., et al. (2011). Angiosperm phylogeny: 17 genes, 640 taxa. *American Journal of Botany, 98*(4), 704–730.

Sonnante, G., & Pignone, D. (2001). Genetic variation in wild and cultivated lentil (Lens culinaris Medik.) investigated by RAPD markers. *Genetic Resources and Crop Evolution, 48*(3), 231–238.

Subbarao, G., Rao, J. Y. D. K., Kumar, J., Johansen, C., & Deb, U. (2001). *Spatial distribution, and quantification of rice-fallows in South Asia potential for legumes. International Crops Research Institute for the Semi-Arid Tropics. Spatial distribution and quantification of rice fallows in South Asia-potential for legumes*. Patancheru, India.

Subroto, K. D., Kishwar, J. S., Hoque, M. I., & Sarker, R. H. (2012). *Agrobacterium*- mediated genetic transformation in lentil (*Lens culinaris* Medik.) followed by in vitro flowering and seed formation. *Plant Tissue Culture and Biotechnology, 22*(1), 13–26.

Sultana, T., & Ghafoor, A. (2008). Genetic diversity in ex-situ conserved black gram (Vigna mungo L. Hepper) germplasm. *Pakistan Journal of Botany, 40*(6), 2347–2354.

Sultana, T., Ghafoor, A., & Ashraf, M. (2006). Genetic diversity in lentil germplasm for botanical descriptors in Pakistan. *Pakistan Journal of Botany, 38*(3), 553–558.

Tahir, M., Sadaqat, H. A., & Tahir, M. H. N. (2012). Genetic variability for yield and related traits in sunflower (Helianthus annuus L.) hybrids under drought stress. *Pakistan Journal of Botany, 44*(1), 177–182.

Tanyolac, B., Ozatay, S., Kahraman, A., & Muehlbauer, F. (2010). Linkage mapping of lentil (*Lens culinaris* L.) genome using recombinant inbred lines revealed by AFLP, ISSR, RAPD and some morphologic markers. *Journal of Agricultural Biotechnology and Sustainable Development, 2*(1), 1–6.

Temel, H. Y., Gol, D., Kahriman, A., & Tanyolac, M. B. (2014). Single nucleotide polymorphism discovery through Illumina-based transcriptome sequencing and mapping in lentil. *Turkish Journal of Agriculture and Forestry, 38*, 1–19.

Toker, C., Yadav, S. S., & Solanki, I. S. (2007). Mutation breeding. In S. S. Yadav, D. L. McNeil, & P. C. Stevenson (Eds.), *Lentil-an ancient crop for modern times* (pp. 209–224). Springer.

Toklu, F., Biçer, B. T., Karaköy, T., & Özkan, H. (2009). Assessment of genetic variation for agronomic traits among lentil (Lens culinaris Medik.) landraces and cultivars. *African Journal of Biotechnology, 8*(17), 4124–4136.

Tomlekova, N. B., Kozgar, M. I., & Wani, M. R. (2014). *Mutagenesis-exploring novel genes and pathways*. Wageningen Academic Publishers.

Tsanakas, G. V., Mylona, P. V., & Katsiotis, A. (2018). Genetic diversity and population structure of Greek lentil (Lens culinaris Medik.) landraces using SSR markers. *Genetic Resources and Crop Evolution, 65*(3), 637–650.

Tullu, A., Buchwaldt, L., Warkentin, T., Taran, B., & Vandenberg, A. (2003). Genetics of resistance to anthracnose and identification of AFLP and RAPD markers linked to the resistance gene in PI 320937 germplasm of lentil (*Lens culinaris* Medikus). *Theoretical and Applied Genetics, 106*, 428–434.

Tullu, A., Bett, K., Banniza, S., Vail, S., & Vandenberg, A. (2013). Widening the genetic base of cultivated lentil through hybridization of *Lens culinaris* "Eston" and *L. ervoides* accession IG 72815. *Canadian Journal of Plant Science, 93*, 1037–1047.

Verma, P., Sharma, T. R., Srivastava, P. S., Abdin, M. Z., & Bhatia, S. (2014). Exploring genetic variability within lentil (*Lens culinaris* Medik.) and a cross related legumes using a newly developed set of microsatellite markers. *Molecular Biology Reports, 41*, 5607–5625.

Williams, D. J., & McHughen, A. (1986). Plant regeneration of the legume *Lens culinaris* Medik. (lentil) *in vitro*. *Plant Cell, Tissue and Organ Culture, 7*, 149–153.

Yüzbaşıoğlu, D., Ünal, F., & Tavukçuoglu, N. (2008). Cytogenetic effects of the herbicide 2,4-D on Allium cepa. *Caryologia, 61*(1), 45–52.

Zamir, D., & Ladizinsky, G. (1984). Genetics of allozyme variants and linkage groups in lentil. *Euphytica, 33*, 329–336.

Zohary, D. (1972). The progenitors of wheat and barley in relation to domestication and agricultural dispersal in the Old World. In P. J. Ucko& G. W. Dimbleby (Eds.), *The domestication and exploitation of plants and animals* (pp. 47–66). Duckworth.

Zohary, D. (1996). *The mode of domestication of the founder crops of the Southwest Asian agriculture and pastoralism in Eurasia* (pp. 142–158). University College London Press.

Zohary, D. (1999). Monophyletic vs. polyphyletic origin of the crops on which agriculture was founded in the Near East. *Genetic Resources and Crop Evolution, 46*(2), 133–142.

Advances in Genetic Modifications for Legume Improvement

Meenal Rathore, Ayushi Tripathi, and Sudhir Kumar Jha

Abstract Legumes are known as smart crops due to their nutritional benefits, climate resilience, and soil-sustaining abilities. However, like any other crop, they are also exposed to several constraints impeding their production and benefits. Though numerous improved lines have been bred using conventional and molecular breeding, modern biotechnological tools for the development of genetically modified (GM) crops were looked upon to aid the improvement process across the species barrier. However, with the success limited to a few legumes only, advanced technologies such as genome editing have been adopted for precise improvement that in sync with speed breeding can enhance the genetic gain in legumes. This chapter brings about a brief of the status of commercialized transgenic legumes and opportunities for legume improvement via genome editing and speed breeding.

Keywords Genome editing · Legumes · Speed breeding · Transgenics · Climate resilience · Molecular breeding · Modern biotechnological tools

1 Introduction

Genetic improvement of pulses is a continuous effort aiming to develop smarter pulses that not only produce more but also produce better, especially in the regime of climate change. While aiming for enhanced productivity is essential, targeting the development of biofortified cultivars is now indispensable. Further, with a change in the dynamics of pest–pathogen–disease complex, holistic development of pulses is inevitable. With the development of precise molecular and analytical tools,

---

M. Rathore (✉) · S. K. Jha
Division of Plant Biotechnology, ICAR-Indian Institute of Pulses Research, Kanpur, Uttar Pradesh, India
e-mail: meenal.rathore@icar.gov.in

A. Tripathi
Amity University, Noida, Uttar Pradesh, India

A. Pratap et al. (eds.), *Integrated Improvement of Food Legumes*,
https://doi.org/10.1007/978-3-031-82629-0_11

the scientific approach for pulse improvement has also advanced. For example, the development of MAGIC populations to introgress multiple alleles and to provide increased recombination and mapping resolution is need of the hour in effort to breed improved pulses. Rapid turnover of generations in breeding by reducing the seed-to-seed cycle aka "speed breeding" is a tool that can immensely amplify the rate of research; enhanced genetic gain per unit time can now be achieved using genomic selection, a strategy of whole genomic prediction-based selection using genome-wide marker profiles. Precision breeding using the innovative CRISPR/Cas genome editing enables efficient targeted modification in most crops, thus promising to accelerate crop improvement.

2 Genetic Modifications

2.1 Transgenics

With complex issues such as recalcitrance, narrow genetic base, self-pollination, and compatibility barriers faced during conventional breeding programs using different species, progress has been limited (Christou, 1994). Therefore, genetic engineering was utilized to genetically transform choice crops to introgress transgenes for desired traits. The development of transgenics in pulses was relatively slow due to its recalcitrant nature; however, *in vitro* regeneration and genetic transformation systems have now developed in major pulse crops such as *Cicer arietinum*, *Cajanus cajan*, *Vigna* species, *Lupinus* spp., *Vicia* spp., *Phaseolus* spp., and *Pisum sativum*, but transgenic pulse crops have not yet been commercially released (Susan Eapen, 2008). Even though the close leguminous oilseed soybean has been extensively used for the development of herbicide-tolerant and/or insect-tolerant transgenics and also released commercially by private industries, transgenics in food legumes still lag behind and are yet to find a place in the market.

2.2 Genetic Transformation in Legumes Using Agrobacterium

Studies have led to a better understanding of the principle of genetic engineering using the popular method of *Agrobacterium*-mediated genetic transformation (Bhattacharya et al., 2010; Christie et al., 2014; Gelvin, 2003, 2010, 2012; Hwang et al., 2017; Lacroix & Citovsky, 2013; Pitzschke, 2013), and methodologies have also been developed across several legumes such as mungbean (Rathore et al., 2023; Yadav et al., 2010, 2012), chickpea (Bhowmik et al., 2019; Sadhu et al., 2022; Tripathi et al., 2013; Yadav et al., 2017), lentil (Polowick & Yan, 2023), blackgram (Kapildev et al., 2016; Sainger et al., 2015; Trishna et al., 2018), soybean (Hada et al., 2018; Soto et al., 2017), *Medicago* (Wen et al., 2019), and cowpea (Bett et al., 2019; Ping Che et al., 2021), yet the list of transgenic legumes moving out of lab is scanty.

2.3 Current Scenario of Commercialized Transgenic Legumes

The International Service for the Acquisition of Agri-biotech Applications (ISAAA) is a nonprofit international organization that shares information on new bioscience technologies and their benefits and advances with the key stakeholders. The GM Approval Database (GMAD) of ISAAA is one of the top sources of information on GM crop approvals, and it features the available information about GM crops that have been approved for planting and importing for food and feed and commercialization. By the middle of the year 2023, the site enlisted 32 crops encompassing a total of 541 GM events (https://www.isaaa.org/gmapprovaldatabase/). Of these 32 crops, only three are legumes, namely common beans (*Phaselous vulgaris*), soybean (*Glycine max* L.), and cowpea (*Vigna unguiculata*), and have only 1, 43, and 1 event to their names (Tables 1 and 2).

The GM common bean viz. BRS FC401 RMD was developed by microprojectile bombardment and was given approval for direct use or processing as food and feed and also for cultivation in 2011 (https://www.isaaa.org/gmapprovaldatabase/event/default.asp?EventID=23&Event=EMBRAPA%205.1). The GM cowpea was developed by *Agrobacterium*-mediated genetic transformation for insect tolerance and was designated for use as food and feed and approved for cultivation in 2019 in Nigeria and in 2022 in Ghana (https://www.isaaa.org/gmapprovaldatabase/event/

Table 1 Details of GM bean and cowpea in GMAD

S. no.	Event name and code	Trade name	Genes used#	GM trait	Developer/country
Phaseolus vulgaris					
1	Name: **EMBRAPA 5.1** Code: **EMB-PVØ51-1**	**BRS FC401 RMD**	BMGV ac1 gene (sense and antisense)	Viral disease resistance	EMBRAPA (Brazil) Brazil
Vigna unguiculata					
2	Name: **AAT709A** Code: **AAT-7O9AA-4**	**Not available**	Synthetic truncated cry1Ab	Lepidopteran insect resistance	African Agricultural Technology Foundation (AATF)

 # genes used:
ac1: viral replication protein (Rep);
cry1Ab : Cry1Ab delta-endotoxin

Table 2 Details of GM soybean in GMAD

S. no.	Event name and code	Trade name	Genes used#	GM trait	Developer/ country
Glycine max L.					
1.	Name: **260-05 (G94-1, G94-19, G168)** Code: **DD-Ø26ØØ5-3**	**Not available**	gm-fad2-1 bla uidA	Modified oil/fatty acid, antibiotic resistance, visual marker	DuPont (Pioneer Hi-Bred International Inc.)
2.	Name: **A2704-12** Code: **ACS-GMØØ5-3**	**Liberty Link® Soybean**	*pat*	Herbicide tolerance	BASF
3.	Name: **A2704-21** Code: **ACS-GMØØ4-2**	**Liberty Link® Soybean**	*pat*	Glufosinate herbicide tolerance	Bayer CropScience (including fully and partly owned companies)
4.	Name: **A5547-127** Code: **ACS-GMØØ6-4**	**Liberty Link® Soybean**			
5.	Name: **A5547-35** Code: **ACS-GMØØ8-6**	**Liberty Link® Soybean**			
6.	Name: **CV127** Code: **BPS-CV127-9**	**Cultivance**	csr1-2	Tolerance to imidazolinone herbicides	BASF
7.	Name: **DAS44406-6** Code: **DAS-444Ø6-6**	**Not available**	aad-12 2mepsps pat	Glufosinate herbicide tolerance, glyphosate herbicide tolerance, 2,4-D herbicide tolerance	Dow AgroSciences LLC
8.	Name: **DAS68416-4** Code: **DAS-68416-4**	**Not available**	aad-12 pat	Glufosinate herbicide tolerance, 2,4-D herbicide tolerance	Dow AgroSciences LLC
9.	Name: **DAS68416-4 x MON89788** Code: **DAS-68416-4 x MON-89788-1**	**Not available**	aad-12 cp4 epsps (aroA:CP4) pat	Glufosinate herbicide tolerance, glyphosate herbicide tolerance, 2,4-D herbicide tolerance	Dow AgroSciences LLC
10.	Name: **DAS81419** Code: **DAS-81419-2**	**Not available**	cry1Ac cry1F pat (selection marker)	Glufosinate herbicide tolerance, lepidopteran insect resistance	Dow AgroSciences LLC
11.	Name: **DAS81419 x DAS44406** Code: **DAS-81419-2 x DAS-444Ø6-6**	**Conkesta Enlist E3™ Soybean**	cry1Ac cry1F pat aad-12 2mepsps	Glufosinate herbicide tolerance, glyphosate herbicide tolerance, lepidopteran insect resistance, 2,4-D herbicide tolerance	Dow AgroSciences LLC

(continued)

Table 2 (continued)

S. no.	Event name and code	Trade name	Genes used#	GM trait	Developer/ country
12.	Name: **DBN9004** Code: **DBN-Ø9ØØ4-6**	**Not available**	cp4 epsps (aroA:CP4) pat	Glufosinate herbicide tolerance, glyphosate herbicide tolerance	INDEAR
13.	Name: **DP305423** Code: **DP-3Ø5423-1**	**Treus™, Plenish™**	gm-fad2-1 (partial sequence) gm-hra	Sulfonylurea herbicide tolerance, modified oil/fatty acid	DuPont (Pioneer Hi-Bred International Inc.)
14.	Name: **DP305423 x GTS 40-3-2** Code: **DP-3Ø5423-1 x MON-Ø4Ø32-6**	**Not available**	gm-fad2-1 (partial seq) gm-hra cp4 epsps (aroA:CP4)	Glyphosate herbicide tolerance, sulfonylurea herbicide tolerance, modified oil/fatty acid	DuPont (Pioneer Hi-Bred International Inc.
15.	Name: **DP305423 x MON87708** Code: **DP-3Ø5423-1 x MON-877Ø8-9**	**Not available**	gm-fad2-1 (partial seq) dmo	Modified oil/fatty acid, dicamba herbicide tolerance	DuPont (Pioneer Hi-Bred International Inc.)
16.	Name: **DP305423 x MON87708 x MON89788** Code: **DP-3Ø5423-1 x MON-877Ø8-9 x MON-89788-1**	**Not available**	gm-fad2-1 (partial seq) dmo cp4 epsps (aroA:CP4)	Glyphosate herbicide tolerance, modified oil/fatty acid, dicamba herbicide tolerance	DuPont (Pioneer Hi-Bred International Inc.)
17.	Name: **DP305423 x MON89788** Code: **DP-3Ø5423-1 x MON-89788-1**	**Not available**	gm-fad2-1 (partial seq) cp4 epsps (aroA:CP4)	Glyphosate herbicide tolerance, modified oil/fatty acid	DuPont (Pioneer Hi-Bred International Inc.)
18.	Name: **DP356043** Code: **DP-356Ø43-5**	**Optimum GAT™**	gm-hra gat4601	Glyphosate herbicide tolerance, sulfonylurea herbicide tolerance	DuPont (Pioneer Hi-Bred International Inc.)
19.	Name: **FG72 (FGØ72-2, FGØ72-3)** Code: **MST-FGØ72-2**	**Not available**	2mepsps hppdPF W336	Glyphosate herbicide tolerance, isoxaflutole herbicide tolerance	BASF
20.	Name: **FG72 x A5547-127** Code: **MST-FGØ72-2 x ACS-GMØØ6-4**	**Liberty Link® GT27™**	2mepsps hppdPF W336 pat	Glufosinate herbicide tolerance, glyphosate herbicide tolerance, isoxaflutole herbicide tolerance	Bayer CropScience (including fully and partly owned companies)

(continued)

Table 2 (continued)

S. no.	Event name and code	Trade name	Genes used#	GM trait	Developer/ country
21.	Name: **GMB151** Code: **BCS-GM151-6**	**Not available**	cry14Ab-1.b hppdPf4Pa	Nematode Resistance, tolerance to HPPD inhibiting herbicides	BASF
22.	Name: **GTS 40-3-2 (40-3-2)** Code: **MON-Ø4Ø32-6**	**Roundup Ready™ Soybean**	cp4 epsps (aroA:CP4)	Glyphosate herbicide tolerance	Monsanto Company (including fully and partly owned companies)
23.	Name: **GU262** Code: **ACS-GMØØ3-1**	**Liberty Link™ Soybean**	pat bla	Glufosinate herbicide tolerance, antibiotic resistance	Bayer CropScience (including fully and partly owned companies)
24.	Name: **HB4** Code: **IND-ØØ41Ø-5**	**Verdeca HB4 Soybean**	Hahb-4	Drought stress tolerance	Verdeca
25.	Name: **HB4 x GTS 40-3-2** Code: **IND-ØØ41Ø-5 x MON-Ø4Ø32-6**	**Not available**	cp4 epsps (aroA:CP4) Hahb-4	Glyphosate herbicide tolerance, drought stress tolerance	INDEAR
26.	Name: **MON87701** Code: **MON-877Ø1-2**	**Not available**	Cry1Ac	Lepidopteran insect resistance	Monsanto Company (including fully and partly owned companies)
27.	Name: **MON87701 x MON89788** Code: **MON-877Ø1-2 x MON-89788-1**	**Intacta™ Roundup Ready™ 2 Pro**	Cry1Ac cp4 epsps (aroA:CP4)	Glyphosate herbicide tolerance, lepidopteran insect resistance	Monsanto Company (including fully and partly owned companies)
28.	Name: **MON87705** Code: **MON-877Ø5-6**	**Vistive Gold™**	atb1-A (sense and antisense segments) fad2-1A (sense and antisense) cp4 epsps (aroA:CP4)	Glyphosate herbicide tolerance, modified oil/fatty acid	Monsanto Company (including fully and partly owned companies)

(continued)

Table 2 (continued)

S. no.	Event name and code	Trade name	Genes used#	GM trait	Developer/ country
29.	Name: **MON87705 x MON87708** Code: **MON-877Ø5-6 x MON-877Ø8-9**	**Not available**	fad2-1A (sense and antisense) fatb1-A (sense and antisense segments) dmo	Modified oil/fatty acid, dicamba herbicide tolerance	Monsanto Company (including fully and partly owned companies)
30.	Name: **MON87705 x MON87708 x MON89788** Code: **MON-877Ø5-6 x MON-877Ø8-9 x MON-89788-1**	**Not available**	fad2-1A (sense and antisense) fatb1-A (sense and antisense segments) cp4 epsps (aroA:CP4) dmo	Glyphosate herbicide tolerance, modified oil/fatty acid, dicamba herbicide tolerance	Monsanto Company (including fully and partly owned companies)
31.	Name: **MON87705 x MON89788** Code: **MON-877Ø5-6 x MON-89788-1**	**Not available**	fad2-1A (sense and antisense) fatb1-A (sense and antisense segments) cp4 epsps (aroA:CP4)	Glyphosate herbicide tolerance, modified oil/fatty acid	Monsanto Company (including fully and partly owned companies)
32.	Name: **MON87708** Code: **MON-877Ø8-9**	**Genuity® Roundup Ready™ 2 Xtend™**	dmo cp4 epsps (aroA:CP4)	glyphosate herbicide tolerance, dicamba herbicide tolerance	Monsanto Company (including fully and partly owned companies)
33.	Name: **MON87708 x MON89788** Code: **MON-877Ø8-9 x MON-89788-1**	**Not available**	dmo cp4 epsps (aroA:CP4)	Glyphosate herbicide tolerance, dicamba herbicide tolerance	Monsanto Company (including fully and partly owned companies)
34.	Name: **MON87708 x MON89788 x A5547-127** Code: **MON-877Ø8-9 x MON-89788-1 x ACS-GMØØ6-4**	**Not available**	pat dmo cp4 epsps (aroA:CP4)	Glufosinate herbicide tolerance, glyphosate herbicide tolerance, dicamba herbicide tolerance	Monsanto Company (including fully and partly owned companies)

(continued)

Table 2 (continued)

S. no.	Event name and code	Trade name	Genes used#	GM trait	Developer/ country
35.	Name: **MON87712** Code: **MON-87712-4**	**Not available**	cp4 epsps (aroA:CP4) bbx32	Glyphosate herbicide tolerance, enhanced photosynthesis/yield	Monsanto Company (including fully and partly owned companies)
36.	Name: **MON87751** Code: **MON-87751-7**	**Not available**	cry1A.105 cry2Ab2	Lepidopteran insect resistance	Monsanto Company (including fully and partly owned companies)
37.	Name: **MON87751 x MON87701 x MON87708 x MON89788** Code: **MON-87751-7 x MON-877Ø1-2 x MON87708 x MON89788**	**Not available**	cry1A.105 cry2Ab2 cry1Ac dmo cp4 epsps (aroA:CP4)	Glyphosate herbicide tolerance, lepidopteran insect resistance, dicamba herbicide tolerance	Monsanto Company (including fully and partly owned companies)
38.	Name: **MON87769** Code: **MON87769-7**	**Not available**	Pj.D6D Nc.Fad3 cp4 epsps (aroA:CP4)	Glyphosate herbicide tolerance, modified oil/fatty acid	Monsanto Company (including fully and partly owned companies)
39.	Name: **MON87769 x MON89788** Code: **MON-87769-7 x MON-89788-1**	**Not available**	Pj.D6D Nc.Fad3 cp4 epsps (aroA:CP4)*	Glyphosate herbicide tolerance, modified oil/fatty acid	Monsanto Company (including fully and partly owned companies)
40.	Name: **MON89788** Code: **MON-89788-1**	**Genuity® Roundup Ready 2 Yield™**	cp4 epsps (aroA:CP4)	Glyphosate herbicide tolerance	Monsanto Company (including fully and partly owned companies)
41.	Name: **SYHT0H2** Code: **SYN-ØØØH2-5**	**Herbicide-tolerant Soybean line**	pat avhppd-03	Glufosinate herbicide tolerance, mesotrione herbicide tolerance	BASF

(continued)

Table 2 (continued)

S. no.	Event name and code	Trade name	Genes used#	GM trait	Developer/ country
42.	Name: **W62** Code: **ACS-GMØØ2-9**	**Liberty Link™ Soybean**	bar	Glufosinate herbicide tolerance	Bayer CropScience (including fully and partly owned companies)
43.	Name: **W98** Code: **ACS-GMØØ1-8**	**Liberty Link™ Soybean**	bar	Glufosinate herbicide tolerance	Bayer CropScience (including fully and partly owned companies)

#genes used:
gm-fad2-1: *Glycine max* fatty acid desaturase-2
bla: beta lactamase
uidA: beta-D-glucuronidase (GUS)
pat: phosphinothricin N-acetyltransferase
csr1-2: acetohydroxyacid synthase large subunit (AtAHASL)
aad-12: aryloxyalkanoate di-oxygenase 12
2mepsps: 5-enolpyruvyl shikimate-3-phosphate synthase enzyme (double mutant version)
cp4 epsps (aroA:CP4): herbicide tolerant form of 5-enolpyruvulshikimate-3-phosphate synthase (EPSPS)
cry1Ac: Cry1Ac delta-endotoxin
cry1F: Cry1F delta-endotoxin
gm-hra: modified acetolactate synthase (ALS)
dmo: dicamba mono-oxygenase
gat4601: glyphosate N-acetyltransferase
hppdPF W336: modified p-hydroxyphenylpyruvate dioxygenase (hppd)
cry14Ab-1.b: Cry14Ab1
hppdPf4Pa: modified p-hydroxyphenyl pyruvate dioxygenase (HPPD)
Hahb-4: Isolated nucleic acid molecule encoding the transcription factor Hahb-4
fatb1-A: FATB enzymes or acyl-acyl carrier protein thioesterases
fatb2-1A: delta-12 desaturase enzyme
bbx32: protein that interacts with one or more endogenous transcription factors to regulate the plant's day/night physiological processes
cry1A.105: Cry1A.105 protein which comprises the Cry1Ab, Cry1F, and Cry1Ac proteins
cry2Ab2: Cry2Ab delta-endotoxin
Pj.D6D: delta 6 desaturase protein
Nc.Fad3: delta 15 desaturase protein
avhppd-03: p-hydroxyphenylpyruvate dioxygenase
bar: phosphinothricin N-acetyltransferase (PAT) enzyme

default.asp.EventID=543&Event=AAT709A). Of the 45 events in soybean, 5 have not been approved by any country for use, while 5 events have been given approvals for use as food and/or feed and approved for cultivation from the year 2020 by any country. For example, the event DAS81419 was given the first approval in 2014 itself, and the latest approval earned was in 2019 for use as food and feed by the Philippines and by the European Union (EU) in 2021. Similarly, Conkesta Enlist E3™ Soybean had been approved for use by the governments of Argentina, Colombia, Mexico, and Taiwan and recently by the EU in 2021. DBN9004 got its latest approval in 2020 by China for use as food and feed and therefore for general cultivation. The event GMB151 has recent approvals for use as food only by Australia and New Zealand in 2020, Canada in 2021, and the USA for cultivation only in 2022. Soybean Verdeca was given its first approval in 2015 for use as food and feed and cultivation in Argentina followed by USA and Brazil, with the latest approvals from the government of Canada for use as food in 2021 and by Australia in 2022. The remaining 35 events have been approved earlier, with the latest approval in 2019 for direct use or processing as food and/or feed and/or also for cultivation (https://www.isaaa.org/gmapprovaldatabase/crop/default.asp?CropID=19& Crop =Soybean).

The data reveal that of the 541 GM crops enlisted, only 37 events from 3 legumes have been approved for use as food and/or feed and/or cultivation. The majority of the events are from soybean, a well-studied legume known to be responsive to *in vitro* regeneration and amenable to genetic transformation.

2.4 Constraints in Legume Transformation

Though a larger number of potential transgenics have been developed in legumes as reported by Pratap et al. (2018), there are related shortcomings or constraints that have slowed down the pace for successful legume transformation. Spontaneous rearrangements during gene transfer, gene silencing (Meyer, 1995), multiple copy insertion, cotransfer of vector backbone sequence (Kohli et al., 2010; Tiwari et al., 2014), somaclonal variations (Jayasankar, 2005), genotype-dependent *Agrobacterium*-mediated transformation and varied transformation efficiencies (Ahmad & Mukhtar, 2017), recalcitrant nature of legumes, narrow genetic base in legumes, public perception of GM crops and biosafety guidelines (Brookes & Barfoot, 2014; Sehaole, 2022) all play a significant role in affecting legume transformation and its further progress.

3 Genome Editing

The past decade is a witness to tremendous developments in the unveiling of genomes through sequencing and structural analysis. Draft genomes and transcriptome data have been generated for many important legume species. These genomic resources provide a platform for novel discovery and deployment of available information for the development of superior cultivars of legume species. With advances in technology, genome editing has emerged as a potential tool to alter crop genomes for desired traits and is free of biosafety guidelines to some extent. This has led to its extensive adoption by the scientific fraternity for both basic and applied research.

Genome editing is based on the utilization of engineered nucleases and cellular DNA repair pathways to make precise, site-specific genome modifications in an organism. The gene-editing methodologies were actually introduced three decades ago with the key discovery that specific double-stranded breaks can be introduced in chromosomes using a meganuclease I-SceI (Choulika et al., 1995; Rouet et al., 1994). However, the application of meganucleases for gene editing was limited due to the low frequency of target sites in most genes. The upscaling and efficiency of gene-editing systems increased with the introduction of programmable zinc finger nucleases (ZFNs; Bibikova et al., 2002) and transcription activator-like effector nucleases (Zhang et al., 2011). However, the major breakthrough in this revolutionary technology came with the successful adaptation of RNA-guided Cas9 nuclease harnessed from the type II prokaryotic clustered regularly interspaced short palindromic repeats (CRISPR) adaptive immune system for genome editing in eukaryotic cells (Cong et al., 2013; Mali et al., 2013).

4 CRISPR/Cas9

Clustered regularly interspaced short palindromic repeats (CRISPR)-associated protein (Cas) system allows precise modification and gene knockouts, creating probable multiplex genome engineering or activating and repressing genes (Arora & Narula, 2017) of the genome using either endogenous nonhomologous end-joining (NHEJ) or homology-dependent repair (HDR) pathway. It can also use the microhomology-mediated end-joining (MMEJ) pathway. The CRISPR/Cas9 system has two main elements: Cas9 protein and guide RNA (gRNA). The Cas9 is an RNA-dependent DNA endonuclease that binds to make a complex with gRNA. Consisting of 20 nucleotides, the gRNA is complementary to the target DNA sequence and serves as a recruiting signal for Cas9. To recognize the target DNA sequence, CRISPR/Cas9 employs RNA–DNA interactions instead of DNA–protein interactions. In 2014, Doudna and Charpentier discovered that target recognition by the Cas9 protein only requires a seed sequence within the gRNA and a conserved protospacer-adjacent motif (PAM) upstream of the gRNA binding site (Jinek et al., 2012). The sequence 5′-NGG-3′ is the PAM for Cas9 from *Streptococcus pyogenes,* which is the most frequently selected and utilized for genome editing (Hsu et al., 2013; Jiang et al., 2013). Cas9 introduces double-stranded breaks (DSBs) in the target DNA, which ultimately turns up into the induction of DNA repair mechanisms. These mechanisms act through nonhomologous end joining (NHEJ) to create gene knockouts and homology-directed repair (HDR) for gene modification and gene insertion (Huang & Puchta, 2019). When NHEJ randomly inserts or deletes a DNA strand within a coding region, a frameshift mutation occurs that results in a gene knockout. CRISPR technology has proven to be a very robust and potential application for gene expression analysis, gain-of-function, and loss-of-function. In agricultural perspective, legume farming can be revolutionized with this technology (Thakur et al., 2021).

4.1 CRISPR/Cas-Mediated Modifications in Legumes

Genome editing has been utilized in legume plants that include alfalfa and lotus and the crop legumes such as soybean, cowpea, and chickpea. The model legume, Medicago truncatula, also known as alfalfa, has been genome edited for biosynthesis of soyasapogenol B wherein 51 CYP93E2 mutant plant lines were developed. The knockout edited plants did not produce soyasapogenols and diverted the metabolic flux toward the production of valuable hemolytic sapogenins (Confalonieri et al, 2021). Alfalafa has also been edited for MtSUP that regulates the floral organ number and its impairment causes defects in floral development and inflorescence architecture in the mtsup mutant allele carrying plants (Rodas et al., 2021). Knockout of the MtPDS coding phytoene dehydrogenase/chromoplastic protein using the hairy root transformation system (Zhang et al., 2020) and the NPD genes for nodulation (Trujillo et al., 2019) are examples of genome editing in legumes. There are many reports in alfalfa, as it is a model legume plant amenable to genetic transformation as is a self-fertilizing legume with a short life cycle and a small diploid genome (Baloglu et al., 2022).

Lotus, another model legume plant, is known for nodulation and has hence been genome edited for nodulation aspects targeting the *Lbs* genes for nodule senescence (Wang et al., 2019) and also *LjCZF1* and *LjCZF2* for root nodule symbiosis (Cai et al., 2018a). It has also been edited for CYP716A51 (triterpenoid production) by Suzuki et al. (2019).

Chickpea (*Cicer arietinum*) is an important crop worldwide, and gene-editing tools can be used to eliminate the associated problems with its yield. The first-ever study by Badhan et al. (2021) reported targeted drought tolerance-associated genes, 4-coumarate ligase (4CL) and Reveille 7 (RVE7), for CRISPR/Cas9 editing in chickpea protoplast, and the knock-out of the *RVE7* gene showed high-efficiency editing in vivo. The results indicate that CRISPR/Cas9 DNA-free gene editing can be used for genes associated with drought tolerance in chickpea.

The most researched-upon legume is soybean, wherein genome editing for many traits has been attempted as described in Table 3.

Most of the studies have been conducted on the model legumes or food legumes. Looking broadly, the minor pulses have not been figured out yet for genome editing. Similarly, the importance of nutritious yet underutilized legumes found in deserts and the possibility to genome edit them for nutraceutical and flavor has been discussed by Joshi et al. (2023).

Table 3 Genome editing in soybean

Parameter	Target gene	Trait of choice	References
Flowering	*GmPRR37*	Change in flowering time	Wang et al. (2020)
	GmFT2a and *GmFT5a*	Flowering time	Cai et al. (2020) Cai et al. (2018b)
	E1	Early flowering	Han et al. (2019)
Morphological	*GmSPL9*	Altered plant architecture	Bao et al. (2019)
	GmAGO7a and *GmAGO7b*)	Controlling leaf pattern	Zheng et al. (2020)
	DCLb	Increased lateral root growth	Curtin et al. (2011)
Biochemical	*GmLox1, GmLox2, GmLox3*	Loss of lipoxygenase activity	Wang et al. (2020)
	KAS1	(Conversion of sucrose to oil)	Virdi et al. (2020)
	GmFAD2-1A, GmFAD2-2A	Increased oleic acid content	Wu et al. (2020) Do et al. (2019) Haun et al. (2014)
	FAD2-2	Increased oleic acid content	Al Amin et al. (2019)
	Conglycinin (7S) and *glycinin (11S)*	Storage proteins—Mutations in three of nine genes	Li et al. (2019a, b)
	GmPDS11 and *GmPDS18*	Albino and dwarf buds	Du et al. (2016)
	FAD3A	Reduced linolenic acid content	Demorest et al. (2016)

4.2 Regulatory Aspects of Genome-Edited Crops

The genetically engineered products (transgenic plants in context here) are regulated by the "Rules 1989" under the Environment (Protection) Act, 1986 in India and notified by the Ministry of Environment, Forest, and Climate Change, Government of India. However, the regulatory considerations for the new technologies have been eased up, and the genome-edited crops falling under the SDN1 and SDN2 categories are exempted from the 1989 regulations of GE products. Yet, all the genome-edited studies will be regulated by the Guidelines for Safety Assessment of Genome Edited Plants, 2022 (Ministry of Science and Technology, New Delhi, 2022).

Though genome editing is an advanced technology being widely used in the scientific community, the need to enhance its precision and efficiency and also to reduce the cost and time involved is yet to be developed. Again, though the genomic resources are in plenty, the need to identify target genes for desired and prioritized traits through functional genomics is indispensable. Another relevant advancement lately has been speed breeding that permits genetic gain in lesser time. Exploitation of speed breeding in combination with genome editing shall definitely aid in legume improvement.

5 Speed Breeding

Speed breeding allows for rapid generation advancement and has been successfully applied in crop plants to achieve quick rates of crop improvement. In contrast to the past, plant breeders now have cutting-edge technologies at their disposal that can assist them in overcoming challenges and enhancing crop varieties. It is a simulated habitat that has been designed to provide longer days in order to manipulate the growth cycles of photo-sensitive crops.

Traditional breeding methods need 8–10 years to develop a new variety; however, speed breeding allows the generation cycle to be sped up by 2–3 times. For a variety of plant crops that act on chromosomes and are given perfect light quality and intensity and correct temperature to promote photosynthesis and increase growth and breeding, speed breeding techniques are available. This procedure helps in attaining at least six generations of unique species. Speed breeding can be carried out in regular settings with some adjustments; specialized labs are not required (Shivakumar et al., 2018). Pyramiding traits, providing transgenic pipelines, and the capacity to speed up backcrossing are just a few of the benefits speed breeding offers over traditional methods. Researchers without access to larger areas can build up smaller speed breeding units and conduct speed breeding in targeted locations (Shimelis & Laing, 2012). Speed breeding has been incorporated with many various techniques to hasten crop development under controlled temperature and light conditions.

5.1 Role of Speed Breeding in Enhancement of Crop Plants

Compared to traditional selection procedures, which only produce 1–2 generations annually, speed breeding produces 3–6 generations annually. Speed breeding enables the quick generational progression and homozygous and stable genotype production, which leads to the creation and dissemination of innovative cultivars (Watson et al., 2018). Speed breeding technology also complements marker-assisted selection and high-throughput phenotyping methods well for multiple trait selection. The target gene or QTL responsible for the desired attribute must be identified in order to perform marker-assisted selection. As a result, when the underlying genetic factors are unknown, marker-assisted selection is less viable for complex traits (Budak et al., 2013).

5.2 Increasing Genetic Gain Through Genomic Selection and Rapid Breeding

Researchers can advance their breeding materials more quickly by reducing the number of breeding cycles needed to generate new varieties as well as the period between crossing and choosing offspring to utilize as parents for the following cross. Recent advancements in breeding methods, such as genetic engineering, genomic selection, and doubled-haploid technology, have shortened breeding cycles and improved genetic gain rates in crops such as wheat, rice, and maize (Hickey et al., 2017). These innovations can have an even greater impact when combined with speed breeding methods, which enable rapid generation advancement by maintaining plant populations under regulated photoperiod and temperature regimes to hasten their growth and development (Wolter et al., 2019).

5.3 Speed Breeding to Accelerate Domestication

Plant domestication is an artificial process of converting wild plant species into crop plants. In this process, early hybridization is followed by a selective breeding strategy. Crops with polyploid genomes are notably related to plant breeding. It is a time-consuming approach; thus to address this problem, speed breeding has been integrated, which reduces the length of time and number of generations of that crop that have been produced. Polyploid plants with rapid breeding, such as bananas and peanuts, are required as plant domestication proof.

5.4 Speed Breeding: A Concept to Rationality

Botanist Pfeiffer (1926) made the discovery that plants can be cultivated in the presence of artificial light almost 100 years ago. Siemens (1880) investigated the effects of continuous supplemental lighting using electric and incandescent lamps. It was subsequently discovered that continuous light promotes early flowering in a variety of plant species, including ornamentals, weed species, weedy plants, vegetables, and grains (Davis & Burns, 2016).

Scientists from the University of Queensland in Australia, the John Innes Centre in the United Kingdom, and collaborators improved the technique for quick plant growth and development and called it "speed breeding" because it was inspired by NASA's methods for growing plants on space stations. This technique has been successfully utilized for a range of crops such as wheat, barley, chickpea, and canola, which use controlled modified glasshouses fitted with light-emitting diode (LED) to grow plants under extended photoperiods to accelerate plant growth, early flowering, and seed maturation, consequently reducing plant generation time by up to 5 times compared to field conditions and by 2.5 times compared to regular greenhouse (Ghosh et al., 2018; Watson et al., 2018). The world is paying close attention to this cutting-edge technology that shortens the plant breeding cycle and quickens breeding research.

5.5 Induction of Early Flowering

Throughout the photoperiod, a higher temperature is maintained, whereas a drop in temperature throughout the night can expedite flowering (Pazos-Navarro et al., 2017). Extended light was used in the instance of amaranth (*Amaranthus* spp. L.) before and after a reduced photoperiod to stimulate flowering and speed up initial vegetative growth (Stetter et al., 2016).

5.6 Drought

According to Berger et al. (2010), one of the main obstacles to the cultivation of sustainable crops is drought. Plants experience water stress when there is insufficient water in the soil. In diverse crops, drought stress has been shown to reduce stem diameter, leaf area, plant height, and plant biomass (Zheng et al., 2016). It is extremely concerning since drought stress is unexpected and heterogeneous, with changes occurring throughout seasons, years, and areas. By adopting speed breeding, breeders can readily overcome the challenges they confront while creating drought-tolerant cultivars (Mitra, 2001). A representative drought stress condition is found using a speed-breeding platform, which also provides uniform screening conditions. Selection of potential individual plants is done by genotype evaluation under controlled conditions. In order to identify superior genotypes in speed breeding, reproducible and exact screening procedures can be used. This eliminates the need to pick lines with a poor correlation between stress tolerance traits and yield. The effects of several shocks on plants are typically recognized; however, speed breeding enables independent research on each stress and clarifies the mechanisms underlying stress tolerance. Given that the majority of research efforts are focused on cereals, speed breeding will aid in extending the study of abiotic stress in legume crops.

6 Conclusion

The past two decades have seen an upsurge in biotechnological advancements, to name whole genome sequencing, transcriptomics, metabolomics, genomic selection, genome editing, and speed breeding as a few. While molecular breeding needs to join hands with genomic selection to utilize the genotyping database effectively, genome editing needs to move ahead with speed breeding to achieve recognizable genetic gains. We have come to a ground where these players need to team up for improving legumes.

References

Ahmad, N., & Mukhtar, Z. (2017). Genetic manipulations in crops: Challenges and opportunities. *Genomics, 109*(5–6), 494–505.

Al Amin, N., Ahmad, N., Wu, N., Pu, X., Ma, T., Du, Y., et al. (2019). CRISPR-Cas9 mediated targeted disruption of FAD2-2 microsomal Omega-6 desaturase in soybean (Glycine max.L). *BMC Biotechnology, 19*, 9. https://doi.org/10.1186/s12896-019-0501-2

Arora, L., & Narula, A. (2017). Gene editing and crop improvement using CRISPR-Cas9 system. *Frontiers in Plant Science, 8*, 193.

Badhan, S., Ball, A. S., & Mantri, N. (2021). First report of CRISPR/Cas9 mediated DNA-free editing of *4CL* and *RVE7* genes in chickpea protoplasts. *International Journal of Molecular Sciences, 22*, 396. https://doi.org/10.3390/ijms22010396

Baloglu, M. C., Celik Altunoglu, Y., Baloglu, P., Yildiz, A. B., Türkölmez, N., & Özden Çiftçi, Y. (2022). Gene-editing technologies and applications in legumes: Progress, evolution, and future prospects. *Frontiers in Genetics, 13*, 859437. https://doi.org/10.3389/fgene.2022.859437

Bao, A., Chen, H., Chen, L., Chen, S., Hao, Q., Guo, W., et al. (2019). CRISPR/Cas9-mediated targeted mutagenesis of *GmSPL9* genes alters plant architecture in soybean. *BMC Plant Biology, 19*, 131. https://doi.org/10.1186/s12870-019-1746-6

Berger, B., Parent, B., & Tester, M. (2010). High-throughput shoot imaging to study drought responses. *Journal of Experimental Botany, 61*, 35193528.

Bett, B., Gollasch, S., Moore, A., Harding, R., & Higgins, T. J. V. (2019). An improved transformation system for cowpea (*Vigna unguiculata* L. Walp) via sonication and a kanamycin-Geneticin selection regime. *Frontiers in Plant Science, 10*, 219. https://doi.org/10.3389/fpls.2019.00219

Bhattacharya, A., Sood, P., & Citovsky, V. (2010). The roles of plant phenolics in defence and communication during agrobacterium and rhizobium infection. *Molecular Plant Pathology, 11*, 705–719.

Bhowmik, S. S. D., Cheng, A. Y., Long, H., Tan, G. Z. H., Hoang, T. M. L., Karbaschi, M. R., Williams, B., Higgins, T. J. V., & Mundree, S. G. (2019). Robust genetic transformation system to obtain non-chimeric transgenic chickpea. *Frontiers in Plant Science, 10*, 524. https://doi.org/10.3389/fpls.2019.00524

Bibikova, M., Golic, M., Golic, K. G., & Carroll, D. (2002). Targeted chromosomal cleavage and mutagenesis in *Drosophila* using zinc-finger nucleases. *Genetics, 161*, 1169–1175.

Brookes, G., & Barfoot, P. (2014). Economic impact of GM crops: The global income and production effects 1996–2012. *GM Crops & Food: Biotechnology in Agriculture and the Food Chain, 5*(1), 65–75.

Budak, H., Kantar, M., & Kurtoglu, K. Y. (2013). Drought tolerance in modern and wild wheat. *Scientifi c World Journal, 2013*, 548246. https://doi.org/10.1155/2013/548246. PMID: 23766697; PMCID: PMC3671283.

Cai, K., Yin, J., Chao, H., Ren, Y., Jin, L., Cao, Y., et al. (2018a). A C3HC4-type RING finger protein regulates Rhizobial infection and nodule organogenesis in *Lotus Japonicus*. *Journal of Integrative Plant Biology, 60*(9), 878–896. https://doi.org/10.1111/jipb.12703. PMID: 30047576.

Cai, Y., Chen, L., Sun, S., Wu, C., Yao, W., Jiang, B., et al. (2018b). CRISPR/Cas9-mediated deletion of large genomic fragments in soybean. *International Journal of Molecular Sciences, 19*(12), 3835. https://doi.org/10.3390/ijms19123835

Cai, Y., Wang, L., Chen, L., Wu, T., Liu, L., Sun, S., et al. (2020). Mutagenesis of *GmFT2a* and *GmFT5a* mediated by CRISPR/Cas9 contributes for expanding the regional adaptability of soybean. *Plant Biotechnology Journal, 18*(1), 298–309. https://doi.org/10.1111/pbi.13199. Epub 2019 Jul 5. PMID: 31240772; PMCID: PMC6920152.

Choulika, A., Perrin, A., Dujon, B., & Nicolas, J. F. (1995). Induction of homologous recombination in mammalian chromosomes by using the I-SceI system of *Saccharomyces cerevisiae. Molecular and Cellular Biology, 15*, 1968–1973.

Christie, P. J., Whitaker, N., & Gonzalez-Rivera, C. (2014). Mechanism and structure of the bacterial type IV secretion systems. *Biochimica et Biophysica Acta, 1843*, 1578–1591.

Christou, P. (1994). *The biotechnology of crop legumes* (Vol. 74, pp. 165–185). Kluwer Academic Publishers.

Confalonieri, M., Carelli, M., Gianoglio, S., Moglia, A., Biazzi, E., & Tava, A. (2021). CRISPR/Cas9-mediated targeted mutagenesis of CYP93E2 modulates the triterpene Saponin biosynthesis in Medicago truncatula. *Frontiers in Plant Science, 12*, 690231. https://doi.org/10.3389/fpls.2021.690231

Cong, L., Ran, F. A., Cox, D., Lin, S., Barretto, R., Habib, N., Hsu, P. D., Wu, X., Jiang, W., Marraffini, L. A., et al. (2013). Multiplex genome engineering using CRISPR/Cas systems. *Science, 339*, 819–823.

Curtin, S. J., Zhang, F., Sander, J. D., Haun, W. J., Starker, C., Baltes, N. J., et al. (2011). Targeted mutagenesis of duplicated genes in soybean with zinc-finger nucleases. *Plant Physiology, 156*(2), 466–473. https://doi.org/10.1104/pp.111.172981

Davis, P. A., & Burns, C. (2016). Photobiology in protected horticulture. *Food Energy Security, 5*(4), 223238.

Demorest, Z. L., Coffman, A., Baltes, N. J., Stoddard, T. J., Clasen, B. M., Luo, S., et al. (2016). Direct stacking of sequence-specific nuclease-induced mutations to produce high oleic and low linolenic soybean oil. *BMC Plant Biology, 16*, 1–8. https://doi.org/10.1186/s12870-016-0906-1

Do, P. T., Nguyen, C. X., Bui, H. T., Tran, L. T. N., Stacey, G., Gillman, J. D., et al. (2019). Demonstration of highly efficient dual gRNA CRISPR/Cas9 editing of the Homeologous GmFAD2-1A and GmFAD2-1B genes to yield a high oleic, low linoleic and α-linolenic acid phenotype in soybean. *BMC Plant Biology, 19*, 311. https://doi.org/10.1186/s12870-019-1906-8

Doudna, J. A., & Charpentier, E. (2014). The new frontier of genome engineering with CRISPR-Cas9. *Science, 346*(6213), S. https://doi.org/10.1126/science.1258096

Du, H., Zeng, X., Zhao, M., Cui, X., Wang, Q., Yang, H., et al. (2016). Efficient targeted mutagenesis in soybean by TALENs and CRISPR/Cas9. *Journal of Biotechnology, 217*, 90–97. https://doi.org/10.1016/j.jbiotec.2015.11.005

Gelvin, S. B. (2003). Agrobacterium-mediated plant transformation: the Biology behind the "Gene-Jockeying" tool. *Microbiology and Molecular Biology Reviews, 67*, 16–37.

Gelvin, S. B. (2010). Plant proteins involved in agrobacterium-mediated genetic transformation. *Annual Review of Phytopathology, 48*, 45–68.

Gelvin, S. B. (2012). Traversing the cell: Agrobacterium T-DNA's journey to the host genome. *Frontiers in Plant Science, 3*, 52.

Ghosh, S., Watson, A., Gonzalez-Navarro, O. E., Ramirez-Gonzalez, R. H., Yanes, L., Mendoza-Suárez, M., Simmonds, J., Wells, R., Rayner, T., Green, P., Hafeez, A., Hayta, S., Melton, R. E., Steed, A., Sarkar, A., Carter, J., Perkins, L., Lord, J., Tester, M., Osbourn, A., Moscou, M. J., Nicholson, P., Harwood, W., Martin, C., Domoney, C., Uauy, C., Hazard, B., Wulff, B. B. H., & Hickey, L. T. (2018). Speed breeding in growth chambers and glasshouses for crop breeding and model plant research. *Nature Protocols, 13*(12), 2944–2963. https://doi.org/10.1038/s41596-018-0072-z

Hada, A., Krishnan, V., Mohamed Jaabir, M. S., et al. (2018). Improved *Agrobacterium tumefaciens*-mediated transformation of soybean [*Glycine max* (L.) Merr.] following optimization of culture conditions and mechanical techniques. *In Vitro Cellular & Developmental Biology – Plant, 54*, 672–688. https://doi.org/10.1007/s11627-018-9944-8

Han, J., Guo, B., Guo, Y., Zhang, B., Wang, X., & Qiu, L.-J. (2019). Creation of early flowering germplasm of soybean by CRISPR/Cas9 technology. *Frontiers in Plant Science, 10*, 1446. https://doi.org/10.3389/fpls.2019.01446

Haun, W., Coffman, A., Clasen, B. M., Demorest, Z. L., Lowy, A., Ray, E., et al. (2014). Improved soybean oil quality by targeted mutagenesis of the fatty acid desaturase 2 gene family. *Plant Biotechnology Journal, 12*, 934–940. https://doi.org/10.1111/pbi.12201

Hickey, L. T., Germán, S. E., Pereyra, S. A., Diaz, J. E., Ziems, L. A., Fowler, R. A., Platz, G. J., Franckowiak, J. D., & Dieters, M. J. (2017). Speed breeding for multiple disease resistance in barley. *Euphytica, 213*(3), 1–14.

Hsu, P. D., Scott, D. A., Weinstein, J. A., Ran, F. A., Konermann, S., Agarwala, V., Li, Y., Fine, E. J., Wu, X., Shalem, O., et al. (2013). DNA targeting specificity of RNA-guided Cas9 nucleases. *Nature Biotechnology, 31*, 827–832.

https://www.isaaa.org/gmapprovaldatabase/. Accessed 27 July 2023.

https://www.isaaa.org/gmapprovaldatabase/crop/default.asp?CropID=19&Crop=Soybean. Accessed 28 July 2023.

https://www.isaaa.org/gmapprovaldatabase/event/default.asp?EventID=23&Event=EMBRAPA%205.1. Accessed 28 July 2023.

https://www.isaaa.org/gmapprovaldatabase/event/default.asp?EventID=543&Event=AAT709A. Accessed 27 July 2023.

Huang, T.-K., & Puchta, H. (2019). CRISPR/Cas-mediated gene targeting in plants: Finally a turn for the better for homologous recombination. *Plant Cell Reports, 38*, 443–453. https://doi.org/10.1007/s00299-019-02379-0

Hwang, H., Manda, Y., & Lai, E. (2017). *Agrobacterium*-mediated plant transformation: Biology and applications. *The Arabidopsis Book, 2017*(15), e0186.

Jayasankar, S. (2005). Variation in tissue culture. In R. N. Trigiano & D. J. Gray (Eds.), *Plant development and biotechnology* (pp. 301–309). CRC Press.

Jiang, W., Bikard, D., Cox, D., Zhang, F., & Marraffini, L. A. (2013). RNA-guided editing of bacterial genomes using CRISPR-Cas systems. *Nature Biotechnology, 31*, 233–239.

Jinek, M., Chylinski, K., Fonfara, I., Hauer, M., Doudna, J. A., & Charpentier, E. (2012). A programmable dual-RNA-guided DNA endonuclease in adaptive bacterial immunity. *Science, 337*, 816–821. https://doi.org/10.3389/fpls.2017.01932

Joshi, T., Sehgal, H., Mandal, S. K., et al. (2023). Nutraceutical and flavor profiles in underutilized desert legumes of India: Gene editing strategies towards sustainable food development. *Journal of Plant Biochemistry and Biotechnology*. https://doi.org/10.1007/s13562-023-00836-2

Kapildev, G., Chinnathambi, A., Sivanandhan, G., Rajesh, M., Vasudevan, V., Mayavan, S., Arun, M., Jeyaraj, M., Alharbi, S. A., Selvaraj, N., & Ganapathi, A. (2016). High-efficient Agrobacterium mediated in planta transformation in black gram (Vigna mungo (L.) Hepper). *Acta Physiologiae Plantarum, 38*, 205.

Kohli, A., Miro, B., & Twyman, R. M. (2010). Transgene integration, expression and stability in plants: Strategies for improvements. In C. Kole, C. H. Michler, A. G. Abbott, & T. C. Hall (Eds.), *Transgenic crop plants* (pp. 201–237). Springer.

Lacroix, B., & Citovsky, V. (2013). The roles of bacterial and host plant factors in Agrobacterium-mediated genetic transformation. *The International Journal of Developmental Biology, 57*, 467–481.

Li, W., Chen, Y., Schnabel, E., Crook, A., & Frugoli, J. (2019a). Comparison of efciency and time to regeneration of *Agrobacterium*-mediated transformation methods in *Medicago truncatula*. *Plant Methods, 15*, 20.

Li, C., Nguyen, V., Liu, J., Fu, W., Chen, C., Yu, K., et al. (2019b). Mutagenesis of seed storage protein genes in soybean using CRISPR/Cas9. *BMC Research Notes, 12*, 176. https://doi.org/10.1186/s13104-019-4207-2

Mali, P., Yang, L., Esvelt, K. M., Aach, J., Guell, M., DiCarlo, J. E., Norville, J. E., & Church, G. M. (2013). RNA-guided human genome engineering via Cas9. *Science, 339*, 823–826.

Meng, Y., Hou, Y., Wang, H., Ji, R., Liu, B., Wen, J., et al. (2017). Targeted mutagenesis by CRISPR/Cas9 system in the model legume *Medicago Truncatula*. *Plant Cell Reports, 36*, 371–374. https://doi.org/10.1007/s00299-016-2069-9

Meyer, P. (1995). Understanding and controlling transgene expression. *Elsevier Science, 13*, 332–337.

Mitra, J. (2001). Genetics and genetic improvement of drought resistance in crop plants. *Current Science, 80*, 758763.

Pazos-Navarro, M., Castello, M., Bennett, R. G., Nichols, P., & Croser, J. (2017). In vitro-assisted single-seed descent for breeding-cycle compression in subterranean clover (Trifolium subterraneum L.). *Crop & Pasture Science, 68*, 958966.

Pfeiffer, N. E. (1926). Microchemical and morphological studies of the effect of light on plants. *Botanical Gazette, 81*(2), 173–195.

Ping Che, Chang, S., Simon, M. K., Zhang, Z., Shaharyar, A., Ourada, J., O'Neill, D., Torres-Mendoza, M., Guo, Y., Marasigan, K. M., Vielle-Calzada, J.-P., Ozias-Akins, P., Albertsen, M. C., & Jones, T. J. (2021). Developing a rapid and highly efficient cowpea regeneration, transformation and genome editing system using embryonic axis explants. *The Plant Journal*. https://doi.org/10.1111/tpj.15202

Pitzschke, A. (2013). Agrobacterium infection and plant defense-transformation success hangs by a thread. *Frontiers in Plant Science, 4*, 519.

Polowick, P. L., & Yan, W. (2023). A protocol for Agrobacterium-mediated genetic transformation of Lens culinaris Medik (lentil). *Plant Cell Tissue and Organ Culture (PCTOC), 152*, 605–618.

Pratap, A., Prajapati, U., Singh, C. M., et al. (2018). Potential, constraints and applications of in vitro methods in improving grain legumes. *Plant Breeding, 00*, 1–15. https://doi.org/10.1111/pbr.12590

Rathore, M., Tripathi, A., Kushwah, N. S., & Singh, N. P. (2023). Multiple shoot regeneration from detached embryonic axis in greengram (*Vigna radiata*) cv.SML 668. *Legume Research*. https://doi.org/10.18805/LR-463

Rodas, A. L., Roque, E., Hamza, R., Gómez-Mena, C., Minguet, E. G., Wen, J., et al. (2021). MtSUPERMAN plays a key role in compound inflorescence and flower development in Medicago Truncatula. *The Plant Journal, 105*, 816–830. https://doi.org/10.1111/tpj.15075

Rouet, P., Smih, F., & Jasin, M. (1994). Introduction of double-strand breaks into the genome of mouse cells by expression of a rare-cutting endonuclease. *Molecular and Cellular Biology, 14*, 8096–8106.

Sadhu, S. K., Jogam, P., Gande, K., Banoth, R., Penna, S., & Peddaboina, V. (2022). Optimization of different factors for an agrobacterium-mediated genetic transformation system using embryo axis explants of chickpea (*Cicer arietinum* L.). *Journal Of Plant Biotechnology, 49*, 61–73.

Sainger, M., Chaudhary, D., Dahiya, S., Jaiwal, R., & Jaiwal, P. K. (2015). Development of an efficient in vitro plant regeneration system amenable to Agrobacterium mediated transformation of a recalcitrant grain legume black gram [*Vigna mungo* (L.) Hepper]. *Physiology and Molecular Biology of Plants, 21*(4), 505–517.

Sehaole, E. K. M. (2022). Genetic transformation in agro-economically important legumes [Internet]. In *Legumes research – Volume 1*. IntechOpen. Available from: https://doi.org/10.5772/intechopen.101262.

Shimelis, H., & Laing, M. (2012). Timelines in conventional crop improvement: Prebreeding and breeding procedures. *Australian Journal of Crop Science, 6*(11), 1542–1549.

Shivakumar, M., Nataraj, V., Kumawat, G., Rajesh, V., Chandra, S., Gupta, S., & Bhatia, V. S. (2018). Speed breeding for Indian agriculture: A rapid method for development of new crop varieties. *Current Science, 115*, 1241–1241.

Siemens, C. W. (1880). III. On the influence of electric light upon vegetation, and on certain physical principles involved. *Proceedings of the Royal Society of London, 30*(200–205), 210–219.

Soto, N., Delgado, C., Hernández, Y., Rosabal, Y., Ferreira, A., Pujol, M., et al. (2017). Efficient particle bombardment-mediated transformation of Cuban soybean (INCASoy-36) using glyphosate as a selective agent. *Plant Cell, Tissue and Organ Culture, 128*, 187–196. https://doi.org/10.1007/s11240-016-1099-x

Stetter, M. G., Zeitler, L., & Steinhaus, A. (2016). Crossing methods and cultivation conditions for rapid production of segregating populations in three grain amaranth species. *Frontiers in Plant Science, 7*, 816.

Susan Eapen. (2008). Advances in development of transgenic pulse crops. *Biotechnology Advances, 26*, 162–168.

Suzuki, H., Fukushima, E. O., Shimizu, Y., Seki, H., Fujisawa, Y., Ishimoto, M., et al. (2019). *Lotus Japonicus* triterpenoid profile and characterization of the *CYP716A51* and *LjCYP93E1* genes involved in their biosynthesis in planta. *Plant Cell Physiology, 60*(11), 2496–2509. https://doi.org/10.1093/pcp/pcz145

Thakur, S., Kushwah, N., Kumar, S., Singh, P., Rathore, M., Pratap Singh, N., & Das, A. (2021). Genome editing using CRISPR/Cas systems in legumes. In *Genome editing in plants* (pp. 89–107). CRC Press. https://doi.org/10.1201/9780367815370-7

Tiwari, V., Chaturvedi, K. A., Mishra, A., & Jha, B. (2014). An efficient method of agrobacterium-mediated genetic transformation and regeneration in local Indian cultivar of groundnut (*Arachis hypogaea*) using grafting. *Applied Biochemistry and Biotechnology, 175*, 436–453.

Tripathi, L., Singh, A. K., Singh, S., Singh, R., Chaudhary, S., Sanyal, I., & Amla, D. V. (2013). Optimization of regeneration and Agrobacteriummediated transformation of immature cotyledons of chickpea (*Cicer arietinum* L.). *Eapen, 113*, 513–527.

Trishna, K., Borah, B., Handique, A. K., Acharjee, S., & Sarmah, B. K. (2018). An efficient in vitro regeneration protocol to generate stable transgenic lines of black gram (*Vigna mungo* L.). *Indian Journal of Genetics and Plant Breeding, 78*(3), 324–332.

Trujillo, D. I., Silverstein, K. A. T., & Young, N. D. (2019). Nodule-specific PLAT domain proteins are expanded in the *Medicago* lineage and required for nodulation. *New Phytologist*. https://doi.org/10.1111/nph.15697

Virdi, K. S., Spencer, M., Stec, A. O., Xiong, Y., Merry, R., Muehlbauer, G. J., et al. (2020). Similar seed composition phenotypes are observed from CRISPR-generated in-frame and knockout alleles of a soybean KASI Ortholog. *Frontiers in Plant Science, 11*, 1006. https://doi.org/10.3389/fpls.2020.01005

Wang, L., Rubio, M. C., Xin, X., Zhang, B., Fan, Q., Wang, Q., et al. (2019). CRISPR/Cas9 knockout of leghemoglobin genes inLotus Japonicusuncovers their synergistic roles in symbiotic nitrogen fixation. *The New Phytologist, 224*, 818–832. https://doi.org/10.1111/nph.16077

Wang, J., Kuang, H., Zhang, Z., Yang, Y., Yan, L., Zhang, M., et al. (2020). Generation of seed lipoxygenase-free soybean using CRISPR-Cas9. *Crop Journal, 8*(3), 432–439. https://doi.org/10.1016/j.cj.2019.08.008. ISSN 2214-5141.

Watson, A., Ghosh, S., Williams, M. J., Cuddy, W. S., Simmonds, J., Rey, M. D., Asyraf Md Hatta, M., Hinchliffe, A., Steed, A., Reynolds, D., & Adamski, N. M. (2018). Speed breeding is a powerful tool to accelerate crop research and breeding. *Nature Plants, 4*(1), 23–29.

Wen, L., Chen, Y., Schnabel, E., Crook, A., & Frugoli, J. (2019). Comparison of efficiency and time to regeneration of *Agrobacterium*-mediated transformation methods in *Medicago truncatula. Plant Methods, 15*, 20. https://doi.org/10.1186/s13007-019-0404-1. Erratum in: Plant Methods. 2019 Aug 19;15:95. PMID: 30858871; PMCID: PMC6394069.

Wolter, F., Schindele, P., & Puchta, H. (2019). Plant breeding at the speed of light: The power of CRISPR/Cas to generate directed genetic diversity at multiple sites. *BMC Plant Biology, 19*(1), 1–8.

Wu, N., Lu, Q., Wang, P., Zhang, Q., Zhang, J., Qu, J., et al. (2020). Construction and analysis of *GmFAD2-1A* and *GmFAD2-2A* soybean fatty acid desaturase mutants based on CRISPR/Cas9 technology. *International Journal of Molecular Sciences, 21*(3), 1104. https://doi.org/10.3390/ijms21031104

Yadav, S. K., Sreenu, P., Maheswari, M., Vanaja, M., & Venkateswarlu, B. (2010). Efficient shoot regeneration from double cotyledonary node explants of green gram [*Vigna radiata* (L.) Wilczek]. *Indian Journal of Biotechnology, 9*, 403–407.

Yadav, S. K., Katikala, S., Yellisetty, V., et al. (2012). Optimization of *Agrobacterium* mediated genetic transformation of cotyledonary node explants of *Vigna radiata*. *Springerplus, 1*, 59. https://doi.org/10.1186/2193-1801-1-59

Yadav, R., Mehrotra, M., Singh, A. K., et al. (2017). Improvement in *Agrobacterium*-mediated transformation of chickpea (*Cicer arietinum* L.) by the inhibition of polyphenolics released during wounding of cotyledonary node explants. *Protoplasma, 254*, 253–269. https://doi.org/10.1007/s00709-015-0940-0

Zhang, F., Cong, L., Lodato, S., Kosuri, S., Church, G. M., & Arlotta, P. (2011). Efficient construction of sequence-specific TAL effectors for modulating mammalian transcription. *Nature Biotechnology, 29*, 149–153.

Zhang, H., Cao, Y., Zhang, H., Xu, Y., Zhou, C., Liu, W., et al. (2020). Efficient generation of CRISPR/Cas9-mediated homozygous/Biallelic *Medicago Truncatula* mutants using a hairy root system. *Frontiers in Plant Science, 11*, 294. https://doi.org/10.3389/fpls.2020.00294

Zheng, M., Tao, Y., Hussain, S., Jiang, Q., Peng, S., Huang, J., et al. (2016). Seed priming in dry direct-seeded rice: Consequences for emergence, seedling growth and associated metabolic events under drought stress. *Plant Growth Regulation, 78*, 167178. Available from: 10.1007/s10725-015-0083-5.

Zheng, N., Li, T., Dittman, J. D., Su, J., Li, R., Gassmann, W., et al. (2020). CRISPR/Cas9-based gene editing using egg cell-specific promoters in Arabidopsis and soybean. *Frontiers in Plant Science, 11*, 800. https://doi.org/10.3389/fpls.2020.00800

Botanical and Economic Significance of African Yam Bean and its Relatives

B. Daniel Adewale, T. E. Konyeme, F. T. Fayose, A. A. Odetoye, J. O. Popoola, B. O. Ojuederie, C. V. Nnamani, B. Ikhajiagbe, and C. O. Aremu

Abstract *Sphenostylis* is a small genus with significant and unmistakable botanical description in the family *Fabaceae*, sub-family *Papilionoideae*, tribe *Phaseoleae* and sub-tribe *Phaseolinae*. It is distinguished for possessing dorsiventrally flattened and bearded-edge stigma and the absence of aril on the seeds. Its nativity is tropical Africa, with seven species, among which only one - *S. stenocarpa* (Hochst Ex. A. Rich. Harms)—African yam bean—is cultivated for the edible aerial grains and subterranean tuber. The other six are *S. angustifolia*, *S. briartii*, *S. erecta*, *S. marginata*, *S. schweinfurthii* and S. *zimbabweensis*. *S. angustifolia* which provide fodder for ruminants, while *S. erecta*, *S. marginata* and *S. schweinfurthii* have food,

---

B. D. Adewale (✉) · A. A. Odetoye
Department of Crop Science and Horticulture, Federal University Oye-Ekiti, Ikole-Ekiti Campus, Ikole-Ekiti, Nigeria
e-mail: daniel.adewale@fuoye.edu.ng

T. E. Konyeme
Department of Biological Sciences, University of Delta, Agbor, Nigeria

F. T. Fayose
Department of Agriculture and Bio-resources Engineering, Federal University Oye-Ekiti, Ikole-Ekiti Campus, Ikole-Ekiti, Nigeria

J. O. Popoola
Department of Biological Sciences, Bowen University, Iwo, Nigeria

B. O. Ojuederie
Department of Biological Sciences, Kings University, Ode-Omu, Nigeria

C. V. Nnamani
Plant Systematics and Conservation Biology Research Unit, Department of Applied Biology, Faculty of Science, Ebonyi State University, Abakaliki, Nigeria

B. Ikhajiagbe
Department of Plant Biology and Biotechnology, University of Benin, Benin City, Nigeria

C. O. Aremu
Department of Crop and Soil Science, Landmark University, Omu-Aran, Nigeria

A. Pratap et al. (eds.), *Integrated Improvement of Food Legumes*,
https://doi.org/10.1007/978-3-031-82629-0_12

medicinal and industrial uses. Specific economic importance of the others is unknown. *S. stenocarpa*, though similar to *Pachyrhizus* spp. is well distinguished. This chapter redresses some subjective assertions about African yam bean: it is an annual crop, endemic only to tropical Africa, the underground product is typical of root tuber, etc. It also unveils the identity, resources and biodiversity of the six wild species. Research on the agricultural and food values of most of them is rare in literature, but each could be hosting promising economic resources and genes for the improvement of the cultivated species. Exploration and conservation of the wild and the cultivated species of *Sphenostylis* is inevitable to reduce further genetic erosion in the genus. The prospect of genetic improvement and the value chain of African yam beans will be enhanced by wider variability within the *Sphenostylis* germplasm.

Keywords *Sphenostylis schweinfurthii* · Genetic resources · Phenotypic traits · Marker-assisted selection · QTLs

1 Introduction

The plant family *Leguminoseae* (*Fabaceae*) provides important genetic resources for food, feed, medicine and nutrition throughout the tropical and temperate world. It is second to the *Gramineae* in economic importance and is the third largest family of flowering plants, with more than 19,500 species and over 750 genera (Heywood, 1985). All pulses species, as well as some other important grain legumes, belong to this family (Snapp et al., 2018). It is the most prominent protein source for the vegetarians. Consumption of leguminous grains enhances human health and provides nutritional benefits (Foyer et al., 2016).

Sphenostylis is a small genus with seven species within the subtribe *Phaseolinae*, tribe *Phaseoleae*, subfamily *Papilionoideae* and family *Fabaceae*. Moteetee and Van Wyk (2012) observed that only three species of *Sphenostylis* are used for food in Africa, including the only cultivated species (*S. stenocarpa*) and the two wild species (*S. erecta* and *S. schweinfurthii*). The seed of the cultivated species is consumed as food in West Africa, while its tuber is cherished above the grains as food in Central and South Africa. The leaves, flowers, pods and seeds of the two wild species are harvested for food (Moteetee & Van Wyk, 2012). In addition, Potter (1992) suggested that the roots of *S. erecta* are utilised as fish poison.

Forests and homesteads serve as important niches for food, feed and medicinal sources. Therefore, many products from the wild are important for food, nutrition and health security, as well as the social and economic welfare of rural communities (Chivenge et al., 2015). It is therefore imperative to study locally available neglected and underutilised crops and their wild relatives for the discovery of hidden natural wealth among them for awareness and optimum utilisation. Many of the wild relatives might have evolved to adapt themselves to adverse environmental conditions (Mabhaudhi et al., 2013; Chivenge et al., 2015) to serve as an important gene pool for the improvement of the cultivated types in the future; their deployment in breeding programmes could resolve many key environmental and genetic issues.

The present chapter descriptively unveils the botanical features of the seven species in the genus—*Sphenostylis*, identifies promising phenotypic traits in the six wild relatives of the African yam bean, redresses misinformation about the cultivated species, compares the African yam bean with the Andean yam bean and presents a way forward through active exploration of the *Sphenostylis* wild relatives and their effective utilisation in the improvement of the cultivated species.

1.1 Sphenostylis

Sphenostylis was evolved to replace initial classification within *Dolichos*, *Macrotyloma* and *Vigna* in the naming of a group of leguminous species, which is distinguished by the style that is wedge-shaped at the tip in a spatulated shape with hairy edges (Harms, 1899). *Sphenostylis* arose from a Greek word, *sphen*, a descriptive term for the dorsi-ventrally flattened and wedge-shaped style that may be bearded (Allen & Allen, 1981; Potter & Doyle, 1994). The level of neglect and poor popularity of the genus in the family *Fabaceae* may have stemmed from its localisation within tropical Africa alone and from a small number of species—only seven, of which one is domesticated.

1.2 Naming of Sphenostylis *and Its Distinguishing Features*

Many initial classifications of this genus were within *Dolichos*, *Macrotyloma* and *Vigna*, such that binomial nomenclature of the seven species bore different generic names associated with either of the above three (Table 1). Harms (1899) uniquely distinguished the genus from other genera by the description of the stigma, which he observed to be 'dorsiventrally cunneate'. This feature is particularly unique to this genus within the Fabaceae. Many legume botanists, including Harvey (1862), Bentham (1865) and Taubert (1894), all cited in Gillett (1966), grouped *Sphenostylis* within *Vigna savi* because of the linear pod characteristics. Even after the discovery of the unique features that guided the renaming of *Sphenostylis* by Harms (1899), some other botanists who probably were not aware of the report still wrongly placed the genus in *Dolichos* and *Macrotyloma* (Wight & Arn.) Verdc. (Wojciechowski et al., 2004); even Gillett (1966) considered it to be closely related to *Lablab*. For instance, *Vigna* has peltate stipules, but stipules in *Sphenostylis* are persistent. However, the absence of appendages on the standard petals of *Sphenostylis* clearly distinguished it from *Vigna, Dolichos* and *Macrotyloma*, whose standard petals have appendages. Furthermore, a molecular phylogenetic study of some genera in the sub-tribe—*Phaseolinae*, including *Sphenostylis*, showed that *Dolichos* is more related to *Sphenostylis* than *Vigna* (Moteetee & Van Wyk, 2012). However, the similarity investigation among the genera did not include *Nesphostylis*. Based on floral features and DNA data, according to Potter and Doyle (1994), it was reported that *Nesphostylis* (a smaller genus containing three species) is the nearest sister genus to *Sphenostylis*.

1.3 Nesphostylis *and* Sphenostylis

Cuneate and dorsiventrally flattened styles are common features in both *Nesphostylis* and *Sphenostylis* (Milne-Redhead & Polhill, 1971; Potter & Doyle, 1994). However, *Nesphostylis* is distinguished from *Sphenostylis* by the presence of aril on seeds, appendages on standard petals, apical dilation on the stamens, a tooth at the base of the vexillary stamen, a large persistent bracteolate and hair below the terminal stigma (Milne-Redhead & Polhill, 1971). Classification of species within these two sister genera had been quite ambiguous because of their botanical closeness. Milne-Redhead and Polhill (1971) and Okigbo (1973) had observed that one species within *Sphenostylis* was identified in India, thus adding up the number of species in *Sphenostylis* to be eight. However, the suspected eighth species was *S. holosericea*, which was reclassified as *Nesphostylis holosericea* (Welw. Ex J.G. Baker) Verdc. (Potter & Doyle, 1994). It is the *Nesphostylis* species for Africa. Its distribution extends from West to South Africa, common growth being observed among tall grasses. Its synonyms include *Sphenostylis holosericea* (Welw. Ex. Baker) Harms, *Sphenostylis calantha* Harms, *Sphenostylis holosericea var. hastifolia* (Baker) Baker f., *Vigna holosericea* Baker and *Sphenostylis kerstingii* Harms. The white-to-yellow hairy and densely pubescent vine twines vigorously, grows on woody rootstock and possesses tuberous roots. It is suspected to be a perennial climbing species. The blue or pale purple flowered species is highly fragrant, and pods are equally densely pubescent (Verdcourt, 1970). A later report (Potter & Doyle, 1994) revealed that there are altogether ten species in *Sphenostylis* and *Nesphostylis* in Africa, India and Burma. Seven of the ten species that are native to tropical Africa were all *Sphenostylis*, and the remaining three are within *Nesphostylis*; one each was found in tropical Africa, India and Burma (Potter & Doyle, 1994).

1.4 Botanical Features of the Sphenostylis

The botanical features of *Sphenostylis* have been described by Milne-Redhead and Polhill (1971), Allen and Allen (1981), Potter (1992), Potter and Doyle (1992, 1994) and Moteetee and Van Wyk (2012). The notable growth habit in *Sphenostylis* includes prostrate, climbing (twining) or erect shrubs. Some species within the genus have been speculated to be perennial herbs, probably because such species possess thick and woody rootstock. Leaves of the *Sphenostylis* are compound, pinnately trifoliate with persistent stipules whose structural shape could be within ovate-lanceolate or ovate-acuminate. The three leaflets may vary in size and shape. Leaflets may be appressed when young and thereafter become glabrescent with aging. The tip/head of the usually long peduncles bears the raceme inflorescence. The calyx has two very short and blunt lobes with the upper lip being partially or entirely connate. Major colours of the corolla in *Sphenostylis* are purple, purplish pink, violet, whitish pink and yellow. At anthesis, the suborbicular standard petal symmetrically twists back with two inflexed auricles to reveal the corolla colour. Prominently, the standard petal does not possess appendages; the wings are obliquely obovate, eared near the base, with a short linear claw. The incurved keel may possess a concave-convex beak with short linear claw. The androecium is diadelphous (i.e., ten filaments of which nine are fused into a tube with only one free stamen (the vexillary)). The filaments bear five basifixed and five dorsifixed anthers alternately at the tip. The ovary is usually narrowly oblong, hosting up to 20 ovules. The ovary tapers with the penicillate style, which holds the dorsiventrally flattened and margin-ciliated stigma. Fruits/pods of species in this genus are linear and compressed with persistent style at the tip. At drying, the pods twist after dehiscence. The minutely papillose seeds exhibit various sizes, shapes, brilliance and colour pattern.

2 Species in the Genus, *Sphenostylis*

Sphenostylis E. Mey. is largely confined to dry forests and wet savannas in tropical Africa. There are two major clades in *Sphenostylis*: clade one has five purple-flowered taxa, and the remaining two (*S. erecta and S. schwenfurthii*) are yellow-flowered taxa (Potter & Doyle, 1994). Among the seven, six exist in the wild, while only two, namely, *S. angustifolia* and *S. marginata*, occur in South Africa and Swaziland. The two species differ in the size, shape and venation of the leaflets as well as the length of the petiole, peduncle and pods (Moteetee & Van Wyk, 2012). Clark et al. (2017) recorded *Sphenostylis erecta* and *Sphenostylis marginata* among the taxa compiled in the Harare Herbarium, and they are endemic species at Nyanga Massif (Manica Highlands, Zimbabwe–Mozambique). The only domesticated species is *S. stenocarpa*, and it is the most important species in the genus. The list of synonyms and geographical regions where the seven species are endemic in Africa is presented in Table 1. Below are some descriptions of the seven species.

2.1 Sphenostylis angustifolia *Sonders*

Sphenostylis angustifolia is commonly called wild sweet-pea. It is a highly branching perennial and a hardy shrub, growing on woody rootstock. The beautiful pink-purple flowers are faintly aromatic. The species name '*angustifolia*' comes from Latin, meaning narrow leaves. Leaves and stems of *S. angustifolia* make good fodder for ruminants. It is endemic to South Africa (Moteetee & Van Wyk, 2012). It has

Table 1 Some descriptive features of the seven *Sphenostylis* species

S/N	Taxa	Botanical synonyms	Status	Flower color	Geographical Distribution	Notable peculiarities/ potentials
1	*S. angustifolia* Sonder	*Vigna angustifolia* Benth.	Wild	Purple/ Pink	Northern province of South Africa and Swaziland	It has the potential to make a very interesting and colourful indigenous garden specimen as an outdoor ornamental around houses. Moreover, leaves, flowers and seeds from the species are food for human in some parts of West Africa
2	*S. briartii* (De Wild) E. G. Baker	*S. wildemaniana* Baker f. *V. briartii* De Wild. *V. homblei* De Wild. *V. matengoana* Harms	Wild	Pale pink or maroon	East and Central Africa (especially in Angola, Burundi, Malawi, Tanzania, Zambia, DRC)	Poorly or lightly researched
3	*S. erecta* (Baker f.) Hutch. Ex. Baker f.	*Dolichos erectus* Baker f. *S. homblei* De Wild. *S. marginata subsp. erecta* (Baker f.) Verdc. *S. erecta subsp. erecta.*	Wild	Yellow	East and Central Africa (especially in Angola, Burundi, DRC, Central African Republic (CAR), Tanzania, Malawi, Mozambique, Zambia and Zimbabwe)	The standard petal of the flower is usually yellow. Leaves, flowers and seeds are food for human in some parts of West Africa
4	*S. marginata* E. Mey	*Vigna marginata* (E. Mey.) Benth. Ex. Harv.)	Wild	Pink/ Purple	South Africa and Swaziland	*S. marginata* is used for food and for medicinal purposes
5	*S. schweinfurthii* Harms	*Sphenostylis gossweileri* Baker f.	Wild	Yellow	West Africa stretching to Ethiopia	Seeds are wild food and the stems and leaves are fodder for livestock

(continued)

Table 1 (continued)

S/N	Taxa	Botanical synonyms	Status	Flower color	Geographical Distribution	Notable peculiarities/ potentials
6	*S. stenocarpa* (Hochst Ex. A. Rich.) Harms	*Sphenostylis ornata* A. Chev. *Sphenostylis congensis* A. Chev. *Dolichos stenocarpus* Hochst. Ex. A. Rich. *Sphenostylis katangensis* (De Wild.) Harms *Vigna katangensis* De Wild. *Vigna ornata* Welw. Ex. Baker.	Cultivated	Purple	Tropical Africa (especially, West and Central Africa)	Both seed and tuber are food and feed. It also have significant medicinal values
7	*S. zimbabweensis* Mithen	*Sphenostylis marginata subsp.* Drummond *Sphenostylis marginata subsp. Obtusifolia,* sensu Verdcourt).	Wild	Purple	Zimbabwe	Short black linear pod

also been reported to be used as a fish poison in southern Zimbabwe. The medicinal or culinary uses of *S. angustifolia* by humans have not been documented. *S. angustifolia* exhibits erect growth habit, reaching up to 50 cm in height. It branches profusely and spreads widely. The trifoliate compound leaves are held by petioles whose length could be up to 20 mm. Leaflet shape is oblong or linear-lanceolate. Inflorescences of *S. angustifolia* are characteristic of sub-umbellate axillary racemes. The usually stout and long peduncles hold up to four flowers at the peak. At anthesis, standard petals of the flowers twist backward to reveal a pink or purple colour with a whitish or yellowish centre. The style is persistently yellow. The pod is linear and holds many seeds whose colour and colour pattern vary extensively. Notable testa colour in *S. angustifolia* includes brown, reddish brown, black, speckled black, etc. (Leistner, 2000; Lewis et al., 2005).

The cladistic analysis by Potter and Doyle (1994) identified a very close affinity between *S. angustifolia* and *S. zimbabweensis* R. Mithen, stating that the two species share resemblance for deciduous bracts and persistent bracteoles. The major difference, however, is that the growth habit of *S. zimbabweensis* is prostrate, and the leaves are broader. This species is distinguished from *S. marginata* by the narrower leaflets and the much shorter petioles, peduncles, and pods (Van Wyk & Malan, 1997).

2.2 **Sphenostylis briartii** *(De Wild.) Baker f.*

Sphenostylis briartii (De Wild.) Baker f. is a native ranging from Tanzania to south tropical Africa, well distributed in Angola, the Democratic Republic of Congo (DRC) and Tanzania (Gillett et al., 1971). It is an annual herb that flourishes in the dry tropical regions.

It is usually found inhabiting the mountainous ecology with high rainfall. It exists in open vegetation of wood and grasslands. It has equally been identified growing in moist and swampy environments and at higher altitudes of 780–1800 m on rocky, sandy soils. This species differs from the others by its unique style, which is bearded below the apical part of the stigma (Milne-Redhead & Polhill, 1971).

S. briartii is a herb with slender and pigmented vines that can either trail or climb. The leaflet shape ranges between ovate to elliptic; the apex could be wide, obtuse or acute, but the base is round. The petiole length could be long, with a length of up to 4 cm. The peduncles, which hold few flowers (1–3), have a length ranging between 0.4 and 7 cm. The standard petal is pale pink or maroon outside; colour gets darker along the middle, and it turns purple at the base with a violet-edged white zone. Mature pods of *S. briartii* are linear and short, reaching up to 5 cm. Seeds, which could be ovoid or oblong, have a relative metric measurement of about 2.5 mm length, 2 mm width and 1.5 mm thickness (Burrows & Willis, 2005; Mackinder et al., 2001; Phiri, 2005).

2.3 **Sphenostylis erecta** *(Baker f.) Hutch. Ex. Baker f.*

This species is an erect sub-shrub that grows on woody rootstock. It is widely distributed throughout Southern Africa and grows in woodlands, open bushlands, furrow edges and riverine vegetation (Chikwanda et al., 2013). It is also suspected to exhibit perennial growth like *S. angustifolia*. The leaflets of the trifoliate compound leaves may be narrowly oblong, ovate or elliptic. The herb or sub-shrub may be prostrate or erect. The plant length from the rootstock widely varies between 0.15 and 6 m long. The woody rootstock produces red exudates. The stem is more pubescent when young. Flowering is often initiated when the stem is almost or entirely leafless. The very long (up to 30 cm) peduncle carries few flowers. This yellow-flowered species bears short (about 15 cm long) linear, hairless and margined pods with minute dot-like scales (Drummond, 1972). For the edible seeds, this species has been domesticated for food in some parts of West Africa, although it is not widely used elsewhere beyond the region (Mackinder et al., 2001; Mapaura & Timberlake, 2004). A report by Moteetee and Van Wyk (2012) documented that the species still exists in the wild, and its leaves, flowers and seeds are harvested for food in West Africa. A sub-species, *S. erecta* subsp. *obtusifolia*, is distinguished by being a trailing, climbing or creeping herb, sometimes forming mats, and the leaflets are also broader (Da Silva et al., 2004; Phiri, 2005; Baumann, 2005). Indigenous uses of *S. erecta* include fish poison; the leaves are forage for herbivores, seeds, flowers and tubers are wild vegetables for human, and some cultures have identified its significance in folklore medicine; it is also a source of dye in some parts of Africa (Potter and Doyle (1992). The use of *S. erecta* as an acaricide was indigenously noted among smallholder ruminant farmers in Zimbabwe (Chikwanda et al., 2013). The investigation by the same authors revealed that root extracts of *S. erecta* were acaricidal against the larvae of *Boophilus decoloratus* tick, stating that toxicity was concentration dependent and directly proportional to the period of exposure. Chikwanda et al. (2013) further revealed that pure root extract gave 94.0% mortality within 24 hours. Moreover, with an increase in duration of treatment, even at low concentration, there was increased toxicity, such that an exposure of the ticks to 20% root extract concentration for 48 hours gave 100% mortality. Primarily, isolation of the active ingredients in *S. erecta* root extracts is necessary as a way forward to optimise its use in veterinary phytotherapy. Moreover, investigation of its residual persistence is equally necessary. The discovered efficacy needs to be tapped for utilisation; industrial awareness and large-scale manufacturing of acaricide from the root of this plant would be a novel organic and environmentally safe product.

2.4 Sphenostylis marginata *E. Mey*

Synonyms of *Sphenostylis marginata* E. Mey. include *Sphenostylis marginata* E. Mey. ssp. *erecta* (Baker f.) Verdc., *Dolichos erectus* Baker f. and *Sphenostylis erecta* (Baker f.) Hutch. ex Baker f. (Table 1). It is otherwise called African yellow pea. *Sphenostylis marginata* differs from *S. angustifolia* in having broader leaflets and longer petioles, peduncles, and pods. *S. marginata* has 2n = 22 (Goldblatt & Davidse, 1977). The species, *S. marginata* exhibits some complexes; the notable two include *S. erecta* (Baker f.) Baker f. and *S. obtusifolia* Harms, respectively, from Central and East Africa.

Verdcourt (1970) divided the complex into three subspecies, namely, *marginata*, *obtusifolia* (Harms) Verdc. and *erecta* (Baker f.) Verdc. However, investigation through morphological and DNA data by Potter and Doyle (1994) corrected the record by limiting the species concept to subspecies, such that there is *S. marginata* subsp. *erecta* and *S. erecta*, which now has a subspecies called *obtusifolia*. *S. marginata* E. Mey. is indigenous to South Africa (especially the KwaZulu-Natal region) and Swaziland. It is a trailing or semi-scandent herb with a glabrous stem that may be densely pubescent when young. The keel, wing and standard are purple; the round or oblate standard petal is usually bi-lobed at the apex. The slightly long pod of *S. marginata* E. Mey. tapers at both ends with the capacity of holding up to eight seeds. This species establishes itself on woody rootstock and exhibits a trailing growth habit. The tubers, flowers and seed of *S. marginata* are edible and are utilised as a food source in some African countries (Malaisse & Parent, 1985).

Earlier reported medicinal values of *S. marginata* include antiseptic activities for pathogens and treatment of abdominal pain, diarrhoea, oedema and fever (Sohni et al., 1994). Gunzinger et al. (1988) equally identified and reported four antifungal pterocarpans isolates (secondary metabolites) from the root bark of *S. marginata*. The antioxidant and scavenging role of reactive oxygen species, detoxification by induced phase II enzyme (quinone reductase (QR)) and inhibition of nuclear factor kappa B (NF-κB) are promising strategies and approaches to both cancer chemotherapy and chemoprevention (Cuendet et al., 2006; Sarkar & Li, 2008). An investigation by Li et al. (2013) discovered that the root bark of *S. marginata* possesses a hydroxyl radical-scavenging role, quinone reductase (QR) induction activities and the sphenostylisin A (1) (isolated from the root bark) was found to be a very potent NF-κB inhibitor. Li et al. (2013) isolated and elucidated the structures of two novel carbon skeletons—sphenostylisins A–C (1–3) and eight new compounds, sphenostylisins D–K (4–11)—in the root bark of *S. marginata*. The biological evaluation of the 11 isolates revealed varied potency for hydroxyl radical-scavenging, QR-induction and NF-κB inhibition in the cell. Root bark of *S. marginata* holds potent naturally occurring compounds that can effectively combat cancer.

2.5 Sphenostylis schweinfurthii *Harms*

This species is an erect shrub, reaching up to 90–150 cm high. Inflorescences are axillary and falsely racemose. Flowers may be few or many, fairly large and yellow. The linear ovary houses many ovules. The style twists and thickens basally. The style terminates with a flattened spatulate feature at the apex, typical of the genus. Most pods are linear and enclose many seeds whose shape can be cylindrical or ellipsoidal (Burkill, 1985; Gillett et al., 1971). *S. schweinfurthii* is found commonly in open savanna from Mali to Northern Nigeria and across sub-Saharan Africa. Hutchinson and Dalziel (1958) reported that they are well distributed in the Wenchi region of Ghana, Jebba and Zaria (Northern Nigeria) and the southwestern area of Nigeria, especially Iseyin, Oyo State, Nigeria. It is an important leguminous forage in Northern Nigeria (Asare et al., 1984).

2.6 Sphenostylis stenocarpa *(Hochst Ex. A. Rich.) Harms*

S. stenocarpa is commonly called African yam bean. It is the only cultivated species in the genus *Sphenostylis*. Nnamani et al. (2020) stated that it is a crop of Africans for the Africans who have culturally and variously utilised it. It is greatly webbed into African historical culture, thus affirming its endemism in tropical Africa, especially the humid tropical forests and the southern savanna edges. Among the seven species in the genus, it has received the most research attention because it is the most economic species.

Inter-specific boundaries within *Sphenostylis* are well defined (Potter & Doyle, 1994), but the intra-specific boundaries are still unclear (Potter, 1992; Potter & Doyle, 1992; Nyananyo & Nyingifa, 2011). This could substantiate the mostly reported high intra-specific genetic variability in the species *S. stenocarpa*. Leaves of the plant are compound trifoliate with long and broader terminal leaflets. Prominent pigmentations may or may not be visible on the vines, branches and petioles (Fig. 1a and b). *S. Stenocarpa* flower buds and flowers are large and complete. The inflorescence is raceme and indeterminate (Adewale & Amazue, 2018). Quantitative information reveals that matured flower buds may reach 2.2–2.7 cm in length. Furthermore, the appearance of a purple line along the joining end of the standard petal is a qualitative indicator of the maturation of *S. Stenocarpa* flower bud (Fig. 1c). The four floral components (calyx, corolla, stamen and pistil) and the prominent floral features (dorsiventrally flattened and margin-ciliated stigma), which delineate the *Sphenostylis*, are presented in Fig. 1d–f. Honey bees are the most frequent visitors of *S. stenocarpa* flowers (Fig. 1g). The foraging by many insects, especially the bee family, must have been responsible for the high outcrossing rate in the species. As revealed by Adewale and Adegbite (2018) and Ojuederie et al. (2016), the proportion of outcrossing in *S. stenocarpa* is 10%. Pods of *S. stenocarpa* are mostly linear. The seed and tubers are the economic products of *S. Stenocarpa*. Depending upon genotype and seed size, pods can hold ≤20 seeds with very significant shapes, testa colour and colour patterns (Fig. 2). Morphologically, African yam bean tuber development is better described as clusters (never single), bearing two to six tuber subsets connecting in diverse patterns to the strings to the main rootstock or other tubers. Shapes of the root tubers differ, ranging from oval to spindle but mostly irregular (Fig. 2). Human selection may have produced two major groups of this domesticated taxon: those cultivated for the seeds in West Africa and those cultivated for the tubers in the Central Africa (Potter & Doyle, 1992; Nyananyo & Nyingifa, 2011).

It is an annual with an obligate climbing habit, usually propagated from the seed in companionship with yam (*Dioscorea* spp.). Its tubers and grains are very rich in carbohydrate and protein. Report on the proximate analysis of the two economic organs of the crop revealed that they are equally rich in some minerals and vitamins; this makes the crop a whole and balanced food when consumed. The nutritional potential from the high-protein pulse grains and the carbohydrate-rich root tubers hosts promises for food and feed for humans and livestock. Despite the level of

Fig. 1 Morphological features of African yam bean: (**a** and **b**)—pigmented and unpigmented vines, branches and petiole, (**c**)—matured flower bud with purple stripe along the point of joint of the wing; a qualitative indicator of floral maturity and readiness to anthesise within 24 hours, (**d**)—anthesised flower, (**e**)—the calyx, holding the diadelphous stamen which encloses the ovary and style terminating with the ciliated and spatulated stigma, (**f**)—the various components of the corolla (S—standard/banner, W—wings (2) and K—keels (2)), (**g**)—foraging by honey bee, (**h**)—newly fertilised pods. (*Photo:* Marvelous Paul, Jacob Popoola, Adefunke Odetoye and Daniel Adewale.)

neglect, its genetic resources are still in the landrace status in the hands of the farmers, who had generally remarked that the crop does not play host to many of the pests and diseases commonly encountered in cowpea and groundnut. Furthermore, shattering tendencies in African yam bean pod is mild, and the crop suffers less damage in storage compared to cowpea (Olayiwola et al., 2012).

Fig. 2 Testa colour and colour patterns in some African yam bean seeds and root tubers (swellings along the root lengths), tubers are in clusters with different shapes. (*Photo*: Daniel Adewale)

The seeds of *S. stenocarpa* are consumed in various culinary forms in Nigeria. In traditional medicine, the powder of the seeds are reported to be used in the management of high blood pressure, diabetes (Ubaka & Ukwe, 2010), high cholesterol and stroke, and also for the induction of milk production in lactating women (Nnamani et al., 2017). The presence of some phytochemical constituents in the seeds and tubers has led to the identification of antimicrobial, antidiabetic, antinflammatory, antioxidant and hepatoprotective activity (Jude et al., 2022). The proximate and nutritional composition of the seed of *S. stenocarpa* fall within the permissible level and are comparable to other regularly consumed legumes within the tropics. *S. stenocarpa* have a potential as a functional food, which could be a better alternative to the synthetic drugs in the treatment and management of such health challenges within the tropical region (Jude et al., 2022; Adewale & Nnamani, 2022).

2.6.1 Anecdotal Information on African Yam Bean

Many reports about the neglected and underutilised species were unverified and highly subjective. This may have prompted Mabhaudhi et al. (2017) to state that neglected and underutilised crops' potential may have remained unattractive because of limited, often incoherent and inconsistent information about them. Information on many of the underutilised crops is generally sketchy, some of them are subjective and also misleading. Moreover, poor statistics and documentation on these crops seem to have encouraged lot of errors, including misnaming, mis-documentation and hence misrepresentation. It is therefore worthwhile to harmonise available information, clean up some misinformation and discard some unfounded details about African yam bean. For example, original research attempts (based on many genotypes) by Ojuederie and Balogun (2019) and Konyeme et al. (2020) on African yam bean tubers reported a harmonious mean for carbohydrate content in the tuber to be within 50–69% as against 86% in a much earlier report (NRC, 1979). It could be inferred that some of the earlier reports were mere speculations or that these were based on limited genetic resources that either did not represent adequate germplasm or their assessment was done with poor facilities. The names of African yam bean within regions, states, tribes, cultures, etc. are numerous; the list of the local names of African yam bean mentioned by Potter (1992), Adewale and Odoh (2013) and Nnamani et al. (2018) may be negligible if a thorough survey for the cultural names of African yam bean is undertaken in Nigeria and Africa. Based on some published reports, the course of botanical progress of naming and renaming the crop led to the coining of many synonyms. This often leads to confusion, poor understanding and awareness of different species at local, national and international levels. To this end, Adewale and Odoh (2013) documented six synonyms for African yam bean, which are *Sphenostylis ornata* A. Chev., *Sphenostylis congensis* A. Chev., *Dolichos stenocarpus* Hochst. Ex. A. Rich., *Sphenostylis katangensis* (De Wild.) Harms, *Vigna katangensis* De Wild. and *Vigna ornate* Welw. Ex. Baker.

Another example of an unfiltered report mentioned much earlier in NRC (1979) and NRI (1987) with the reference mentioned in many subsequent publications stated that the tuber of African yam bean possesses perennating ability, and hence it is a vegetative propagule. Some reports even called African yam bean a perennial crop. In another version, Klu et al. (2001) remarked that the underground tubers also serve as organs of perennation in the wild. Researchers are in dearth of information on the gene pool of this crop, and our access to the wild germplasm is very limited. However, it is noteworthy to state here that the domesticated and cultivated African yam bean is a root tuber that does not perennate; this is in conformity with the report of Akande (2008) that the African yam bean is an annual arable crop and its propagation is from the seeds and not from the tubers. Generally, tubers are of two types: the stem and root tubers. Stem tubers are formed as thickened underground stems at the side of parent plants and are located near the surface of the soil. Tubers possess buds, nodes or eyes from where new shoots grow out. Root tubers are modified lateral roots that swell to form storage organs. The enlarged area of the root may be one end of the root or the entire root of the plant. They are insensitive to light and do not possess buds, nodes or eyes; hence, they do not regenerate (Crang et al., 2018). The later description matches the tuber type in African yam bean—root tuber. On the floral status of the crop, Popoola et al. (2011) reported that the crop is an obligate selfer. Some contrary reports by Ojuederie et al. (2016) and Adewale and Adegbite (2018) negated this, as they revealed up to 10% outcrossing in the species. The identified outcrossing (Ojuederie et al., 2014; Adewale et al., 2015; Shitta et al., 2016) has been advocated to be responsible for notable heterozygosity within the species.

Snapp et al. (2018) suggested that there are also errors in the definition of underutilised crops, which are reflected in their production statistics and documentation. The report by Agunbiade and Longe (1999) and Anya and Ozung (2019) identified *Sphenostylis stenocarpa* as 'yam bean', while Adekayode (2011) called it 'long yam bean'. These common names are misleading and capable of leading to confusion within the scientific domain. Grain legume taxonomists, agricultural and rural sociologists could collaboratively pitch on this to clean up probable misleading facts so as to improve the level of awareness of the crop.

2.7 **Sphenostylis zimbabweensis** *Mithen*

S. zimbabweensis is a near-endemic species in the Afromontane area within the Chimanimani-Nyanga sub-centre of the Manica province of Mozambique. It is a herbaceous climbing perennial that is also endemic in Zimbabwe (Darbyshire et al., 2019). The status of *Sphenostylis zimbabweensis* has been controversial before the formal description by Mithen (1993). Verdcourt (1970) had placed it within *S. marginata* subsp. *obtusifolia*, although with a comment that it might deserve a reclassification. Drummond (1972) listed it among the undescribed subspecies of *S. marginata*. It differs from *S. margínata* and *S. erecta*, which have numerous bright flowers. Furthermore, fewer flowers and shorter and broader pods distinguished it from *S. stenocarpa*. Mithen (1993) described *S. zimbabweënsis* as a species with a small number of purple flowers on each inflorescence. It has almost glabrous black seeds and narrower black pods in addition to persistent bracts, bracteoles, small mucronate leaves and a deeply emarginate standard petal. This taxon is only and most endemic in the Eastern Highlands of Zimbabwe (Mithen, 1993).

This species is well distributed in the montane vegetation of Mozambique and Zimbabwe. It equally thrives in grassland and woodland areas with scattered trees and at altitudes of 1200–1700 m. The climbing or procumbent herb grows on a woody rootstock. Compound leave are trifoliate, whose shape could be ovate or elliptic with a rounded base and acute apex. Peduncles that hold the axillary inflorescence are short, carrying few (2–3) brightly purpled flowers. Short linear pods are usually black in colour at maturity. Seed colour in *S. zimbabweensis* is mostly black or speckled with an oblong or cylindrical shape (Mackinder et al., 2001).

3 Yam Beans and Tuberization Inconsistencies in African Yam Bean

Yam beans are leguminous crops that produce both aerial pods (containing the pulse grain) and the subterranean swollen root. This group of crops combines the features of pulses and tuberous crops. The uses of the root tubers of legumes are many, including food, medicinal, insecticidal, flavouring, etc. (Saxon, 1981). Among the identified tuberous legumes (Table 2), the utility of some is unknown, probably due to neglect and under-exploration.

Among these, two yam beans are most notable in popularity: African yam bean (*S. stenocarpa* (Hochst Ex. A. Rich.) Harms), also called 'yam pea', and yam bean (*Pachyrhizus* spp.), also called 'potato bean'. By the report of Zanklan et al. (2018), *Pachyrhizus* DC., in the subtribe *Glycininae*, has three edible root crop species (*P. erosus* (L.) Urban - Mexican yam bean, *P. tuberosus* (Lam.) Spreng.—Amazonian yam bean and *P. ahipa* (Wedd.) Parodi)—Andean yam bean. Andean, Amazonian and Mexican yam beans are, however, more popular in research, cultivation and utilisation than African yam beans, such that in most literature the earlier were called yam beans. The three *Pachyrhizus* spp. and *S. stenocarpa* share some similarities: both belong to the tribe Phaseoleae, they are annual crops, produce edible tubers, and profusely fix atmospheric nitrogen in the root nodules to improve soil fertility. However, some features that distinguish the two are presented in Table 3.

Research focus on the tuber of the African yam bean has been very poor because it is less cherished compared to the grains, especially in West Africa. Much of the information is lacking on African yam bean tuber, and the available information is mostly based on assumptions unsupported by scientific evidence, some of which have degenerated to ideological complexes needing to be unknotted. Most striking is the issue of inconsistent tuberization among accessions/landraces of African yam bean in time and space. Ojuederie and Balogun (2019) suggested that this could be due to the photoperiodic sensitivity of the species, earlier hinted by Okpara and Omaliko (1995). No report has yet confirmed the right photoperiod that favour African yam bean tuberization. However, in the three *Pachyrhizus* spp., short day length induced tuberization. The photoperiod, according to Vaillant and Desfotaines (1995), initiated the development of a very strong sink leading to the export of much assimilate for tuber development. However, a large-scale screening programme to test a large number of African yam bean accessions under short-day photoperiods for tuberization may provide meaningful information.

Table 2 List of leguminous genera hosting some of the species which produce tubers

S. No.	Genera	Species	Family and tribe	Global, regional or national nativity
1	*Atylosia*	1	Tribe—Phaseoleae, Subtribe—Cajaninae	Australia
2	*Dolichos*	2	Tribe—Phaseoleae, subtribe—Phaseolinae	Mexico, Namibia, Africa
3	*Eriosema*	2	Tribe—Phaseoleae, Subtribe—Cajaninae	S.E. Asia and Africa
4	*Flemingia*	2	Tribe—Phaseoleae, Subtribe—Cajaninae	South America
5	*Glycine*	1	Tribe—Phaseoleae, Subtribe—Glycininae	Australia
6	*Hardenbargia*	1	Tribe—Phaseoleae, Subtribe—Kennediinae	–
7	*Macrotyloma*	1	Tribe—Phaseoleae, subtribe—Phaseolinae	Australia, Zimbabwe
8	*Pachyrhizus*	3	Tribe—Phaseoleae, Subtribe—Glycininae	Central and South America
9	*Phaseolus*	5	Tribe—Phaseoleae, subtribe—Phaseolinae	Central and South America, North America
10	*Psophocarpus*	2	Tribe—Hedysareae	West Africa and New Guinea
11	*Psoralea*	2	Tribe—Psoraleeae	Australia and North America
12	*Pueraria*	3	Tribe—Phaseoleae, subtribe—Glycineae	Kenya
13	*Sphenostylis*	4	Tribe—Phaseoleae, subtribe—Phaseolinae	Ethiopia, Zambia
14	*Tylosema*	2	Tribe—Bauhinieae	
15	*Vigna*	5	Tribe—Phaseoleae, subtribe—Phaseolinae	India, Australia, Ethiopia
16	*Derris*	2	Tribe—Tephrosieae	Namibia, Africa
17	*Lonchocarpus*	2	Tribe—Millettieae	Tanzania
18	*Otoptera*	1	Tribe—Phaseoleae	–
19	*Rynchosia*	1	Tribe—Phaseoleae, subtribe—Cajaninae	Australia
20	*Apios*	2	Tribe—Phaseoleae	North America
21	*Austrodolichos*	1	Tribe—Phaseoleae	Australia
22	*Clitoria*	1	Tribe—Desmodieae	Australia
23	*Dunbaria*	1	Tribe—Phaseoleae, subtribe—Cajaninae	Australia
24	*Periandra*	1	Tribe—Phaseoleae, subtribe—Clitoriinae	Brazil
25	*Physostigma*	1	Tribe—Phaseoleae, subtribe—Phaseolinae	–
26	*Psophocarpus*	2	Tribe—Hedysareae	Zambia

Table 3 Differences between yam bean and African yam bean

S. No.	Features	Yam bean (*Pachyrhizus spp.*)	African yam bean (*Sphenostylis stenocarpa*)	References
1	Number of cultivated species	*Three:* Andean, Amazonian and Mexican yam bean	*One:* African yam bean	Zanklan et al. (2018), Potter and Doyle (1992)
2	Centre(s) of diversity	Central and South America	Central, East, West and South Africa	Zanklan et al. (2018), Potter and Doyle (1994)
3	Near relative genus/species	*Glycine max* and *Puereria phaseoloides*	*Nesphostylis* and to some extent: *Dolichos* and *Vigna*	Zanklan et al. (2018), Potter and Doyle (1994)
4	Outcrossing ratio	8%	10%	Sorensen (1990), Adewale (2011), Ojuederie et al. (2016)
5	Subtribe	Subtribe—*Glycininae*	Subtribe—*Phaseolinae*	
6	Number of wild sister species	Two: *P. ferrugineus* and *P. panamensis*	Six: *S. angustifolia, S. briartii, S. erecta, S. marginata, S. schweinfurthii* and *S. zimbabweensis*	Zanklan et al. (2018), Milne-Redhead and Polhill (1971), Potter and Doyle (1994)
7	Growth habit	Grow as erect, bushy or spreading (needing no staking)	An obligate climber (needs to be staked)	Gruneberg (2016), Okpara and Omaliko (1997)
8	Most economic food product	The tubers	Seed and tuber	Ramos-de-la-Pen˜a et al.(2013), Okigbo (1973)
9	Seed utilisation	Industrial use as biodegradable insecticides due to high rotenone content	Purely for food	Zanklan et al. (2018), Okigbo (1973), Rachie (1973)
10	Shape(s) of tubers	Round, oval, elliptic, obovate, oblong, etc.	Sometimes spindle and mostly irregular in shape	NRC (1979, 2006) Adewale and Dumet (2011)
11	Global level of awareness	Fairly good	Poor	
12	Gestation period	3–4 months	6–8 months	Zanklan et al. (2018), Adewale et al. (2012)
13	Cropping system	Mostly mono-cropping	Mostly intercropping	Zanklan et al. (2018), Ibeawuchi et al. (2007)
14	Moisture content in tubers	Higher moisture content (≥82%)	Lower moisture content (≤10%)	Zanklan et al. (2018), Ojuederie and Balogun (2019)
15	Protein and carbohydrate content in tubers	Low protein and starch: 1.5% and 10% respectively	High protein and starch: ≤16% and ≤67% respectively	Zanklan et al. (2018), Ojuederie and Balogun (2019)

(continued)

Table 3 (continued)

S. No.	Features	Yam bean (*Pachyrhizus spp.*)	African yam bean (*Sphenostylis stenocarpa*)	References
16	Sugar content in tubers	Sucrose is 15% and soluble sugars (sucrose, glucose and fructose) is 60% of the tuber dry weight	No record	Vaillant and Desfotaines (1995)
17	Taste of tuber flesh	Sweet, crisp and mildly flavoured	Quite insipid	NRC (2006)
18	Mode of tuber consumption	Mostly as fruity vegetables	As cooked or processed food	
19	Vitamin C content	A good source of ascorbic acid	No record	Zanklan et al. (2018)
20	Number of chromosomes (2n)	2n = 22	Reported 2n ranges between 18 and 24 (22 is most common)	Kumar and Hymowitz (1987), Sørensen (1988), Adesoye and Nnadi (2011), Popoola et al. (2011)
21	Tuber initiation	Short day length highly aided tuber initiation in *P. erosus*	No record yet on tuber sensitivity to photoperiodism	Vaillant and Desfotaines (1995)

The African yam bean is a class above the mainline root crops, containing more than twice the protein in sweet and Irish potatoes, yams and more than ten times that is obtainable in cassava (NRC, 2006). Crude protein content in *Pachyrhizus spp* and Marama bean (*Tylosema esculentum*) is 1.5% and 9%, respectively (Cullis et al., 2018; Zanklan et al., 2018), as against ≥15% in African yam bean (Ojuederie & Balogun, 2019). The content of crude protein in African yam bean tubers is no match to any other tuberous crops or legumes. Therefore, this neglected product (the tuber) deserves a renewed focus in research as it promises to significantly support food and nutritional security in the poor economic regions of the central, east, and west African.

4 Intra-Generic Diversity: A Reservoir for Livelihood Improvement

Some reports have documented successful association of a few phenotypic traits, such as nutritional traits (Oluwole et al., 2020), seed shape, tuber flesh colour, seed, and tuber yield (Aina et al., 2021), seed size (Olomitutu et al., 2022) and 27 other phenotypic traits (Shitta et al., 2022), with some of the genetic markers. Studies on association mapping have reliably suggested that the markers could be useful for marker-assisted selection in African yam bean improvement, as in the case of chickpea (Pratap et al., 2017; Bharadwaj et al., 2022). Olomitutu et al. (2022) identified 12 SNP markers that were significantly associated with seed size traits in African yam bean, supporting the observation of Adewale et al. (2010) that seed metric traits of African yam bean could be very genetically reliable because the influence of the environment was not significant on the traits. In a preliminary GWAS study by Oluwole et al. (2020) for AYB seed protein and oil, some QTLs were found on chromosome 1, aligning with the reference genome of *Phaseolus vulgaris*, although the alignment of the other QTLs, especially for starch, was unknown. Although the unveiling of the reference genome of African yam bean is underway, studies of Shitta et al. (2022), Olomitutu et al. (2022), Aina et al. (2021) and Oluwole et al. (2020) lead the path to the possibility of markers-assisted breeding in African yam bean, because identification of the regions of the QTLs could lead to its fixation for particular breeding interest.

The study of Shitta et al. (2022) involved 1789 SNPs and 27 phenotypic traits, most of which did not align with the genomic data. Absence of a strong association between the phenotypic and genotypic data could be because the SNP data are not associated with the phenotypic traits evaluated or probably because molecular markers generally detect the non-adaptive types of variation (Shitta et al., 2022). Adewale et al. (2021) had earlier remarked that reliable phenotypic data generated from genetically traceable lines (not landraces) in space and time remains the best matching genetic materials to link QTLs. Furthermore, in African yam bean research, more phenotypic traits need to be studied from wider genetic resources. Different markers have confirmed the presence of high polymorphism across many populations of African yam bean, all indicating high genetic diversity in the germplasm of the species. The initially discovered high outcrossing ratio in the species (Ojuederie et al., 2016; Adewale & Adegbite, 2018) stands most significant for the notable heterozygosity in the germplasm.

Wild plants grow spontaneously in self-maintaining natural and/or semi-natural ecosystems independent of direct human action (Heywood, 2007). Most of them are usually dismissed as 'weeds'. They are, in fact, valuable genetic resources, especially with respect to the improvement of their cultivated species. Their direct or indirect uses as reported by Cook (1995) and Heywood (2007) include food (for human, livestock and wild animals, invertebrates), food additives, industrial materials (wood, fibre resins, gum, etc.), ornamental/environmental uses (advertisement and foraging), vertebrate and non-vertebrate poison, medicine, cosmetics and perfumery, etc. The six wild species of the *Sphenostylis* also host promising utility for *S. stenocarpa*. As they are linked to the biodiversity of the genus *Sphenostylis* to provide meal diversity, improve dietary nutrition, provide backup candidate genes for subsequent disease or pest outbreaks, plant growth habit/architecture, industrial and ornamental quality, etc. For instance, the micronutrient and vitamin content of the wild species may be superior to that of the domesticated taxon (Msuya et al., 2010). Therefore their consumption may improve the dietary diversity and nutritional quality of rural diets, thus contributing to the global food security (Harris & Mohammed, 2003; Bharucha & Pretty, 2010). Wild populations containing higher genetic variability, having survived in a wide range of habitats without human selection (Huang & Han, 2014); may also possess some desirable traits such as biotic and abiotic stress resistances, high productivity and special nutritional values (Gockowski et al., 2003; Li et al., 2016; Asif et al., 2013, Pratap et al., 2021). The potential they hold makes them suitable candidates for the improvement of the cultivated relatives.

Commercial cultivation of African yam bean would be considerable if cultivars with erect growth habits can be developed. This chapter described some important qualities of *S. angustifolia*, *S. biartii*, and *S. magrinata*, which include non-climbing and erect growth habits, wide spreading of branches with stout peduncles, wide adaptability to different ecologies, etc. A breeding programme that can develop suitable varieties of African yam bean with a non-climbing growth habit would enhance its commercial value. Plant breeding and crop improvement greatly acknowledges the importance of wild genetic resources (Hawkes, 1977; Tanksley & McCouch, 1997; Baydoun et al., 2017; Ojelel et al., 2019, Pratap et al., 2021, 2023). The direct utilisation of the wild crop relatives through gene introgression has led to groundbreaking historical successes in many crops (Hawkes, 1977; Tanksley & McCouch, 1997; Mackill et al., 2012; Bernal-Galeano et al., 2020). However, their deployment is dependent upon several factors, including crossability status (Bortolotto et al., 2015). An understanding of the mechanisms of unique adaptations as seen in Marama bean (Cullis et al., 2018); identification of rich nutraceutical properties as obtained in *Canavalia*, *Mucuna*, and *Sesbania* (Bhat & Karim, 2009); etc. could set the pace for research into the wild relatives of African yam bean. Subsequent exploration for wild crop relatives should involve detailed documentation of the utility of collected genepools to be conserved.

5 Conclusions

This chapter, beyond an update on the genus *Sphenostylis* and all its species, unveiled significant potential characteristics capable of being exploited to improve the cultivated species *S. stenocarpa*. The wild species of *Sphenostylis* might be holding several genes capable for improving African yam bean against many of the notable constraints, such as growth habit, staking, long maturity duration, hardness of the seed coat, climate resilience, photoperiodic sensitivity, etc. Therefore, assembly of many morphs of the six other species of *Sphenostylis* from the wild should be followed by genotypic and throughput phenotypic and biochemical evaluation. This primary investigation will reveal the potential of these taxa for different economic traits that will provide information on candidate genes for the genetic improvement of African yam bean.

There is also a need to carry out further phytochemical analysis to isolate and characterise bioactive constituents of those species that have medicinal value to promote their industrial uses.

The potential impact of underutilised crops on food security is globally evident, but comprehensive policy on this group of crops is rare in any of the national agricultural programmes. In Ghana, for example, like in most countries in Africa, interventions and programmes aimed at promoting effective utilisation of underutilised crop species have also not been well coordinated (Baa-Poku, 2018). It would be needful to support researchers and/or institutions that have shown a demonstrated record of accomplishments and commitment to advancing the status of neglected and underutilised crops (Mabhaudhi et al., 2017).

References

Adekayode, F. O. (2011). Comparative effectiveness of long yam bean (*Sphenostylis stenocarpa)* and organic and inorganic fertilizers in improving soil nutrient status and yield of maize in Southwest Nigeria. *Journal of Soil Science and Environmental Management, 2*(7), 198–205.

Adesoye, A. I., & Nnadi, N. C. (2011). Mitotic chromosome studies of some accessions of African yam bean *Sphenostylis stenocarpa* (Hochst. Ex. A. Rich.) Harm. *African Journal of Plant Science, 5*(14), 835–841. https://doi.org/10.5897/AJPS10.100

Adewale, B. D. (2011). Genetic diversity, stability and reproductive biology of African yam bean, *Sphenostylis stenocarpa* (Hochst. ex A. Rich.) Harms. Thesis, Federal University of Agriculture, Abeokuta, Nigeria.

Adewale, B. D., & Nnamani, C. V. (2022). Introduction to food, feed, and health wealth in African yam bean, a locked-in African indigenous tuberous legume. *Frontiers in Sustainable Food Systems, 6*, 726458. https://doi.org/10.3389/fsufs.2022.726458

Adewale, B. D., & Adegbite, E. A. (2018). Investigation of the breeding mechanism of African yam bean [Fabaceae] (*Sphenostylis stenocarpa* Hochst. Ex. A. Rich) Harms. *Notulae Scientia Biologicae, 10*(2), 199–204.

Adewale, B. D., & Amazue, E. U. (2018). Floral maturation indices of African yam bean (*Sphenostylis stenocarpa* Hochst Ex. A. Rich) Harms (Fabaceae). *Notulae Scientia Biologicae, 10*, 102–106.

Adewale, B. D., & Dumet, D. (2011). Descriptors for African yam bean, *Sphenostylis stenocarpa* (Hochst ex. A. Rich.) Harms. http://old.iita.org/cms/articlefiles/1488–ayb descriptors.

Adewale, B. D., Kehinde, O. B., Aremu, C. O., Popoola, J. O., Dumet, D. J. (2010). Seed metrics for genetic and shape determinations in African yam bean Fabaceae (Sphenostylis stenocarpa Hochst. Ex. A. Rich) harms. *African Journal of Plant Science, 4*, 107–115.

Adewale, B. D., Dumet, D. J., Vroh-Bi, I., Kehinde, O. B., Ojo, D. K., Adegbite, A. E., & Franco, J. (2012). Morphological diversity analysis of African yam bean and prospects for utilization in germplasm conservation and breeding. *Genetic Resources and Crop Evolution, 59*, 927–936.

Adewale, B. D., Vroh-Bi, I., Dumet, D. J., Nnadi, S., Kehinde, O. B., Ojo, D. K., Adegbite, A. E., & Franco, J. (2015). Genetic diversity in African yam bean accessions based on AFLP markers: Towards a platform for germplasm improvement and utilization. *Plant Genetic Resources: Characterization and Utilization, 13*(2), 111–118. https://doi.org/10.1017/S1479262114000707

Adewale, B. D., Aremu, C. O., Alake, C. O., & Ige, S. A. (2021). Multi-year genetic estimates of quantitative traits and intraspecific diversity of some *Sphenostylis stenocarpa* accessions for reliable selection and characterization. *Agricultural Research, 10*(2), 165–174. https://doi.org/10.1007/s40003-020-00496-x

Adewale, B. D., & Odoh, N. C. (2013). A review on genetic resources, diversity and agronomy of African yam bean (*Sphenostylis stenocarpa* (Hochst. ex A. Rich.) Harms): A potential future food crop. *Sustainable Agriculture Research, 2*, 32–43.

Agunbiade, S. O., & Longe, O. G. (1999). Essential amino acid composition and biological quality of yam bean *Sphenostylis stenocarpa* (Hoechst Ex. A. Rich.) Harms. *Nahrung, 43*, 22–24.

Aina, A., Garcia-Oliveira, A. L., Ilori, C., Chang, P. L., Yussuf, M., Oyatomi, O., Abberton, M., & Potter, D. (2021). Predictive genotype-phenotype relations using genetic diversity in African yam bean (Sphenostylis stenocarpa (Hochst. ex. A. Rich) Harms). *BMC Plant Biology, 21*, 547.

Akande, S. R. (2008). Germplasm characterization of African yam bean (*Sphenostylis stenocarpa*) from South-West Nigeria. *Plant Genetic Resources Newsletter, 154*, 25–29.

Allen, O. N., & Allen, E. K. (1981). *The leguminous. A source book of characterization, uses and nodulation*. Macmillan Publishers' Ltd.

Anya, M. I., & Ozung, P. O. (2019). Proximate, mineral and anti-nutritional compositions of raw and processed African yam bean (*Sphenostylis stenocarpa*) seeds in Cross River state, Nigeria. *Global Journal of Agricultural Sciences, 18*, 19–29.

Asare, E. O., Shehu, Y., & Agishi, E. A. (1984). Preliminary studies on indigenous species for dry season grazing in the northern Guinea savanna zone of Nigeria. *Tropical Grasslands, 18*, 148–152.

Asif, M., Rooney, L. W., Ali, R., & Riaz, M. N. (2013). Application and opportunities of pulses in food system: A review. *Critical Reviews in Food Science and Nutrition, 53*, 1168–1179.

Baa-Poku, F. (2018). Review on the role of underutilized crops in achieving food security in Ghana: Implications for policy. *Ghana Journal of Agricultural Sciences, 52*, 113–120.

Baumann, G. (2005). *Photographic guide to wildflowers of Malawi*. Wildlife and Environmental Society of Malawi.

Baydoun, S. A., Kanj, D., Raafat, K., Aboul Ela, M., Chalak, L., & Arnold-Apostolides, N. (2017). Ethnobotanical and economic importance of wild plant species of Jabal Moussa Bioreserve, Lebanon. *Journal of Ecosystem Ecography, 7*, 245–252 https://doi.org/10.4172/2157-7625.1000245

Bernal-Galeano, V. N. G., Ellis, D., Anglin, N. L., Hareau, G., Smale, M., Jamora, N., Alwang, J., & Pradel, W. (2020). The effect of Andean potato diversity conserved in the International Potato Center genebank helps develop agriculture in Uganda: The example of Victoria. *Food Security, 12*(4), 959–973. https://doi.org/10.1007/s12571-020-01037-8

Bharadwaj, C., Jorben, J., Rao, A., Roorkiwal, M., Patil, B. S., Jayalakshmi, Ahammed, S. K., Saxena, D. R., Yasin, M., Jahagirdar, J. E., Sontakke, P. L., Pithia, M. S., Chudasama, M. K., Swarup, I., Singh, R. K., Nitesh, S. D., Chitikineni, A., Singh, S., Singh, I., Pratap, A., Dixit, G. P., Srivastava, A. K., & Varshney, R. K. (2022). Development of high yielding Fusarium wilt resistant cultivar by pyramiding of "genes" through marker-assisted backcrossing in Chickpea (*Cicer arietinum* L.). *Frontiers in Genetics*. https://doi.org/10.3389/fgene.2022.924287

Bharucha, Z., & Pretty, J. (2010). The roles and values of wild foods in agricultural systems. *Philosophical Transactions of the Royal Society B: Biological Sciences, 365*(1554), 2913–2926.

Bhat, R., & Karim, A. A. (2009). Exploring the nutritional potential of wild and underutilized legumes. *Comprehensive Reviews in Food Science and Food Safety, 8*, 305–331.

Bortolotto, I. M., de Mello Amorozo, M. C., Neto, G. G., Oldeland, J., & Damasceno-Junior, G. A. (2015). Knowledge and use of wild edible plants in rural communities along Paraguay River, Pantanal, Brazil. *Journal of Ethnobiology and Ethnomedicine, 11*, 46. https://doi.org/10.1186/s13002-015-0026-2

Burkill, H. M. (1985). *The useful plants of west tropical Africa*. Royal Botanic Gardens.

Burrows, J. E., & Willis, C. K. (2005). *Plants of the Nyika Plateau*. Southern African Botanical Diversity Network (SABONET) Report No. 31, Pretoria.

Chikwanda, D., Mapanda, T., Chikosi, V. E., & Chikwanda, A. T. (2013). The efficacy of *Sphenostylis erecta* root extracts in controlling Blue tick, *Boophilus decoloratus* Infestation. *Biopesticide International, 9*(2), 195–203.

Chivenge, P., Mabhaudhi, T., Modi, A., & Mafongoya, P. (2015). The potential role of neglected and underutilised crop species as future crops under water scarce conditions in Sub-Saharan Africa. *International Journal of Environmental Research and Public Health, 12*(6), 5685–5711.

Clark, V. R., Timberlake, J. R., Hyde, M. A., Mapaura, A., Coates Palgrave, M., Wursten, B. T., Ballings, P., Burrows, J. E., Linder, H. P., Mcgregor, G. K., Chapano, C., Plowes, D. C. H., Childes, S. L., Dondeyne, S., Müller, T., & Barker, N. P. (2017). A first comprehensive account of floristic diversity and endemism on the Nyanga Massif, Manica highlands (Zimbabwe–Mozambique). *Kirkia, 19*(1), 1–53.

Cook, F. E. M. (1995). *Economic botany—Data collection standard. Prepared for the international working group on taxonomic databases for plant sciences (TDWG)*. Royal Botanic Gardens, UK.

Crang, R., Lyons-Sobaski, S., & Wise, R. (2018). *Plant anatomy: A concept-based approach to the structure of seed plants*. Springer Nature.

Cuendet, M., Oteham, C. P., Moon, R. C., & Pezzuto, J. M. (2006). Quinone reductase induction as a biomarker for cancer chemoprevention. *Journal of Natural Products, 69*, 460–463.

Cullis, C., Chimwamurombe, P., Barker, N., Kunert, K., & Vorster, J. (2018). Orphan legumes growing in dry environments: Marama Bean as a case study. *Frontiers in Plant Science, 9*(1199), 1–7. https://doi.org/10.3389/fpls.2018.01199

Darbyshire, I., Timberlake, J., Osborne, J., Rokni, S., Matimele, H., Langa, C., Datizua, C., de Sousa, C., Alves, T., Massingue, A., Hadj-Hammou, J., Dhanda, S., Shah, T., & Wursten, B. (2019). The endemic plants of Mozambique: Diversity and conservation status. *PhytoKeys 136*, 45–96. https://doi.org/10.3897/phytokeys.136.39020

Da Silva, M. C., Izidine, S., & Amude, A. B. (2004). *A preliminary checklist of the vascular plants of Mozambique*. Southern African Botanical Diversity Network (SABBONET), Report No. 30, Pretoria.

Drummond, R. B. (1972). A list of Rhodesian Legumes. *Kirkia, 8*(2), 227–242.

Foyer, C. H., Lam, H. M., Nguyen, H. T., Siddique, K. H., Varshney, R. K., Colmer, T. D., Cowling, W., Bramley, H., Mori, T. A., Hodgson, J. M., Cooper, J. W., Miller, A. J., Kunert, K., Vorster, J., Cullis, C., Ozga, J. A., Wahlqvist, M. L., Liang, Y., Shou, H., Shi, K., Yu, J., Fodor, N., Kaiser, B. N., Wong, F. L., Valliyodan, B., & Considine, M. J. (2016). Neglecting legumes has compromised human health and sustainable food production. *Nature Plants, 2*(8), 1–10. https://doi.org/10.1038/NPLANTS.2016.112

Gillett, J. B. (1966). Notes on Leguminosae (Phaseoleae). *Kew Bulletin, 20*, 103–111.

Gillett, J. B., Polhill, R. M., & Verdcourt, B. (1971). *Flora of tropical East Africa*. The Whitefriabs Press Ltd.

Gockowski, J., Mbazo'o, J., Mbah, G., & Moulende, T. F. (2003). African traditional leafy vegetables and the urban and peri-urban poor. *Food Policy, 28*(3), 221–235.

Goldblatt, P., & Davidse, G. (1977). Chromosomes numbers in legumes. *Annals of the Missouri Botanical Garden, 64*(1), 121–128.

Gruneberg, W. J. (2016). *Enhancing the nutrient-rich yam bean (Pachyrhizus spp.) storage roots to improve food quality and availability and sustainability of farming systems in Central and West Africa*. International Potato Center Report No. 978-92-9060-474-7. https://goo.gl/ovVCNL, https://doi.org/10.4160/9789290604747, p. 267.

Gunzinger, J., Msonthi, J. D., & Hostettmann, K. (1988). New pterocarpinoids from Dolichos marginata ssp. *erecta*. *Helvetica Chimica Acta, 71*, 72–76.

Harms, H. (1899). Leguminosae Africanae II. *Botanische Jahrbücher für Systematik, 26*, 308–310.

Harris, F. M. A., & Mohammed, S. (2003). Relying on nature: Wild foods in Northern Nigeria. *Ambio, 32*(1), 24–30.

Hawkes, J. (1977). The importance of wild germplasm in plant breeding. *Euphytica, 26*, 615–621.

Heywood, V. H. (1985). *Flowering plants of the world*. Beckenham.

Heywood, V. H. (2007). The use and economic potential of wild species: An overview. In N. Maxted, B. V. Ford-Lloyd, S. P. Kell, J. M. Iriondo, M. E. Dulloo, & J. Turok (Eds.), *Crop wild relative conservation and use* (pp. 581–600). CAB International.

Huang, X. H., & Han, B. (2014). Natural variations and genome-wide association studies in crop plants. *Annual Review of Plant Biology, 65*, 531–551.

Hutchinson, J., & Dalziel, J. M. (1958). *Floral of west tropical Africa. Vol.1 part 2*. Agents for Oversea Governments and Administration, Crown.

Ibeawuchi, I. I., Dialoke, S. A., Ogbede, K. O., Ihejirika, G. O., Nwokeji, E. M., Chigbundu, I. N., Adikuru, N. C., & Oyindo, P. O. (2007). Influence of yam/cassava based intercropping systems with legumes in weed suppression and disease/pest incidence reduction. *Journal of American Science, 3*, 49–59.

Jude, A. I., Omeh, R. C., & Chekwube, A. C. (2022). Ethnobotanical, trado-medicinal uses, pharmacological investigations and chemical constituents of *Sphenostylis stenocarpa* (Hochst Ex. A. Rich) Harms seed. A review. *International Journal of Botany Studies, 7*(1), 522–527.

Klu, G. Y. P., Amoatey, H. M., Bansa, D., & Kumaga, F. K. (2001). Cultivation and use of African yam bean (*Sphenostylis stenocarpa*) in the Volta Region of Ghana. *The Journal of Food Technology in Africa, 6*, 74–77.

Konyeme, T. E., Nyananyo, B. L., & Tanee, F. B. G. (2020). Diversity in proximate analysis of tubers of some African yam bean (*Sphenostylis stenocarpa*) (Hochst Ex. A. Rich.) Harms (Fabaceae) accessions. *Journal of Applied Sciences & Environmental Management, 24*(10), 1787–1793. https://doi.org/10.4314/jasem.v24i10.12

Kumar, P. S., & Hymowitz, T. (1987). Where are the diploid (2n = 2x = 20) genome donors of Glycine Willd. (Leguminosae, Papilionoidae)? *Euphytica, 40*(3), 221–226.

Leistner, O. A. (2000). *Seeds plants of southern Africa: Families and genera. Strelitzia 10*. National Botanical Institute.

Lewis, G., Schrire, B., Mackinder, B., & Lock, M. (2005). *Legumes of the world*. Royal Botanic Gardens, Kew.

Li, M. W., Muñoz, N. B., Wong, C. F., Wong, F. L., Wong, K. S., Wong, J. W., Qi, X., Li, K. P., Ng, M. S., & Lam, H. M. (2016). QTLs regulating the contents of antioxidants, phenolics, and flavonoids in soybean seeds share a common genomic region. *Frontiers in Plant Science, 7*, 854.

Li, J., Pan, L., Deng, Y., Muñoz-Acuña, U., Yuan, C., Lai, H., Chai, H., Chagwedera, T. E., Farnsworth, N. R., de Blanco, E. J. C., Li, C., Soejarto, D. D., & Kinghorn, A. D. (2013). Sphenostylisins A-K, bioactive modified Isoflavonoid constituents of the root bark of *Sphenostylis marginata* ssp. *erecta*. *Journal of Organic Chemistry, 78*(20). https://doi.org/10.1021/jo401573h

Mabhaudhi, T., Chimonyo, V. G. P., Chibarabada, T. P., & Modi, A. T. (2017). Developing a roadmap for improving neglected and underutilized crops: A case study of South Africa. *Frontiers in Plant Science, 8*, 2143. https://doi.org/10.3389/fpls.2017.02143

Mabhaudhi, T., Modi, A., & Beletse, Y. (2013). Response of taro (*Colocasia esculenta* L. Schott) landraces to varying water regimes under a rainshelter. *Agricultural Water Management, 121*, 102–112. https://doi.org/10.1016/j.agwat.2013.01.009

Mackill, D. J., Ismail, A. M., Singh, U. S., Labios, R. V., & Paris, T. R. (2012). Development and rapid adoption of submergence-tolerant (Sub1) rice varieties. *Advances in Agronomy, 115*, 299–352.

Mackinder, B., Pasquet, R., Polhill, R., & Verdcourt, B. (2001). Papilionoideae. *Floral Zambesica, 3*(5), 69–73.

Malaisse, F., & Parent, G. (1985). Edible wild vegetable products in the Zambesian woodland area: A nutritional and ecological approach. *Ecology of Food and Nutrition, 18*, 43–82.

Mapaura, A., & Timberlake, J. (2004). *A checklist of Zimbabwean vascular plants*. Southern African Botanical Diversity Network (SABONET), Report No. 33, Pretoria and Harare.

Milne-Redhead, E., & Polhill, R. M. (1971). *Flora of Tropical East Africa*. Crown Agents for Oversea Governments and Administrations, Millbank.

Mithen, R. (1993). A new species of *Sphenostylis* (Leguminosae – Papilionoideae) from Zimbabwe. *Kirkia, 14*(1), 19–23.

Moteetee, A. N., & Van Wyk, B. E. (2012). Revision of the genus *Sphenostylis* (Fabaceae: Phaseoleae) in South Africa and Swaziland. *Bothalia, 42*(1), 1–6.

Msuya, T. S., Kideghesho, J. R., & Mosha, T. C. (2010). Availability, preference, and consumption of indigenous forest foods in the Eastern Arc Mountains, Tanzania. *Ecology of Food and Nutrition, 49*(3), 208–227.

National Research Council (NRC). (1979). *Tropical legumes: Resources for the future*. National Academy of Sciences Press.

National Research Council (NRC). (2006). *Lost crops of Africa. Volume II: Vegetables*. The National Academies Press. http://nap.edu/11763

Nnamani, C. V., Adewale, D. B., Oselebe, H. O., & Atkinson, C. J. (2020). African yam bean the choice for climate change resilience: Need for conservation and policy. In *African Handbook of Climate Change Adaptation* (pp. 1–17). https://doi.org/10.1007/978-3-030-42091-8_203-1

Nnamani, C. V., Ajayi, S. A., Oselebe, H. O., Atkinson, C. J., Igboabuchi, A. N., and Ezigbo, E. C. (2017). Sphenostylis stenocarpa (ex. A. rich.) harms. a fading genetic resource in a changing climate: Prerequisite for conservation and sustainability. *Plants, 6*, 30. https://doi.org/10.3390/plants6030030

Nnamani, C. V., Ajayi, S. A., Oselebe, H. O., Atkinson, C. J., Adewale, B. D., Igwe, D. O., & Akinwale, R. O. (2018). Updates on nutritional diversity in Sphenostylis stenocarpa (Hoechst ex. A. Rich.) Harms, for food security and conservation. *American Journal of Agricultural and Biological Sciences, 13*, 38–49. https://doi.org/10.3844/ajabssp.2018.38.49

NRI. (1987). *Root crops*. D:/cd3wddvd/NoExe/.../meister13.htm. pp. 1–15.

Nyananyo, B. L., & Nyingifa, A. L. (2011). Phytochemical investigation on the seed of *Sphenostylis stenocarpa* (Hochst ex. A. Rich.) Harms (Family: Fabaceae). *Journal of Applied Sciences and Environmental Management, 15*(3), 419–423.

Ojelel, S., Mucunguzi, P., Katuura, E., Kakudidi, E. K., Namaganda, M., & Kalema, J. (2019). Wild edible plants used by communities in and around selected forest reserves of Teso-Karamoja region, Uganda. *Journal of Ethnobiology and Ethnomedicine, 15*(3), 1–14. https://doi.org/10.1186/s13002-018-0278-8

Ojuederie, O. B., Balogun, M. O., & Abberton, M. T. (2016). Mechanism for pollination in African yam bean. *African Crop Science Journal, 24*, 405–416.

Ojuederie, O. B., & Balogun, M. O. (2019). African Yam Bean (*Sphenostylis Stenocarpa*) tubers for nutritional security. *Journal of Underutilized Legumes, 1*(1), 56–68.

Ojuederie, O. B., Balogun, M. O., Fawole, I., Igwe, D. O., & Olowolafe, M. O. (2014). Assessment of the genetic diversity of African yam bean (*Sphenostylis stenocarpa* Hochst ex. A Rich. Harms) accessions using amplified fragment length polymorphism (AFLP) markers. *African Journal of Biotechnology, 13*(18), 1850–1858. https://doi.org/10.5897/AJB2014.13734

Okigbo, B. N. (1973). Introducing the yam bean (*Sphenostylis stenocarpa*) (Hochst ex. A. Rich.) Harms. In *Proceedings of the first IITA grain legume improvement workshop Ibadan, Nigeria*, 29 October—2 November 1973.

Okpara, D. A., & Omaliko, C. P. E. (1995). Effects of staking, nitrogen and phosphorus fertilizer rates on yield and yield components of African yam bean (*Sphenostylis stenocarpa*). *Ghana Journal of Science, 28*, 23–28.

Okpara, D. A., & Omaliko, C. P. E. (1997). Response of African yam bean *(Sphenostylis stenocarpa)* to sowing date and plant density. *Indian Journal of Agricultural Sciences, 67*, 220–221.

Olayiwola, O. A., Latona, D. F., & Oyeleke, G. O. (2012). Evaluation of the macronutrients composition of soil, leaves and seeds of African yam bean (*Sphenostylis sternocarpa* Harms). *Journal of Applied Chemistry, 1*(1), 13–17.

Olomitutu, O. E., Paliwal, R., Abe, A., Oluwole, O. O., Oyatomi, O. A., & Abberton, M. T. (2022). Genome-wide association study revealed SNP alleles associated with seed size traits in African Yam Bean (Sphenostylis stenocarpa (Hochst ex. A. Rich.) Harms). *Genes, 13*, 2350.

Oluwole O. O., Olomitutu O. E., & Paliwal R., et al. (2020). Preliminary assessment of the association between DArT-SEQ SNP and some nutritional traits in african yam bean. *Tropical Journal of Natural Product Research, 4*, 877–879.

Phiri, P. S. M. (2005). *A checklist of Zambian vascular plants*. Southern African Botanical Diversity Network (SABONET), Report No. 32, Pretoria.

Popoola, J. O., Adegbite, A. E., & Obembe, O. O. (2011). Cytological studies on some accessions of African yam bean (AYB) (*Sphenostylis stenocarpa* Hochst. Ex. A. Rich. Harms). *International Research Journal of Plant Science, 2*(8), 249–253.

Potter, D. (1992). Economic botany of *Sphenostylis* (Leguminosae). *Economic Botany, 46*, 262–275.

Potter, D., & Doyle, J. J. (1992). Origin of African yam bean (*Sphenostylis stenocarpa*, Leguminosae): Evidence from morphology, isozymes, chloroplast DNA and linguistics. *Economic Botany, 46*, 276–292.

Potter, D., & Doyle, J. J. (1994). Phylogeny and systematics of *Sphenostylis* and *Nesphostylis* (Leguminosae: Phaseoleae) based on morphological and Chlooplast DNA data. *Systematic Botany, 19*(3), 389–406.

Pratap, U. P., Hima, L., Priyanka, H. P., & ThyagaRajan, S. (2017). Noni (Morinda citrifolia L.) fruit juice reverses age-related decline in neural-immune interactions in the spleens of old F344 rats. *Journal of Ethnopharmacology, 198*, 363–371. https://doi.org/10.1016/j.jep.2017.01.034

Pratap, A., Somta, P., Smýkal, P., & Gupta, S. (2023). Editorial: Genetics of domestication and diversification towards evolution of crop plants. *Frontiers in Genetics, 14*, 1175931. https://doi.org/10.3389/fgene.2023.1175931

Pratap, A., Das, A., Kumar, S., & Gupta, S. (2021). Current perspectives on introgression breeding in food legumes. *Frontiers in Plant Science*. https://doi.org/10.3389/fpls.2020.589189

Rachie, K. O. (1973). Highlight of Grain Legume Improvement at IITA 1970–73. In *Proceedings of the first IITA Grain Legume Improvement Workshop*, Ibadan, Nigeria, 29 October—2 November 1973.

Ramos-de-la-Pen˜a, A. M., CMGC, R., Wicker, L., & Contreras-Esquivel, J. C. (2013). Advances and perspectives of *Pachyrhizus spp.* in food science and biotechnology. *Trends in Food Science and Technology, 29*, 44–54.

Sarkar, F. H., & Li, Y. (2008). NF-kappaB: A potential target for cancer chemoprevention and therapy. *Frontiers in Bioscience, 13*, 2950–2959. https://doi.org/10.2741/2900

Saxon, E. C. (1981). Tuberous legumes: Preliminary evaluation of tropical Australian and introduced species as fuel crops. *Economic Botany, 35*, 163–173.

Shitta, N. S., Abberton, M. T., Adesoye, A. I., Adewale, B. D., & Oyatomi, O. (2016). Analysis of genetic diversity of African yam bean using SSR markers derived from Cowpea. *Plant Genetic Resources: Characterization and Utilization, 14*(1), 50–56. https://doi.org/10.1017/S1479262115000064

Shitta, N. S., Unachukwu, N., Edemodu, A. C., Abebe, A. T., Oselebe, H. O., & Abtew, W. G. (2022). Genetic diversity and population structure of an African yambean (Sphenostylis stenocarpa) collection from IITA GenBank. *Scientific Reports, 12*, 4437.

Snapp, S., Rahmanian, M., & Batello, C. (2018). *Pulse crops for sustainable farms in sub-Saharan Africa*. FAO.

Sohni, Y. R., Mutangadura-Mhlanga, T., & Kale, P. G. (1994). Bacterial mutagenicity of eight medicinal herbs from Zimbabwe. *Mutation Research, 322*, 133–140.

Sørensen, M. (1988). A taxonomic revision of the genus Pachyrhizus (Fabaceae—Phaseoleae). *Nordic Journal of Botany, 8*, 167–192.

Sørensen, M. (1990). Observations on distribution, ecology and cultivation of the tuber-bearing legume genus *Pachyrhizus* Rich. ex DC. (Fabaceae: Phaseoleae). *Agricultural University Wageningen papers, 90*(3), 38–78.

Tanksley, S. D., & McCouch, S. R. (1997). Seed banks and molecular maps: Unlocking genetic potential from the wild. *Science, 277*, 1063–1066.

Ubaka, C. M., & Ukwe, C. V. (2010). Antidiabetic effect of the methanolic seed extract of *Sphenostylis stenocarpa* (Hoechst ex A. Rich. Harms) in rats. *Journal of Pharmacy Research, 3*, 2192–2194.

Vaillant, V., & Desfotaines, L. (1995). Assimilate partitioning in *Pachyrhizus erosus* tubers under short day. *Physiologia Plantarum, 93*, 558–562.

Van Wyk, A. E., & Malan, S. (1997). *Field guide to the wildflowers of the Highveld*. Struik.

Verdcourt, B. (1970). Studies in the Leguminosae-Papilionoideae for the Flora of Tropical East Africa IV. *Kew Bulletin, 24*, 507–569.

Wojciechowski, M. F., Lavin, M., & Sanderson, M. J. (2004). A phylogeny of legumes (Leguminosae) based on analysis of the plastid *matK* gene resolves many well-supported subclades within the family. *American Journal of Botany, 91*, 1846–1862.

Zanklan, A. S., Becker, H. C., Sørensen, M., Pawelzik, E., & Grüneberg, W. J. (2018). Genetic diversity in cultivated yam bean (*Pachyrhizus* spp.) evaluated through multivariate analysis of morphological and agronomic traits. *Genetic Resources and Crop Evolution, 65*, 811–843. https://doi.org/10.1007/s10722-017-0582-5

Postharvest Technology of Pulses

Prasoon Verma and Man Mohan Deo

Abstract Globally, the vegetarian population depends heavily on pulses as a major source of proteins. The dietary system for the vegetarian population requires a balanced combination of cereals and pulses to provide essential amino acids that the human body cannot synthesize. Besides proteins, pulses are rich in carbohydrates, soluble and insoluble fibers, and several minerals and vitamins. Nonetheless, the supply of pulses has always been shorter than the demand. Efforts are being made to develop insect-pest and disease resistant and climate resilient varieties to increase production and productivity through crop improvement techniques. Availability of pulses is also negatively affected by the losses that take place at various postharvest handling, transport and processing stages. If the postharvest losses can be contained, more pulses can be made available for human consumption. Pulses attract insects from the field itself, which get carried to the storage go-downs after harvesting and threshing. During the congenial environmental conditions, huge losses take place due to increased insect activities, and quantitative as well as qualitative losses take place. Minimal processing of pulses before consumption is required to minimize antinutritional substances, improve culinary properties, and enhance the bioavailability of nutrients. Dehusking and splitting, referred to as pulse milling, is the most prevalent method for reducing the anti-nutritional compounds. Pulses can also be used in powder form for making secondary and tertiary processed products. Milling byproducts obtained from pulses also contain proteins, fiber, and phenols; hence, they can effectively be utilized for edible purposes in combination with cereals and other ingredients.

Keywords Pulses, Crop improvement, Pulse milling, Dehusking, Processing units, Primary, secondary, tertiary processing

P. Verma (✉) · M. M. Deo
Division of Crop Production, ICAR-Indian Institute of Pulses Research, Kanpur, Uttar Pradesh, India

A. Pratap et al. (eds.), *Integrated Improvement of Food Legumes*,
https://doi.org/10.1007/978-3-031-82629-0_13

1 Introduction

Most of the vegetarian population around the world depends upon pulses to meet the requirements of essential proteins that cannot be synthesized by the human body. Belonging to Leguminosae family, pulses are consumed as whole grains, sprouts, dehusked splits, immature seeds, and in powdered forms. Lysine-rich pulses complement the methionine-rich cereals to make the combined diet a complete food. The presence of higher proportions of proteins (21–25%) (Tiwari & Singh, 2012; Singh, 2017) makes pulses considered as poor man's meat and therefore, an integral part of human diets. In combination with cereals, the protein complex thus formed is complementary to each other and quite comparable with the animal proteins. Pulses are also a good source of minerals and vitamins, along with complex carbohydrates, as well as soluble and insoluble fibers (Singh & Pratap, 2015). Incorporation of pulses in crop rotations is good for soil health due to the property of fixing atmospheric nitrogen in roots, resulting in a lesser requirement of exogenous application of nitrogenous fertilizers in the field. Low water requirements of the leguminous crops make them suitable to be grown in rain-fed areas by the resource-poor farmers. Pulses, being the cheapest source of proteins, play an important role in food and nutritional security for malnourished populations. However, the availability of pulses is quite low as compared to the recommended level of per capita consumption. Efforts are being made to enhance the production and productivity of pulses through systematic breeding programs to combat biotic and abiotic stresses. Rich in proteins, pulses start attracting insects (bruchids or pulse beetles) from the field itself, which are carried to the storage go-downs after harvesting and threshing. These insects multiply during hot and humid environmental conditions, conducive to insect growth. Losses can be very high if these insects are not monitored properly during storage. Since larvae of insects thrive on cotyledons, by the time the external insect activity is observed, grain has either been lost or it is rendered unfit for human consumption. Postharvest losses further reduce the availability of pulses for human consumption. Therefore, one of the ways to bridge the gap between demand and supply is to reduce postharvest losses, which are quite high, especially during storage and milling. It has been estimated that the postharvest losses in pulses can be as high as 20–30% (Lal & Verma, 2007), if not managed properly. For major pulses, namely, pigeonpea, chickpea, greengram, and blackgram, storage losses range between 4.28% and 6.04% and storage insects have been identified as main cause for the storage losses (Nanda et al., 2012). Evidently, curbing these losses will make more pulses available for human needs. Pulses also contain a wide range of anti-nutritional substances, such as oligosaccharides, lectins, inhibitors, polyphenols, phytic acid, etc. (Ramadoss & Shunmugam, 2014), which restricts the bioavailability and absorption of nutrients. This makes processing of pulses essential prior to consumption. As of now, much of the processing, except of the major pulses, is an age-old traditional method adopted at the domestic and cottage scale to improve digestibility and culinary properties and needs to be upgraded to the modern level. Germination, soaking, cooking, fermentation, milling, etc. are various processing techniques adopted for the pulses.

2 Postharvest Value Chain of Pulses

The postharvest chain of pulses includes all stakeholders involved right from their production till consumption. Pods in pulses are harvested after attaining the proper physiological maturity, and usually the harvesting operation is performed at higher grain moisture to minimize shattering losses. After harvest, the crop is sun-dried for ease of threshing. Grains are extracted by threshing crops either manually or with the help of commercial threshers. For mechanical threshing, since pulse grains of different crops are of different sizes, threshers need to be modified in terms of clearance between cylinder and concave, number of beaters, and concave openings. Grains are collected by the aggregators after cleaning, grading, and packing. These aggregators bring them to the procurement agencies for storage or millers for further processing. It is a prerequisite to clean, grade, and dry the grains as per the norms and requirements of procurement agencies to fetch better market prices. Grains can either be stored in bag storage structures or in silos for bulk storage. Grains can also be stored at farm and household levels for local and domestic consumption. The stored grains are further transported to small, medium, and large-scale commercial processors, where pulses are stored till the time of processing. Whole grains or dehusked splits are also stored in secondary and tertiary processing units depending upon their production capacity. The finished product is stored in the production units until it is supplied to the consumers through the wholesaler and retailer marketing network. Evidently, right from procurement till consumption, agricultural commodities undergo a complex postharvest value chain. Since it involves multiple stakeholders from farmers to sellers and consumers, an ideal value chain system ensures a win-win situation for all the stakeholders. Pulses being more prone to insect-pest infestation, these require extra precaution to minimize losses at every postharvest stage. Still, some losses are inevitable, and it has been observed that handling, threshing, transport, storage, milling, etc. are the critical stages where even minor interventions can reduce the losses.

3 Postharvest Losses

Sizeable quantitative and qualitative losses take place during harvesting, threshing, winnowing, transportation, storage, and processing, thereby adversely affecting the overall availability of pulses. These reductions in grain quality and quantity ultimately result in monetary losses to the stakeholders. These losses can be attributed to the poor management practices adopted at each stage. The extent of loss varies with the method of handling of each pulse crop in the value chain. Weather conditions, varieties, location of storage, and processing units affect the postharvest losses. Furthermore, it has been observed that the maximum losses occur during storage and milling. This is mainly because the storage protocols followed for pulses are basically the adoption of methods used for the storage of wheat and paddy. Furthermore, pulse milling methods are also the scaled-up versions of traditional pulse milling. Therefore, implementation of pulse-specific storage protocols and improved milling machinery can minimize the storage and milling losses. Gunny bags, the prevalent bag storage system, are pervious to atmospheric conditions, allowing grain moisture to increase or decrease according to the environmental conditions. Grain moisture increases during hot and humid ambient environments, creating a conducive environment for the growth of mold and insects. If not monitored properly, it leads to heavy storage losses. Controlling initial field-level infestation prior to storage and using packaging materials impervious to the external environment can help in the long-term safe storage of pulses. The development of improved pre-milling treatments to loosen the seed coat can help in increased *dal* recovery, thereby reducing losses. For example, in abrasive dehusking devices, scarification of the precious protein-rich outer layer of cotyledons takes place, which in turn results in lesser *dal* recovery. Therefore, there is a need to develop improved machinery to minimize milling losses. Out of total production of pulses, about 75% produce is milled to be consumed as dehusked splits (*dal*) and other processed products. According to Majumdar and Parpia (1976), about 25% of food losses takes place in the field itself. Storage contributes to 15% losses, whereas the contribution of handling processing is 7%, and 3% loss is due to other factors. Farm-level losses are determined by harvesting techniques, crop maturity, crop moisture content at the time of harvest, transportation mode, threshing technique, cleaning/winnowing procedure, and type of threshing floor. Postharvest losses are caused by a variety of factors, including birds, insects, mold, fungi, transportation, storage conditions, rats, spills, and pilferage (DMI, 2002). It has been reported that about 8.6–16.6% losses take place during pigeonpea processing (Goyal et al., 2005). According to a study conducted on harvest and postharvest losses in four major pulses, namely, chickpea, pigeonpea, greengram, and blackgram, ranges between 6.36% and 8.41% (Jha et al., 2015). Considering postharvest losses reported by various researchers, it can be as high as 25–30% (Acharya et al., 2019).

4 Interventions to Reduce Postharvest Losses

Postharvest losses in pulses can be reduced significantly by adopting the following practices:

(i) Harvesting crops at proper grain maturity and comparatively higher moisture content
(ii) Using appropriate threshers with suitable adjustments in machine parameters
(iii) Drying and fumigation of grains prior to their storage

(iv) Using pilferage-free packaging materials during transportation and storage
(v) Minimal polishing to reduce milling losses

5 Unit Operations in Pulse Value Chain

5.1 Harvesting

After grain maturity, the pulse crops need to be harvested at slightly higher moisture content in order to minimize shattering losses. Usually, the harvest time is determined by the duration of the plant growth cycle, and it usually takes place at physiological maturity. Suitable moisture content for harvesting is between 15% and 20% of grain moisture and when pods and leaves start turning black. In the case of pulses, harvesting is usually done manually with the help of sickles. Genetic improvement efforts are being made to develop combinable varieties of different pulses. Few combines have been developed and commercially made available for harvesting and threshing of pulse crops, except pigeonpea.

5.2 Threshing and Winnowing

After harvesting, the pulses are either threshed manually or with the help of threshers to separate grains from the plants. The crop also needs to be dried up to a moisture level of 10–12%, congenial for threshing and ease of grain separation from the pods. Furthermore, low grain moisture is also important for the safe storage of pulse grains. Hand and animal pounding and treading with tractors are quite common practices of threshing. Commercial threshers are usually designed for wheat, and therefore, before their deployment in the threshing of pulses, these require settings in machine parameters due to varied grain sizes and plant structure for each pulse crop in order to reduce breakage of grains. Chaff and light impurities are blown away with the help of blowers. For threshing of pulse crops, threshers with chopper and double blowers may be preferred due to the heavy and sturdy plant stalk.

5.3 Cleaning and Grading

After threshing, grains need to be pre-cleaned and graded in order to remove physical impurities and to meet the procurement norms and standards prior to their storage or supply to the processors. For long-term safe storage of pulses, the grains should be free from foreign materials, damaged, infested, or shriveled grains.

5.4 Drying and Fumigation

For long-term storage of pulses, grain moisture needs to be brought under the safe limits (10–12%). Sun drying and heated air drying are the common methods of grain drying. To control initial field-level infestation, it is imperative to fumigate the grains using suitable fumigant prior to storage. In storage go-downs, use of aluminum phosphide is quite prevalent.

5.5 Storage

Improper and inefficient storage methods cause huge quantitative and qualitative losses. Infestations by insects, rodents, birds, etc. are the main causes of physical grain loss, whereas their excreta and microbial activities cause qualitative losses during grain storage. Traditional cottage or rural household structures, namely, mud bins, bamboo and mud plastered bins, paddy straw mud bins, etc., and local methods for storage of pulses are not adequate for a safe and longer storage of pulses. Unfortunately, lack of farm-level storage facilities compels farmers to sell their produce during production peaks at prices much below the support prices. In such a scenario, the use of metal bins may extend the facility's ability to store grains by preventing external infestation and maintaining functional airtightness for fumigation.

5.6 Processing

Traditionally, pulse milling has been a domestic or cottage-scale industry for dehusking and splitting of whole grains to convert them into *dal*. Water soaking, roasting, parching, etc. were the prevailing pre-milling treatments adopted at cottage scale to loosen the seed coat before manually milling pulses in stone *chakkies*. Gradually, pulse milling has grown into the third largest processing industry after wheat and rice milling. In commercial mills, abrasive dehusking devices are used for removal of husk after pre-milling treatment. Oil and water, and thermal treatment are commonly adopted in commercial-level pulse milling. In the abrasive dehusking method, a significant quantity of the outer layer of cotyledons gets scratched and mixed with the outer seed coat. *Dal* recovery is always lesser than the potential *dal* recovery. Milling is also performed to reduce anti-nutritional compounds present in whole grains and improve cooking and culinary properties.

5.7 *Pulse Milling*

Pulse milling is the third largest food processing industry after wheat and rice. Unlike many other crops, pulses are consumed after minimal processing. Pulses are mostly consumed in the form of dehusked splits, i.e., *dal* as about 80% of total pulse production in India is consumed in the form of *dal*. There are about 15,000 dal mills across the country where 75% of total produce is processed (Ali, 2004). There are about 2000 mills with running capacity of 10–20 tons/day (MoFPI, 2009). Flour of different pulses is used in making various traditional food recipes. The remaining 20% is utilized as whole grain or kept for seed purposes. Depending upon the pulse type, variety, milling process, and machineries used, up to 5% variation is observed in milling recoveries, with an average *dal* recovery of 65–70% against the potential recovery of 88–89% (Lal & Verma, 2007).

Dehusking of pulses is an age-old practice that started from domestic level in the initial days to the present day with large-capacity modern mills. Nonetheless, the major steps involved in dehusking, either at the cottage or commercial levels are same. The process includes (i) Pre-milling treatment—to loosen the seed coat, (ii) Dehusking or husk removal—to add culinary properties, and (iii) Splitting—to convert dehusked whole (*gota*) into dal. Dehusking and splitting processes take place simultaneously in the milling units. Dehusking is an essential process for pigeonpea, whereas other pulses can be consumed as whole, splits with husk, and dehusked splits. In most of the secondary and tertiary processed products, the powder form of dehusked grains is used. Removal of husk not only reduces fiber but also improves appearance, texture, cooking quality, storability, palatability, and digestibility (Singh, 1995; Narasimha et al., 2003). Dehusked splits offer better resistance to insect activities in comparison to whole grains without any pesticide application if external infestation is prevented. The presence of gummy substances between the seed coat and cotyledons makes husk removal difficult during milling. According to the quality and quantity of gum present, pulses are categorized into easy-to-mill (lentil, chickpea, fieldpea, etc.) or difficult-to-mill (arhar, urdbean, mungbean, etc.) kinds of pulses.

6 Unit Operations in Commercial Pulse Milling

The milling process essentially involves cleaning, grading, pitting, treatment, milling, and polishing operations. Each unit operation requires specific machinery to perform the individual task. Usually pre-milling treatments and processes were developed for the toughest-to-mill pulse grains, i.e., pigeonpea. The major steps of pulse milling are described as under:

6.1 Cleaning and Grading

Pre-cleaning and grading of raw material is an essential step in milling. It involves removing dust, dirt, and foreign materials, as well as off-sized, immature, and damaged grains, followed by grading. At the first step, raw material is passed through a cleaner-cum-grader. The quality of raw material used for milling determines the quality of the end product, i.e., *dal*. Cleaners are used to remove impurities, foreign matter, and damaged and shriveled grains. Depending upon the size and shape of the grain and gravity, raw material is graded to get uniform quality of raw grains. Commonly, reel machines are used in the commercial mills for size grading. Destoner is used to remove stones and pebbles. Lighter impurities are removed with the help of aspirators, fans, or pneumatic separators.

6.2 Pitting

It is a common practice in commercial *dal* mills that are using emery-coated rollers for abrasive action. Whole pulses are passed through an abrasive roller machine for scratching the coat of the seeds. Cracked seed coat facilitates entry of oil/water applied to the grain during pre-milling treatment. Ideally, cracks should be so small that these are not visible to the naked eyes; however, during the pitting process, some dehusking and splitting takes place.

6.3 Pre-milling Treatments

Depending upon the scale of milling, wet or dry pre-milling treatments can be adopted. Wet milling is adopted at domestic or cottage-scale milling, whereas dry pre-treatment is commonly practiced for commercial milling.

6.3.1 Cottage-Scale Pulse Milling (Wet Milling)

At the cottage scale, whole grain is soaked overnight or for a stipulated time period, followed by sun drying and tempering for 2–3 days to attain a 10–12% moisture level before its dehusking and splitting in hand-operated wooden or stone *chakkies*. Husk is removed by manual winnowing. Several other pre-treatments, namely, red earth application, iron pan roasting with or without sand, oil smearing, boiling, etc., are also being practiced in different parts of the country. However, at the cottage scale, pulse milling is a time-consuming, weather–dependent, and labor-intensive process. Furthermore, despite pre-treatment, it is hard to achieve 100% dehusking in traditional *chakkies*, which diminishes the physical appearance of *dal* and reduces its market value. However, of late, a section of the affluent market segment is developing a preference for cottage-scale unpolished *dals*.

6.3.2 Large-Scale Milling (Dry Milling)

At commercial-scale pulse milling, efforts were made to overcome constraints of traditional cottage-scale mills. Problems of drudgery, time consumed, weather dependency, and improper dehusking were addressed in commercial pulse mills. To minimize soaking and drying time, oil and water application, often known as dry treatment, is quite popular. Linseed oil is applied at the rate of 0.7% on pitted grains. Alternate daytime heating under the sun and heaping and cooling during the nights for 3–4 days loosens the seed coat. Thus, treated grains are milled in emery roller mills for dehusking. However, this process also depends upon weather and the intensity of solar radiation for sun drying.

6.3.3 Thermal Treatment

Pulse milling pre-treatments at both the cottage level and industry scale are laborious, time-consuming and weather-dependent processes. A weather-independent thermal process was developed by CFTRI, Mysore. This process also completely eliminates the use of oil. In this process, grains are subjected to high temperature (166 °C) using heated air, followed by tempering in insulated bins for 6 h to maintain the grain temperature. The process is repeated twice for proper conditioning. This is a weather-independent process. However, it is not being adopted widely due to more investment in machinery and high operational expenses. At GBPUA&T, Pantnagar, India, sodium bicarbonate, vinegar, and enzyme treatments were evaluated. However, none of the treatments could reach up to the level of commercial acceptance.

6.3.4 Tempering

After water soaking or oil application, pulse grains are heaped and covered with gunny bags for 12–18 h for uniform conditioning of the entire grain mass. This process allows penetration of oil/water beneath the seed coat and helps in loosening of gummy substances.

6.3.5 Drying

Sun-drying of treated and tempered grains is quite common in commercial mills in the Indian subcontinent. For year-round operation and reduced weather dependency, electricity-based mechanical dryers are being used on a limited scale. Alternate sun drying in thin layer during daytime (heating) and heaping at night (cooling) for 2–3 days results in the loosening of the seed coat. Nevertheless, the sun drying process is time-consuming and a labor-intensive operation. Drying in mechanical dryers takes 2–3 h, thus saving both time and energy.

6.3.6 Dehusking and Splitting

After thorough pre-treatment, grains are exposed to abrasive surfaces in commercial mills to remove the husk. The operation of dehusking and splitting is performed in emery roller mills. The emery coating, also referred to as carborundum, is made of silicon carbide (carbon + crystalline alumina). The grit size of carborundum affects the dehusking efficiency. Different grit sizes are used for different kinds of pulses. Most of the mills use emery rollers for dehusking and splitting. Some millers use disk shellers for splitting of dehusked grains (*gota*), whereas dehusking is performed in emery rollers only. Milled fractions, namely, unhusked whole, dehusked whole, unhusked *dal*, dehusked *dal*, and breakage, are separated to achieve a quality product. Byproduct from the milling industry (mixture of husk and powder) commonly goes for cattle feeding.

6.3.7 Polishing

Polishing is required to give a uniform appearance to the milled product and increase the appeal to consumers. Certain fractions of milled products include *dal* with husk cover. Polishers help in the removal of husk from such *dal*. Different kinds of polishes, such as nylon polish, oil/water polish, leather, and *makhmal* polish, are applied to different kinds of *dals*. Oil and edible color polish are commonly adopted in commercial mills. With increasing awareness of adulteration and the use of non-edible compounds, consumers are developing a preference for unpolished *dal*.

6.4 Processing

6.4.1 Primary Processing

Prior to final use, the raw grains are cleaned and graded to remove impurities, and immature and damaged grains so that the grain quality does not affect the quality of the end product. Primary processing needs to be carried out at the farm level so that quality grain is kept for storage and for secondary level processing.

6.4.2 Secondary Processing

Whole grains are converted into dehusked *dal* or *dal* flour (*besan*) so that it can be directly utilized for human consumption. Germination, roasting, flaking, cooking, boiling, frying, etc. are a few examples of secondary processing.

6.4.3 Tertiary Processing

Tertiary processing is referred to as the conversion of products obtained from secondary processing into ready-to-eat or cooked kinds of value-added products.

7 Mini *Dal* Mills

Like most of the food processing industries, high-capacity pulse processing plants are generally located in cities. Whole grains are mostly procured by middlemen at cheaper rates, and value-added product, *dal*, is sold back to the consumers via retail outlets at higher prices. This results in sometimes doubled the cost of the finished product than the price of the raw material. In forward and backward linkages, ultimately, the procurement agencies, wholesaler, and retailer network take the maximum advantage. The actual growers and rural population are at a loss as they have to shell out more money for processed *dal*, and profit margins move to the urban areas. Keeping this in view, it was realized that pulse processing can be adopted among rural entrepreneurs to strengthen the rural economy, and a concept of small-capacity mini *dal* mills was visualized by several research organizations. Mini *dal* mills developed by CFTRI, Mysore; Central Institute of Agricultural Engineering, Bhopal; Punjab Rao Deshmukh Krishi Vidyapeeth, Akola; and ICAR-IIPR, Kanpur, are popular in different pockets of the country. Mini *dal* mills in rural areas will reduce the expenditure made on both ways of transportation and make cheaper *dal* available to the rural population. Furthermore, it is easy to control milling losses in small-scale processing machines. Therefore, such units give higher *dal* recovery. The establishment of processing units in the rural catchments has the potential to promote rural entrepreneurship, generate employment opportunities, enhance rural income, and ultimately, contain migration to the cities. Details of some of the mini *dal* mills are given below.

7.1 CIAE Mini Dal *Mill*

An emery roller-based low-cost dehusking and splitting unit was developed at CIAE, Bhopal. The machine consists of an abrasive carborundum roller rotating inside the cylindrical sieve. Treated grains are passed through the roller unit. In this

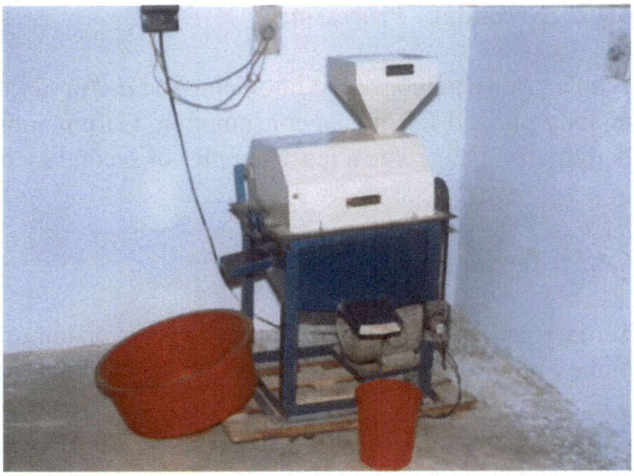

Fig. 1 CIAE Mini *Dal* Mill

machine, dehusking and splitting take place simultaneously. It can easily be operated with semi-skilled workers. Operated by a 2-hp three-phase motor, the mill gives a capacity of 100–150 kg/h for different pulses. Water soaking or oil treatments can be applied prior to milling in the unit (Fig. 1).

7.2 PDKV Mini Dal *Mill*

This is a complete mill, where all the pulse milling operations, namely, dehusking, splitting, sieving, aspiration, grading, etc., can be performed in a single unit. This simple mill offers ease of operation and maintenance. It consists of a tapered horizontal roller covered with a coating of emery rotating inside a cylindrical sieve. Husk and powder mixture obtained from dehusking are removed with the help of an aspirator. *Dal* is graded using an oscillating sieve set. The capacity of this mill is about 300 kg/h (Fig. 2).

7.3 CFTRI Mini Dal *Mill*

CFTRI mini *dal* mill gives better control over dehusking of pulses. The vertical emery cone rotates inside the conical sieve screen. The gap between the screen and the cone surface can be increased or decreased depending upon the grain size. It is available in both manual and motorized operation modes. Hand-operated manual unit has a milling capacity of 40–60 kg/h, whereas the motorized unit has a milling capacity of 100–150 kg/h. The motorized *dal* mill consists of a conical dehusking and splitting unit, an aspiration system to remove husk and powder dust, and a grading unit to separate *dal* from the milled fractions (Fig. 3).

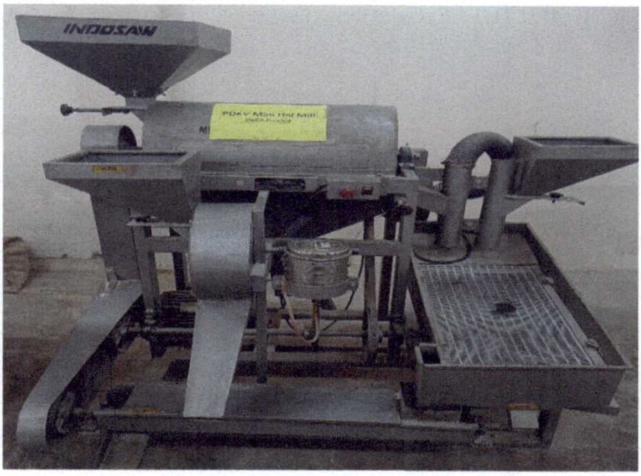

Fig. 2 PDKV Mini *Dal* Mill

Fig. 3 CFTRI Mini *Dal* Mill

7.4 IIPR Dal Chakki *and Mini* Dal *Mill*

IIPR, Kanpur, initially developed a *dal chhaki*, using a rubber and steel discs mechanism, for dehusking of pulses. A good number of farmers and small entrepreneurs used IIPR *Dal Chakki*. However, based on the feedback received from the users of the machine and prospective entrepreneurs, this *dal chakki* was upgraded into a mini *dal* mill. This mill has the provision of an emery roller attachment to make the pitting process easier and enable the unit for the production of dehusked whole, i.e.,

Fig. 4 IIPR *Dal Chakki*

Fig. 5 IIPR Mini *Dal* Mill

gota, also like *malka masoor*, which was not possible in the *chakki*. Additionally, a grading system was attached for cleaning and grading of raw grains and finished products, mounted on the same frame. Incorporation of these units in an upgraded model has made this mill a complete *dal* mill, wherein grading of raw grain, pitting of grains, milling of all types of pulses like dehusked splits (pigeonpea, chickpea, fieldpea, lathyrus, etc.), unhusked splits (mungbean and urdbean), and dehusked *gota* (*malka masoor*), cleaning and separation of husk, etc., and grading of finished products (*dal*) are executed in the same machine, and all the operations can be performed simultaneously (Figs. 4 and 5).

8 Value-Added Products

Pulses are consumed in a myriad of ways, namely, whole grain, dehusked splits, milled, mixed with cereals, roasted, puffed, salted, sweetened, etc. Immature pulse grains are also utilized in green, roasted, boiled, fried, crushed, and cooked forms.

After maturity, the same pulse is consumed as a dried grain, soaked, sprouted, boiled, steamed, fried, or cooked as *dal*. *Dal* or *sambar* is prepared from dehusked cotyledons on a daily basis at almost every household. Dehusked splits are also converted into flour to make various delicacies throughout the country. Soaking, cooking, roasting, puffing, extrusion, germination, fermentation, etc., are the essential processes involved in the preparation of different recipes from pulses.

9 Utilization of Milling Byproduct of Pulses

In the abrasive dehusking process of pulse milling, a significant quantity of cotyledons gets scoured and mixed with the seed coat to produce a milling byproduct. In the case of pigeonpea, *dal* recovery is around 70% against the potential *dal* recovery of 85%. Therefore, about 15% of the protein-rich outer layer of cotyledons is lost to the byproduct, which is mostly utilized as low-value animal or cattle feed currently. The milling byproducts obtained from different pulses were evaluated for bioactive compounds, and it was observed that they are also equally rich in protein, phenol, antioxidant, and fiber, which can be utilized for edible purposes. Therefore, a number of traditional and commercial recipes were developed from the pulses milling byproducts as elaborated in Fig. 6.

10 Future Research Thrusts

1. Variety Improvement: In order to minimize milling losses in pulses, there is a need to develop varieties with lower gummy substances between the seed coat and the cotyledons. Such varieties will require shorter duration and milder pre-milling treatments for loosening of husk.
2. Pre-milling Treatments: Efforts are required to be intensified to develop suitable pre-milling treatments to loosen the seed coat for husk removal at minimum force application.
3. Weather Independency: Pulse milling at cottage or large scale is weather-dependent process. During the rainy and winter seasons, pulse milling plants mostly remain idle. In order to use the plant year-round, weather-independent processes and machines need to be developed.
4. Dehusking Mechanism: To achieve better dehusking, pulse mills need to be designed using suitable emery grit size, roller-sieve clearance, diameter-length ratio, etc. Machine parameters need to be standardized to minimize powdering losses for different kinds of pulses.
5. Composite Milling Plant: In the pulse milling unit, a lot of material handling is involved. There is a need for a fully automated plant that can perform pre-treatment in continuous mode.

Fig. 6 Value-added products from pulses milling byproduct

6. Utilization of Milling Byproducts: Broken from the pulse milling industry are sold at lower prices. It can be utilized for secondary and tertiary processing to generate more income. Milling byproduct husk and powder can also be used for developing low-protein, high-fiber edible products and cuisines.

11 Conclusions

With a vast majority of the vegetarian population, pulses are an integral part of the Indian dietary system. Furthermore, there has been an ever-increasing global population that is having an inclination toward 'vegan' food. Pulse protein in combination with that of cereals makes a balanced vegetarian diet. Although India is the largest producer of pulses, imports are inevitable to meet the domestic requirement. Shifts in areas toward rice-wheat cropping systems show a decline in the interest of farmers toward pulses. Efforts are underway to increase the domestic production and productivity of pulses through genetic enhancement and crop management so that the country becomes self-reliant in the near future. Simultaneously, the international scientific community is also advocating the consumption of pulses for a healthy life. Substantial quantitative and qualitative losses take place during various postharvest stages. There is a need to minimize postharvest losses, which especially occur during storage and milling. This can contribute substantially to the availability of pulses to the masses. In order to minimize storage losses, care must be taken from the pre-storage stage itself. By adopting appropriate methods and structures for pulse storage, a significant amount of storage losses can be averted. Similarly, on the milling front, it is easy to curb losses in the smaller pulse processing units. Various research organizations, namely, CFTRI, Mysuru; CIAE, Bhopal; PDKV, Akola; and IIPR, Kanpur, have developed mini dal mills to cater to the milling requirements in production areas. The establishment of such units at rural thresholds will not only reduce postharvest milling losses but also minimize costs involved in material handling. Primary, secondary, and tertiary level processing of pulses can generate income and employment opportunities for farm families and unemployed rural youth. The profit margin taken away by a chain of middlemen, processors, wholesalers, retailers, etc. will remain with the growers and consumers, which in turn will help in boosting the rural economy.

References

Acharya, B., Mukherjee, B., Lalramhlimi, B., Dutta, S., Chakraborty, I., & Nath, R. (2019). Value addition to pulses by postharvest technology and processing. In I. Chakraborty, I. Riadh, B. Vikram, O. Sujayashree, & A. Mani (Eds.), *Recent trends & advances in food science & postharvest technology* (pp. 51–86). Satish Serial Publishing House.

Ali, N. (2004). Postharvest technology and value addition in pulses. In M. Ali, B. B. Singh, S. Kumar, & V. Dhar (Eds.), *Pulses in a new perspective* (pp. 530–543). Indian Society of Pulses Research and Development, Indian Institute of Pulses Research.

Directorate of Marketing and Inspection. (2002). *Estimation of marketable surplus and postharvest losses of food grains in India (1996–1999)*. Department of Agriculture and Cooperation, Government of India. Https://Dmi.Gov.In/Documents/Foodgrains.Pdf

Goyal, R. K., Wanjari, O. D., Ilyas, S. M., Vishwakarma, R. K., Manikantan, M. R., & Mridula, D. (2005). *Pulse Milling Technologies* (Technical Bulletin, Ciphet/Pub/2005). Central Institute of Postharvest Engineering and Technology.

Jha, S. N., Vishwakarma, R. K., Ahmad, T., Rai, A., & Dixit, A. K. (2015). *Report on assessment of quantitative harvest and postharvest losses of major crops and commodities in India*. Icar-All India Coordinated Research Project on Postharvest Technology, ICAR-CIPHET.

Lal, R. R., & Verma, P. (2007). *Postharvest management of pulses*. ICAR-Indian Institute of Pulses Research.

Majumdar, S. K., & Parpia, H. B. (1976). *Prevention of food losses and food potential* (pp. 1–11). NISI Symposium of Science and India's Food Problem.

MOFPI. (2009). *Annual Report 2009*. Ministry of Food Processing Industries, Govt. of India.

Nanda, S. K., Vishwakarma, R. K., Bathla, H. V. L., Rai, A., & Chandra, P. (2012). *Harvest and postharvest losses of major crops and livestock produce in India*. All India Coordinated Project on Postharvest Technology (ICAR). Mofpi.Nic.In/Sites/Default/Files/Phl_2010_Ciphet.Pdf.

Narasimha, H. V., Ramakrishnaiah, N., & Pratape, V. M. (2003). Milling of pulses. In A. Chakraverty, A. S. Mujumdar, G. S. V. Raghavan, & H. S. Ramaswamy (Eds.), *Handbook of post harvest technology* (pp. 427–454). Marcel Dekker, Inc.

Ramadoss, B. R., & Shunmugam, A. S. K. (2014). Anti-dietetic factors in legumes – local methods to reduce them. *International Journal of Food and Nutritional Sciences., 3*(3), 84–89.

Singh, U. (1995). Methods for dehulling of pulses: A critical appraisal. *Journal of Food Science and Technology, 32*(2), 81–93.

Singh, N. (2017). Pulses: An overview. *Journal of Food Science and Technology, 54*, 853–857.

Singh, N. P., & Pratap, A. (2015). Towards achieving self sufficiency in pulses. In *Agriculture Year Book 2015* (pp. 98–104).

Tiwari, B. K., & Singh, N. (2012). *Pulse chemistry and technology*. Royal Society of Chemistry.

Index

A. Pratap et al. (eds.), *Integrated Improvement of Food Legumes*, https://doi.org/10.1007/978-3-031-82629-0

The manufacturer's authorised representative in the EU is Springer Nature Customer Service Centre GmbH, Europaplatz 3, 69115 Heidelberg, Germany. If you have any concerns regarding our products, please contact ProductSafety@springernature.com

Printed and bound by CPI Group (UK) Ltd, Croydon, CR0 4YY
30/09/2026
02225596-0009